ISSN 1020-1203

Water Reports 9

IRRIGATION IN THE NEAR EAST REGION IN FIGURES

FOOD AND AGRICULTURE ORGANIZATION OF THE UNITED NATIONS
Rome, 1997

Reprinted 2003

The designations employed and the presentation of material in this information product do not imply the expression of any opinion whatsoever on the part of the Food and Agriculture Organization of the United Nations concerning the legal or development status of any country, territory, city or area or of its authorities, or concerning the delimitation of its frontiers or boundaries.

ISBN 92-5-103969-0

Foreword

With increasing pressure on natural resources in the Near East, and considering the regional dimension of water resources development and management, the need for reliable and systematic information on water, its quality and its use is increasingly recognized as being of prime importance for sustainable economic development in the region.

In 1993, the Food and Agriculture Organization of the United Nations initiated a programme, known as AQUASTAT, to meet the demand for information on water resources and its use at regional level, with particular emphasis on irrigation. The objectives of the AQUASTAT programme are to collect and present, in a systematic way, data on rural water use at country level.

The Near East Region has a concentration of most of the complex water allocation issues of the world. With a population of about 560 million inhabitants, or 10% of the world's population, increasing at a rate of 2.7% per year, it receives only 3.5% of the world's precipitation and has as little as 2.2% of its renewable water resources. Considering that more than 90% of the water withdrawal in the Near East is used for irrigation, and in view of the increasing importance of joint management of water resources, reliable information on agricultural water use at regional level is of critical importance. This is why FAO decided to launch a special AQUASTAT survey for the Near East Region.

This publication presents a description of the rural water situation in the 29 countries represented by FAO's Regional Office for the Near East, with relevant tables and maps, and a regional synopsis emphasizing the sub-regional characteristics of this large and diverse region. It is hoped that the information provided here will be of use to all those working for sustainable water resources development in the region.

Acknowledgements

This report was prepared and written with the collaboration of a large number of individuals. It is a joint activity of FAO's Land and Water Development Division (AGL) and its Regional Office for the Near East (RNE).

The whole project was coordinated and managed by Karen Frenken, acting as main author and editor of the survey, in collaboration with Jean-Marc Faurès and Mathieu Bousquet, from the AQUASTAT team of the Land and Water Development Division of FAO in Rome. Coordination of the project at the Regional Office for the Near East was carried out by Selim Sarraf, Regional Water Resources and Management Officer.

Country surveys were carried out in cooperation with national experts: Waleed Al-Zubari, Mohamed Abdelmonem El-Nahal, Karim Shiati, Hazim el Naser, Khalid Masabahi, Adib Salim Geadah, Tony Misfud, Asadullah Taqi, George Soumi, Huseyin Yavuz, Abdelkader Hamdane, Mohamed Saqr Al-Asam and Nasser Mohamed Nasser.

The authors wish to acknowledge the assistance provided by Jean Margat and Philippe Pallas in reviewing all the information related to water resources. Valuable contributions were also received from Azad Azad, and colleagues from the Land and Water Development Division of FAO. The data processing software was developed by Dario Berardi and Mauro de Castro. Data processing was performed by Marco Tagliaferri.

The publication has been edited by Jennifer Ann Parkinson and prepared for printing by Chrissi Smith Redfern, Edoardo Buricchi and Franco Buricchi.

Table of contents

Page

List of tables

List of figures

UNITS

$1\ km^3 = 1 \times 10^9\ m^3 = 1\ 000 \times 10^6\ m^3$

$1\ km^2$ = 100 hectares
1 dunum = 0.10 hectare
1 feddan = 0.42 hectare
1 acre = 0.4046 hectare

The information presented in this publication is collected from a variety of sources. It reflects FAO's best estimates, based on the most accurate and up-to-date information available at the date of printing.

Presentation of the survey

In 1993, FAO started to develop a programme named AQUASTAT, which is an information system on water use for agriculture and rural development. Its aim is to collect information at country and sub-country level and to make it available, in a standard format, to users interested in global, regional and national perspectives. In 1994-95, the 53 African countries were surveyed, which resulted in FAO Water Report 7 *'Irrigation in Africa in Figures'*. The present survey concentrates on the country members of FAO's Regional Office for the Near East[1].

The purpose of the survey is twofold:

- to provide a clear picture of the situation of rural water resources management on a country basis, with emphasis on irrigation, and featuring major characteristics, trends, constraints and perspectives;

- to help support continental and regional analyses by providing systematic, up-to-date and reliable information on water for agriculture and rural development, and to serve as a tool for large-scale planning and forecasting..

In order to obtain information that was as reliable as possible, the survey was developed and carried out for each country as follows:

1. Review of literature and existing information on the country and the sub-region.
2. Data collection by means of a detailed questionnaire.
3. Data processing and critical analysis of the information, with the assistance of the AQUASTAT data processing software and selection of the most reliable information.
4. Preparation of a country profile.

Lastly, regional and sub-regional tables and maps and a general summary were prepared. Cross-checking of the information took place wherever possible.

Country profiles

Each country profile describes the situation regarding water resources and use in the country, with special attention to the water resources, irrigation and drainage sub-sectors. Its aim is to emphasize the particularities of each country, as well as the problems encountered in rural water management and irrigation. It also summarizes the trends of irrigation in the countries, as described in the available literature. It was a deliberate choice to try and standardize the country profiles as much as possible. All profiles follow the same pattern, organized in six sections:

[1] There were 29 country members of FAO's Regional Office for the Near East as at May 1996: Afghanistan, Algeria, Bahrain, Cyprus, Djibouti, Egypt, Iran, Iraq, Jordan, Kuwait, Kyrgyzstan, Lebanon, Libya, Malta, Mauritania, Morocco, Oman, Pakistan, Qatar, Saudi Arabia, Somalia, Sudan, Syria, Tajikistan, Tunisia, Turkey, Turkmenistan, United Arab Emirates, Yemen.

- geography and population;
- climate and water resources;
- irrigation and drainage development;
- institutional environment;
- trends in water resources management;
- main sources of information.

Standardized tables are used for all country profiles. When information is not available, this is indicated by a dash (-). As most of the data are available for a limited number of years only, the most recent reliable data are used in the tables, with an indication of the year to which they refer.

Data collection, processing and reliability

The main sources of information were:

- national water resources and irrigation master plans;
- national yearbooks, statistics and reports;
- reports from FAO or other projects;
- international surveys;
- results from surveys made by national or international research centres.

In total, 65 variables were selected and are presented in the tables attached to each country profile. They are grouped into categories corresponding to the profile sections. A detailed definition of each variable is given below.

In most cases, a critical analysis of the information was necessary to ensure consistency between the different data collected for a given country. When several sources give different or contradictory figures, preference was always given to information collected at national or sub-national level and, unless proved wrong, to official rather than unofficial sources. In the case of shared water resources, a comparison between countries was made to ensure consistency at river-basin level.

Nevertheless, the accuracy and reliability of the information vary greatly between regions, countries and categories of information, as does the year in which the information was gathered. These considerations are discussed in the country profiles.

Definition of the main variables

The main variables presented in the tables in the country profiles are defined as follows and in the order in which they appear in the tables:.

Area of the country: (ha) The total area of the country, including area under inland water bodies. Data in this category are obtained from the United Nations Statistical Office, except for Bahrain where the area has increased due to sea reclamation. The total area of Oman is being revised by the UN. In the meantime, the official government figure has been retained, as agreed with the UN.

Cultivable area: (ha) Area of land potentially fit for cultivation. This term may or may not include part or all of the forests and rangeland. (see the note on cultivable land in the next section).

Cultivated area: (ha) Area under temporary (annual) and permanent crops. This refers to the physical area actually cultivated and does not include land which is temporarily fallow (see the note on arable versus cultivated land in the next section).

Annual crops: (ha) Physical area under temporary (annual) crops.

Permanent crops: (ha) Area cultivated with crops that occupy the land for long periods that do not need to be replanted after each harvest. This does not include woodland and forests.

Total population: (inhabitants) The figures are the UN estimates for 1995.

Rural population: (%). The figures are the UN estimates for 1995.

Water supply coverage: (%) The percentage of urban and rural population with access to safe drinking water (criteria may vary between countries).

Average precipitation: (mm/yr and km^3/yr) Double average over space and time of water falling on the country in a year.

Internal renewable water resources: (km^3/yr) Average annual flow of rivers and recharge of groundwater generated from endogenous precipitation.

Total (actual) renewable water resources: (km^3/yr) The sum of internal renewable water resources and incoming flow originating outside the country, taking into consideration the quantity of flows reserved to upstream and downstream countries through formal or informal agreements or treaties. This gives the maximum theoretical amount of water actually available for the country (see note on the computation of water resources in the next section).

Dependency ratio: (%) That part of the total renewable water resources originating outside the country.

Total dam capacity: (10^6 m^3) The total cumulative capacity of all dams.

Water withdrawal: (10^6 m^3/yr) Gross amount of water extracted from the resources for a given use. It includes conveyance losses, consumptive use and return flow (see note on water withdrawal in the next section).

Agricultural water withdrawal: (10^6 m^3/yr) Annual quantity of water withdrawn for agricultural purposes including irrigation and livestock watering (see note on agricultural water withdrawal in the next section).

Domestic water withdrawal: (10^6 m^3/yr) Annual quantity of water withdrawn for domestic purposes. It is usually computed as the total amount of water withdrawn by public distribution networks, and usually includes the withdrawal by those industries connected to public networks.

Industrial water withdrawal: (10^6 m^3/yr) This sector usually refers to self-supplied industries not connected to any distribution network.

Total water withdrawal: (10^6 m^3/yr) Annual quantity of water withdrawn for agricultural, industrial and domestic purposes. It does not include other withdrawals (see below).

Other water withdrawal: (10^6 m^3/yr) This includes all other sectors: energy, mining, recreation, navigation, fisheries and environment. These sectors usually have a very low consumption rate.

Average groundwater depletion: (10^6 m^3/yr) Annual amount of water withdrawn from aquifers and which is not replenished (mining of rechargeable aquifers or extraction from fossil aquifers).

Produced wastewater: (10^6 m^3/yr) Annual quantity of wastewater produced in the country. This does not include agricultural drainage water.

Treated wastewater: (10^6 m^3/yr) Annual quantity of wastewater which is treated.

Reused treated wastewater: (10^6 m^3/yr) Annual quantity of treated wastewater which is reused.

Desalinated water: (10^6 m^3/yr) Installed capacity of desalination plants.

Irrigation potential: (ha) Area of land suitable for irrigation development, taking into account land and water resources. It includes land already under irrigation. The definition may vary among countries (see note on irrigation potential in the next section).

Full or partial control irrigation: equipped area: (ha) Irrigation schemes carried out and managed either by government, private estates or farmers, and where a full or partial control of the water is achieved. Gardening is included in this category.

Surface irrigation: (ha) Part of the full or partial control area under surface irrigation: furrow, border, basin, and flooded irrigation of rice.

Sprinkler irrigation: (ha) Part of the full or partial control area irrigated by aspersion (sprinkler).

Micro-irrigation: (ha) Part of the full or partial control area irrigated by micro-irrigation.

Percentage of area irrigated from groundwater: (%) Part of the full or partial control area irrigated from wells (shallow wells and deep tubewells) or springs.

Percentage of area irrigated from surface water: (%) Part of the full or partial control area irrigated from rivers or lakes (reservoirs, pumping or diversion).

Percentage of area irrigated from non-conventional sources: (%) Part of the full or partial control area irrigated from non-conventional sources of water such as treated wastewater or desalinated water.

Percentage of equipped area actually irrigated: (%) Part of the full or partial control area which is actually irrigated. Often, the whole area equipped is not irrigated for various reasons, such as lack of water, absence of farmers, damage, organizational problems, and so forth. It concerns the actual physical areas. Irrigated land that is cultivated twice a year is counted once.

Spate irrigation area: (ha) Area of land equipped for spate irrigation.

Equipped wetlands and inland valley bottoms: (ha) Part of cultivated wetlands and inland valley bottoms which have been equipped with water control structures (intake, canals, etc.). Developed mangroves are included in this category.

Total irrigation: Area equipped to provide water to the crops. It includes areas equipped for full and partial control irrigation, spate irrigation areas, and equipped wetland and inland valley bottoms. It does not include flood recession cropping areas. In the text it is also referred to as *irrigated area* or *area under irrigation*.

Flood recession cropping area: (ha) Area along rivers where cultivation occurs in the areas exposed as floods recede. The special case of floating rice is included in this category.

Total water managed area: (ha) The sum of full or partial control irrigation equipped areas, spate irrigation areas, equipped wetland and inland valley bottoms and flood recession cropping areas. It does not include water harvesting areas. Other cultivated wetland and inland valley bottoms (i.e. wetland and inland valley bottoms which have not been equipped with water control structures but are used for cropping) are included in this definition, but are not reported present in the Near East Region.

Power irrigated area as percentage of water managed area: (%) Part of the water managed area where pumps are used for water supply. It does not include areas where water is pumped with human- or animal-driven water lifting devices.

Full or partial control schemes: (ha) Areas of large, medium and small schemes, following the criteria used by the country, with the criteria given.

Total number of households in irrigation: Total number of households living directly on earnings from full or partial controlled irrigation schemes.

Total irrigated grain production: (tons) The total quantity of cereals harvested annually in the water managed area.

Harvested crops under irrigation: (ha) Total harvested irrigated area for the crop for the given year. Areas under double cropping should be counted twice. For permanent crops data generally refer to the total planted area.

Drained area: (ha) The area equipped with subsurface or open drains. It can be broken down into *drained areas in full or partial control irrigated areas; drained areas in equipped wetland and inland valley bottoms;* and *other drained areas*, equipped for drainage but not equipped for irrigation. Flood recession cropping areas are not considered as being drained.

Flood-protected area: (ha) Area of land equipped with flood control structures.

Area salinized by irrigation: (ha) Total irrigated area affected by salinization as a result of irrigation. This does not include naturally saline areas.

Other definitions and conventions

Notwithstanding the detailed description of each variable, some problems persist due to the fact that it is not always clearly indicated in the available literature which definition has been used in computing the figures. The most frequent problems encountered in computing the figures are listed below. In addition, a number of terms used in the country profiles but not in the tables are also explained below.

Arable versus cultivated land: The official definition of arable land is 'land under temporary crops, temporary meadows for mowing or pasture, land under market and kitchen gardens, and land temporarily fallow'. In this study, 'land cultivated with annual crops' has been preferred to 'arable land' and refers to the physical area actually cultivated with temporary crops, excluding the temporarily fallow land. However, the way multiple cropping and inter-cropping are accounted for is not always clearly explained in the literature.

Cultivable land: The notion of cultivable land in arid areas is closely linked to the capacity to provide water to the crops. Assumptions made in assessing cultivable land vary from country to country. In this survey, national figures have been used, whenever available, despite possible large discrepancies in computation methods.

Irrigation potential: Assumptions made in assessing irrigation potential vary from country to country. In most cases it is computed on the basis of available land and water resources, but economic and environmental considerations may also have been taken into account. Some countries include the possible use of non-conventional sources of water for irrigation. Except in a few cases, no consideration is given to the possible double counting of shared water resources. Wetland and flood plains are usually, but not systematically included in irrigation potential.

Water withdrawal: This term includes all the water used for agricultural, domestic and industrial purposes. The use of desalinated and treated wastewater is thus included, although in the strict sense this should be called 'water use' rather than 'water withdrawal'. Desalinated water and treated wastewater are also referred to as *non-conventional sources of water*.

Agricultural water withdrawal: Methods for computing agricultural water withdrawal vary from country to country. The figure was reviewed for each country on the basis of crop water requirements and irrigated areas, and comments were added in the country profiles to explain the figure when necessary.

Livestock water withdrawal: By default, livestock water withdrawal is accounted for in agricultural water withdrawal. However, some countries include it in domestic water withdrawal.

Computation of water resources: The following terms have been used in the computation of water resources: *Internal renewable water resources* is the average annual flow of rivers and recharge of groundwater generated from endogenous precipitation. A critical review of the data was made to ensure that double counting of surface water and groundwater was avoided. *Total natural renewable water resources* is the sum of internal renewable water resources and natural incoming flow originating outside the country. Special rules are used to take into account the flow of border rivers. *Natural incoming flow* is the average annual amount of water which would flow into the country in natural conditions, i.e. without human influence. This figure is not time dependent. *Total actual renewable water resources* is the sum of internal renewable water resources and incoming flow originating outside the country, taking into account the quantity of flow reserved to upstream and downstream countries through formal or informal agreements or treaties and reduction of flow due to upstream withdrawal. It corresponds to the maximum theoretical amount of water actually available for a country at a given moment. The figure may vary with time. *Manageable water resources*, or *development potential*, refer to that part of the resources which is considered to be available for development under specific economic conditions. This figure considers factors such as the dependability of the flow, floods, extractable groundwater, minimum flow required for non-consumptive uses etc.

Safe yield of groundwater: The definition varies between countries and is explained in the country profiles. It may or may not include fossil water. For example, in Malta, the safe yield of groundwater is considered less than the renewable groundwater potential in order not to deplete the storage capacity of the main aquifer without causing salt water intrusion. In Qatar, it includes fossil water from the aquifers underlying a large part of the Arabian Peninsula.

Kareze or Qanat: Unlined tunnel in the hillside, bringing water by free flow from underground aquifers to the surface.

Shat or Chott: Salty depression.

Sabkhas: Evaporative areas.

General summary

The Near East region (grouping the countries represented in the FAO Regional Office for the Near East) extends from the Atlantic Ocean (Mauritania and Morocco) in the west to Pakistan and Kyrgyzstan in the east and from Turkey and Kyrgyzstan in the north to Somalia in the south. It falls between longitudes 17° west and 80° east and latitudes 43° north and 2° south. In May 1996, it comprised 29 countries with a total area of 18.5 million km^2 (Table 11), which is about 14% of the total area of the world (including the interior lakes).

For the purpose of this study, the 29 countries have been grouped in five sub-regions based primarily on geographic conditions and, as far as possible, on hydro-climatic homogeneity. These sub-regions, presented in Figure 1, are here referred to as Maghreb, North-eastern Africa, Arabian Peninsula, Middle East and Central Asia[1].

Climate, geography and population

Because of the aridity prevailing in the region, the Near East is the poorest region in the world in terms of water resources, globally and per inhabitant, even when considering the contribution of rivers flowing from the bordering and more humid regions of tropical Africa (the Nile) or Himalayan Asia (the Indus).

However, the water resources distribution within this vast area, extending over three continents, is far from being uniform: land relief, location with respect to the sea, latitude and resulting hydro-climatic conditions, diversity in hydrographic and geological structures, and matching or mismatching of the river basins with the national territories, all give rise to extremely different water situations.

Many countries in the region are characterized by long coastal boundaries. The coasts are located on the North Atlantic Ocean, the Mediterranean Sea, the Red Sea, the Gulf of Aden, the Persian Gulf, the Gulf of Oman and the Arabian Sea. In addition, several important interior seas can be found in the region: the Black Sea to the north of Turkey, the Caspian Sea to the north of Iran and to the west of Turkmenistan and the Dead Sea to the west of Jordan. The Aral Sea is not bordering the Near East region, but the northern part of Central Asia is located within the Aral Sea basin. Only three countries, Afghanistan, Kyrgyzstan and Tajikistan, are landlocked.

Several international rivers cross the Near East Region (Figure 2). The most important rivers are the Nile in North-eastern Africa, which originates outside the region in the Equatorial Lake

[1] Distribution of the 29 Near East countries over the five sub-regions:
Maghreb: Algeria, Libya, Mauritania, Morocco, Tunisia
North-eastern Africa: Djibouti, Egypt, Somalia, Sudan
Arabian Peninsula: Bahrain, Kuwait, Oman, Qatar, Saudi Arabia, United Arab Emirates, Yemen
Middle East: Cyprus, Iraq, Jordan, Lebanon, Malta, Syria, Turkey
Central Asia: Afghanistan, Iran, Kyrgyzstan, Pakistan, Tajikistan, Turkmenistan

Plateau and the highlands of Ethiopia, the Euphrates and the Tigris in the Middle East, the Amu Darya, the Syr Darya and Indus in Central Asia (the latter originating outside the region in the Himalayas). Smaller rivers, like the Jordan and the Orontes in the Middle East, also play a fundamental role in international relations regarding water resources.

The area extending between the Atlantic Ocean and the Persian Gulf is covered with vast deserts. The Great Desert (the Sahara), extending between the Atlantic Ocean and the Red Sea, comprises large parts of the Maghreb and North-eastern Africa. The area extending between the Red Sea and the Persian Gulf includes the Rub Al Khali (Empty Quarter) desert in the south of the Arabian Peninsula and the Badiat-El-Sham desert in the north of the Arabian Peninsula and the south of the Middle East. In Central Asia, a large part of Iran is covered with desert, as is the case for the southern part of Afghanistan and the south-western part of Pakistan. The Kara Kum desert comprises 80% of the total area of Turkmenistan.

Many mountain ranges can be found in the Near East Region. In the centre of the Great Desert, several mountains appear, with the highest peak at an elevation of 3 000 metres at Tebetsy Mountain in Libya. In addition to these internal mountains, there are mountain ranges running parallel to the coasts of the Mediterranean and Red Seas. The Atlas mountain range occupies the north-western corner of the Maghreb, with its peak in Morocco at an elevation of 4 165 metres. In Lebanon, the Lebanon mountain chain has its highest crest at just over 3 000 metres. In Yemen, there is a mountain peak at an elevation of 3 268 metres; in south-eastern Turkey, at an elevation of 4 135 metres. The peak of the Zagros mountain chain in Iran reaches 4 432 metres. The highest peak of the Hindu Kush range on the border between Afghanistan and Pakistan reaches 7 690 metres (Tirichmir). In the north-eastern part of Central Asia, the Himalayan mountain ranges have peaks reaching almost 8 000 metres.

The total population of the region was about 561 million in 1995, of which almost 52% rural. Population increased by 2.7% between 1994 and 1995, ranging from 2.2% in the Maghreb and North-eastern Africa to 3.1% in Central Asia. During the last 10 years (1985-1995), average annual population growth was 3.0% in the Near East Region as against 1.7% in the world. Over 43% of the population of the region is concentrated in Central Asia, which occupies 21% of the total area of the Near East; 25% of the total population lives in Pakistan alone, which occupies only 4% of the total area. Population density is lowest in the Maghreb, with 12 inhabitants per km^2, and highest in the Middle East, with 71 inhabitants per km^2. The least densely populated country is Mauritania, with 2 inhabitants per km^2. Malta and Bahrain are the most densely populated countries, with 1 158 and 811 inhabitants per km^2 respectively, followed far behind by Lebanon and Pakistan, with 288 and 177 inhabitants per km^2 respectively. All the remaining countries have less than 100 inhabitants per km^2 (Table 12).

Water resources

In this study, a distinction has been made between renewable and non-renewable water resources. The following indicators have been compiled or estimated for each country. *Internal renewable water resources (IRWR)* is that part of the water resources generated from endogenous precipitation. It is computed by adding up surface runoff and groundwater recharge occurring inside the countries' borders. Special care is taken to avoid double counting of their common part. *Total renewable water resources* refers to the sum of IRWR and incoming flow originating outside the countries' borders. A distinction is made between *natural* flow (NRWR), computed by assessing the long-term yearly average of flow without any human-induced abstraction, and *actual* flow (ARWR), which is the

TABLE 1
Sub-regional distribution of the renewable water resources

Region	Area	Population '95	Annual precipitation		Annual internal renewable water resources			
	thousand km^2	thousand inhabitants	mm	km^3	mm	km^3	as % of precipitation	m^3 per inhabitant
Maghreb	5 777	71 544	86	495	8	48	9.8	677
North-eastern Africa	4 168	100 856	306	1 275	10	43	3.4	427
Arabian Peninsula	3 103	39 110	79	246	2	8	3.1	197
Middle East	1 512	106 635	421	637	162	245	38.4	2 294
Central Asia	3 926	243 316	304	1 195	138	541	45.3	2 226
Total Near East	18 486	561 461	208	3 848	48	885	23.0	1 577
World	134 223	5 716 407	820	110 000	298	40 000	36.4	7 000
N. East as % of world	13.8	9.8		3.5		2.2		22.5

Note: Rainfall figures have been estimated for Afghanistan, Tajikistan and Turkmenistan

maximum theoretical amount of water actually available for a country. Actual flow takes into account abstraction in upstream countries and the volumes allocated through formal or informal agreements or treaties between countries. The IRWR figures are the only water resources figures that can be added up for regional assessment and they have been used for this purpose.

While the Near East covers 14% of the total area of the world and contains 10% of its population, its water resources are only about 2% of the total renewable water resources of the world. Further to this, large differences exist between the five sub-regions, as is shown in Table 1. The Maghreb, North-eastern Africa and the Arabian Peninsula have very limited water resources, with less than 10 mm annually on average and suffer severe water scarcity, with values per inhabitant varying between 200 and 700 m^3/year. In contrast, the Middle East and Central Asia show much higher values, mostly thanks to the abundant flows generated in the mountainous areas of Turkey and in the Himalayas.

The internal renewable water resources per inhabitant in the Near East are among the lowest in the world. The average for the region is 1 577 m^3/inhabitant per year, as against over 7 000 m^3/year per inhabitant for the whole world. It ranges from near 0 for Kuwait, which has practically no internal renewable water resources, to about 10 000 m^3/inhabitant per year for Tajikistan and Kyrgyzstan (Table 13). For 16 out of the 29 countries the internal renewable water resources per inhabitant are below 500 m^3/year and for 11 of them even the total actual renewable water resources are below 500 m^3/year (Figures 3 and 4 and Table 2).

For only four countries (Turkey, Kyrgyzstan, Tajikistan and Afghanistan) the internal renewable water resources per inhabitant are above 2 000 m^3/year and three of them act as 'water towers' for the region, with large amounts of water flowing to downstream countries. They are: Turkey (the Euphrates and the Tigris rivers mainly), Kyrgyzstan and Tajikistan (the Amu Darya and the Syr Darya). Two countries, Syria and Sudan, are intermediate countries in that they depend to a large extent, around 80%, on upstream countries for their renewable water resources (mainly the Euphrates from Turkey and the Nile from Ethiopia), but on the other hand they are located upstream from other countries depending on the same rivers (Iraq and Egypt respectively). Five countries depend for over 90% on other countries for their renewable water resources: Turkmenistan, Egypt and Mauritania for surface water and Kuwait and Bahrain for groundwater. To a lesser extent, but still over 50% dependent on other countries are Somalia and Iraq (Figure 5 and Table 3).

Non-renewable groundwater

Several countries, that have few renewable water resources, overlie important non-renewable (fossil) groundwater basins, partly shared with neighbouring countries. In several countries (Saudi Arabia, Libya or the United Arab Emirates) by far the largest part of the total water withdrawn is fossil water, as seen in the next section (Table 6). However, although groundwater reservoirs may allow storage of huge quantities of water accumulated during the pluvial periods of Quaternary, its development cannot be considered sustainable in the long term, as the lack of present recharge would result in the slow depletion of the aquifers. Moreover, the water level decline and the resulting increase of the cost of pumping, as well as the deterioration of the water quality in some areas may also make the abstraction of fossil water less attractive with time.

TABLE 2
Countries with IRWR below 500 m^3/inhab. per year (1995)

Country	Internal renewable water resources per inhabitant m^3/year	Actual renewable water resources per inhabitant m^3/year
Countries with IRWR < 500 and ARWR < 500:		
Kuwait	0	13
Bahrain	7	206
Malta	42	42
United Arab Emirates	79	79
Qatar	93	96
Libya	111	111
Jordan	124	161
Saudi Arabia	134	134
Yemen	283	283
Tunisia	396	463
Oman	455	455
Countries with IRWR < 500 and ARWR > 500:		
Egypt	29	926
Mauritania	176	5 013
Turkmenistan	244	17 321
Syria	477	1 791
Algeria	498	512

TABLE 3
Countries with a renewable water resources dependency ratio above 50%

Country	Internal Renewable Water Resources million m^3/year	Actual Renewable Water Resources million m^3/year	Dependency ratio * %	Main source of incoming water
Kuwait	0	20	100.0	Groundwater from Saudi Arabia
Turkmenistan	1 000	71 000	98.6	Amu Darya river
Egypt	1 800	58 300	96.9	Nile river
Bahrain	4	116	96.6	Groundwater from Saudi Arabia
Mauritania	400	11 400	96.5	Senegal river
Syria	7 000	26 260	80.3	Euphrates, Tigris rivers
Sudan	35 000	88 500	77.3	Nile river
Somalia	6 000	15 740	61.9	Shebelli, Juba rivers
Iraq	35 200	75 420	53.3	Euphrates, Tigris rivers

* *The dependency ratio is equal to the part of the renewable water resources which originates outside the country*

Water withdrawals

Table 4 shows the distribution of water withdrawal by region between the three major sectors of water use: agriculture (irrigation and livestock), communities (domestic water supply) and industries. Water requirements for navigation, fisheries, mining, environment and recreation, although they may represent a significant part of the water resources, have a very low net consumption rate and are computed in very different ways by the different countries. For these reasons, they are not included in the computation of regional water withdrawal.

TABLE 4
Sub-regional distribution of water withdrawal

Region	Water withdrawal by sector							% by region	m^3/year per inhabitant (1993)
	agricultural		domestic		industrial		total		
	km^3 per year	% of total	km^3 per year	% of total	km^3 per year	% of total	km^3 per year		
Maghreb	21.1	85	2.5	10	1.2	5	24.8	4.8	363
North-eastern Africa	65.0	88	3.9	5	4.8	7	73.7	14.4	764
Arabian Peninsula	21.2	87	2.6	11	0.5	2	24.3	4.8	659
Middle East	77.7	85	7.7	8	6.0	7	91.4	17.8	907
Central Asia	282.9	95	8.3	3	7.0	2	298.2	58.2	1 302
Total Near East	467.9	91	25.0	5	19.5	4	512.4	100.0	964
World 1990	2 235.6	69	259.2	8	745.2	23	3 240.0		660
N. East as % of world	20.9		9.6		2.6		15.8		146

In the Near East, 91% of the water withdrawal is directed towards agriculture. Central Asia has the highest level of water withdrawal for agriculture (95%). This is the sub-region where the largest part of the cultivated area (80%) is irrigated, as compared to 16% in the Maghreb (Table 8). Afghanistan is the country with the largest percentage of water withdrawal directed to agriculture (99%), Malta, with 12%, is the country with the lowest percentage (Table 14). In Malta by far the largest part of the total water withdrawal, over 87%, is directed to the communities, due to the extremely high population density (1 158 inhabitants/km^2) and the extensive development of tourism in the country, among other reasons. In absolute terms, Central Asia represents over 58% of the total water withdrawal for the Near East, Pakistan alone withdrawing 30%. The water withdrawal per inhabitant varies from 20 m^3/year in Djibouti (1985) to over 6 000 m^3/year in Turkmenistan (1989). At sub-regional level, the water withdrawal per inhabitant in Central Asia (1 300 m^3/year) is on average 3.6 times the water withdrawal per inhabitant in the Maghreb (360 m^3/year; Table 4).

Use of non-conventional sources of water

Water withdrawal, expressed as a percentage of *internal renewable water resources* is an indicator of the sub-region's or country's capacity to rely on its own, renewable, sources of water. Values above 100% indicate that either renewable water flowing into the sub-region or country from outside, or fossil, or non-conventional sources of water are used in addition to the internal renewable water resources (Figure 6 and Table 14). In North-eastern Africa, where the percentage is 171% (Table 5), water transfer plays a crucial role (the Nile flows into the region from upstream countries outside the region), while in the Arabian Peninsula, with a percentage

TABLE 5
Sub-regional water withdrawal as % of IRWR

Region	Water withdrawal	
	total km^3/year	as % of IRWR
Maghreb	24.8	51
North-eastern Africa	73.7	171
Arabian Peninsula	24.3	317
Middle East	91.4	37
Central Asia	298.2	55
Total	512.4	58

of 317%, the deficit is mainly made good by the use of fossil and non-conventional sources of water.

Water withdrawal, expressed as a percentage of *actual renewable water resources* is a good indicator of the pressure on the renewable water resources. Roughly, it can be considered that pressure on water resources is high when this value is above 25%. In the Near East, only five countries are below the 25% limit (Figure 7 and Table 14). Values above 100% indicate that the country relies at least partly on non-conventional water sources (desalinated water, treated wastewater) or is mining its groundwater resources. In 14 countries, annual water withdrawal is greater than the *internal* renewable water resources (Figure 6), but five countries benefit from rivers flowing in from upstream countries, resulting in an annual water withdrawal that is lower than the *actual* renewable water resources (Table 6). This is the case for Egypt (the Nile river), Turkmenistan (the Amu Darya), Mauritania (the Senegal river), Syria and Iraq (the Euphrates and Tigris rivers). The remaining nine countries are those of the Arabian Peninsula (except Yemen), Libya, Jordan and Malta. In these countries, water withdrawal is greater than the total actual renewable water resources and they have to rely on non-conventional water sources and on fossil water to satisfy water demand (Table 15).

TABLE 6

Countries with total water withdrawal greater than the internal renewable water resources and contribution of other sources of water to total water withdrawal

Country	Water withdrawal		Use of non-conventional water and groundwater depletion as % of total water withdrawal		
	in % of IRWR	in % of ARWR	non-conventional	groundwater depletion	total
Countries with water withdrawal above IRWR, but below ARWR:					
Egypt	3 061	95	0.4	2.5	2.9
Turkmenistan	2 280	32	-	-	-
Mauritania	408	14	0.1	-	-
Syria	206	55	2.6	12.6	15.2
Iraq	122	57	-	-	-
Countries with water withdrawal above IRWR and above ARWR:					
Kuwait	IRWR negligible	2 690	52.6	46.5 *	91.1
Bahrain	5 981	206	21.8	40.2	62.0
United Arab Emirates	1 405	1405	23.4	70.9	94.3
Libya	767	767	3.7	90.0 *	93.7
Saudi Arabia	709	709	5.5	79.7	85.2
Qatar	559	538	43.5	14.9	58.4
Malta	359	359	59.2	32.2	91.4
Jordan	145	112	5.3	17.5	22.8
Oman	124	124	4.9	-	-

* *Figures have been estimated*

It can be seen in Table 14 that Pakistan, Iran, Sudan and Afghanistan have high rates of use of their internal renewable water resources (between 48 and 63%), but benefit from important internal resources as well as from incoming water. Tunisia and Yemen also use most of their internal renewable water resources (87 and 72% respectively), but they cannot benefit from important internal resources or incoming water. It is likely that these two countries will have to rely increasingly on alternative sources of water in the future.

TABLE 7

Countries using the largest quantities of desalinated water and treated wastewater

Country	Use of non-conventional sources of water								
	desalinated water 10^6 m^3/yr	as % of total withdr.	as % of withdrawal of country	treated wastewater 10^6 m^3/yr	as % of total withdr.	as % of withdrawal of country	total non-conventional 10^6 m^3/yr	as % of total withdr.	as % of withdrawal of country
Saudi Arabia	714	41.3	4.20	217	18.1	1.28	931	31.8	5.48
United Arab Emirates	385	22.3	18.26	108	9.0	5.12	493	16.8	23.38
Kuwait	231	13.4	42.94	52	4.3	9.67	283	9.7	52.61
Syria	-	-	-	370	30.8	2.57	370	12.6	2.57
Egypt	25	1.5	0.05	200	16.7	0.36	225	7.7	0.41
Other 24 countries	372	21.5	0.09	253	21.1	0.06	625	21.4	0.15
Total Near East	1 727	100.0	0.34	1 200	100.0	0.23	2 927	100.0	0.57

The total use of desalinated water in the Near East is estimated at 1 727 million m^3/year. In absolute terms, three countries, Saudi Arabia, the United Arab Emirates and Kuwait, are by far the largest users of desalinated water with 77% of the total for the region, with Saudi Arabia alone accounting for 41% (Table 7).

The total quantity of reused treated wastewater in the Near East is estimated at 1 200 million m^3/year. Syria, Saudi Arabia and Egypt are the largest users of treated wastewater in absolute terms, accounting for almost 66% of all the wastewater reused in the region, with Syria alone accounting for almost 31%.

Considering the use of both desalinated water and treated wastewater, the above five countries account for almost 80% of the total for the Near East (Table 7).

The Arabian Peninsula is the sub-region using the largest quantity of desalinated water and treated wastewater: 1 953 million m^3/year or almost 67% of the total in the Near East Region. It is also the sub-region where the contribution of non-conventional sources of water to total water withdrawal is greatest (8%). Central Asia uses the smallest quantity of desalinated water and treated wastewater: 12 million m^3/year or only 0.4% of the total in the Near East Region. The contribution of non-conventional sources of water to water withdrawal is also lowest in this sub-region (0.004%). In two countries (Malta and Kuwait), the contribution of non-conventional sources of water to total water withdrawal is over 50%. In three countries (Qatar, the United Arab Emirates and Bahrain) it is between 20 and 50% (Figure 8 and Table 15). For the remaining countries it is less than 10%.

Irrigation potential

Methods used by countries to estimate their irrigation potential vary, with significant influence on the results. In computing water available for irrigation, some countries only consider renewable water resources, while others, especially arid countries, include the availability of fossil or non-conventional water. For this reason, comparison between countries should be made with caution. In the case of international rivers, calculation by the individual countries of their irrigation potential in the same river basin may lead to double counting of part of the shared water resources. It is thus not possible to systematically add up country figures to obtain regional estimates of irrigation potential.

Already at present, as shown above, most countries of the Near East have to rely on fossil and non-conventional water. This means that, for those countries, any extension of existing irrigation would require more fossil or non-conventional water if no improvement in water use efficiency is made.

By far the largest irrigation potential is concentrated in Pakistan and Iran and is based only on renewable water resources (Table 11). One country, Cyprus, estimates that its irrigation potential is lower than the area equipped for irrigation at present, even including the future availability of non-conventional water (Table 16). The reason for this is the increasing demand for water for domestic and industrial purposes and the groundwater depletion already taking place.

Arid countries, where no agriculture is possible without irrigation, tend to consider the cultivable area as the irrigation potential area, for the development of which they would certainly have to rely on the use of fossil groundwater and non-conventional sources of water. In the United Arab Emirates, where the irrigated area is reported to be equal to the irrigation potential (which is in turn equal to the cultivable area), the land is usually developed by transporting suitable soil to areas where water (renewable, fossil or non-conventional) is available. For this reason the terms 'cultivable land' and 'irrigation potential' are somewhat relative and subject to changes over time.

Water managed areas and irrigation

Different types of water management have been distinguished. The areas on which water, other than direct rainfall, is used for the purpose of agricultural production have been called *water managed areas* in the text. The term *irrigation* refers to that part of the water managed areas equipped to provide water to the crops and includes areas equipped for full and partial control irrigation, spate irrigation areas and equipped wetland or inland valley bottoms (Table 8).

TABLE 8
Sub-regional distribution of water management methods

Region	Irrigation						Flood recession cropping	Water managed area
	full or partial control	spate irrigation	equipped wetl./ivb*	total irrigation	as % of total	as % of cultivated		
	ha	ha	ha	ha			ha	ha
Maghreb	2 412 900	305 000	0	2 717 900	6	16	64 000	2 781 900
N-eastern Africa	5 196 674	196 200	0	5 392 874	11	46	0	5 392 874
Arabian Peninsula	2 139 887	98 320	0	2 238 207	5	80	0	2 238 207
Middle East	8 801 127	393	115 164	8 916 684	19	30	0	8 916 684
Central Asia	27 067 534	1 402 448	0	28 469 982	59	75	1 240 552	29 710 534
Total Near East	45 618 122	2 002 361	115 164	47 735 647	100	48	1 304 552	49 040 199
	93%	4%	-	97%			3%	100%
World				246 408 529				
NE as % of world				19.4%				

* *ivb = inland valley bottom*

Irrigation covers 47.7 million hectares in the Near East. Central Asia represents 59% of this total, although it covers only 21% of the total area of the region. Pakistan alone, covering a little over 4% of the region, accounts for 33% of the irrigated areas. By adding Iran, Turkey, Iraq and Egypt, 72% of the areas under irrigation are controlled by five of the 29 countries, covering only 25% of the Near East (Table 16).

The part of the irrigated areas in national agricultural land varies considerably between the sub-regions and between the countries. In the Arabian Peninsula as a whole, 80% of the cultivated area is under irrigation. In all of the seven countries within this sub-region, except Yemen, the whole cultivated area is under irrigation. In Central Asia, 75% of the cultivated area is equipped for irrigation, playing a crucial role in the production of cereals (especially wheat) and cotton. The part of the cultivated areas under irrigation is less important for the other regions as a whole, but is crucial for some countries within the regions, like Egypt and Djibouti, where the whole cultivated area is under irrigation, and Iraq, where almost 95% of the cultivated area is under irrigation (Figure 9 and Table 16). On the contrary, less than 20% of the cultivated area is under irrigation in Morocco, Algeria, Tunisia and Malta.

Full or partial control irrigation is by far the most widespread type of water management, covering 93% of the area (Table 8). In relative terms, spate irrigation is most important in the Maghreb (11% of the water managed area). In absolute terms, spate irrigation occupies by far the largest area in Pakistan, accounting for 70% of the spate irrigation in the whole Near East and 8% of the water managed area in Pakistan. Equipped wetland and inland valley bottoms (ivb) are reported only in Turkey and flood recession cropping is practised in Pakistan (with over 94% of the flood recession cropping area of the whole region), in Mauritania (5%) and Iran (< 1%), as shown in Table 16.

In four of the 22 countries, for which information was available, (Cyprus, the United Arab Emirates, Somalia and Syria) over 80% of the irrigation potential is at present already equipped for irrigation (Figure 10 and Table 16).

Irrigation techniques

Detailed information on the irrigation techniques used in full or partial control irrigation schemes was available for only 18 out of the 29 countries and the figures below refer to these countries (Table 17). Surface irrigation is by far the most widely used technique, practised on 87.6% of the total area. Sprinkler irrigation is practised on 11.0% and micro-irrigation on 1.4% of the total area. In Libya and Saudi Arabia, sprinkler irrigation is by far the most predominant, while in Cyprus, Malta, Jordan and the United Arab Emirates, micro-irrigation is the most widely used technique, being practised on over half of their full and partial control irrigation areas. In Kuwait and Lebanon sprinkler irrigation and micro-irrigation techniques together are practised on more than 37% and 39% of their full and partial control irrigation area respectively. In particular the arid countries, without large rivers, choose to develop more intensively the micro-irrigation and sprinkler irrigation techniques to save water.

Origin of irrigation water

There are five possible sources of irrigation water: surface water, renewable groundwater, fossil water, treated wastewater and desalinated water. Spate irrigation areas, equipped wetland and flood recession cropping areas are all irrigated by surface water. Table 9 shows the origin of irrigation water for the full or partial control irrigation equipped areas.

It was not possible to make a distinction between renewable and non-renewable groundwater use, but a large part of the groundwater used is fossil water, especially in the more arid sub-regions and countries. For the Near East as a whole the origin of irrigation water consists mainly of surface

TABLE 9

Origin of irrigation water by sub-region

Region	Full or partial control irrigation: area equipped in hectares						
	surface water	% of total	groundwater (renewable and fossil)	% of total	non-conventional water	% of total	total area
Maghreb	1 037 649	43.0	1 355 251	56.2	20 000	0.8	2 412 900
North-eastern Africa	4 969 800	95.6	222 674	4.3	4 200	0.1	5 196 674
Arabian Peninsula	51 000	2.4	2 066 874	96.6	22 013	1.0	2 139 887
Middle East	7 198 091	81.8	1 598 856	18.2	4 180	(-)	8 801 127
Central Asia	17 865 132	66.0	9 202 402	34.0	0	0.0	27 067 534
Total	31 121 672	68.2	14 446 057	31.7	50 393	0.1	45 618 122

Note: Figures for Algeria, Libya, Somalia and Turkmenistan have been estimated on the basis of sub-regional trends

water, but there are significant differences between the sub-regions. The large contribution of surface water in North-eastern Africa and the Middle East reflects the fact that these regions' hydrology is dominated by the presence of large rivers: the Nile, the Euphrates and the Tigris. The Indus, the Amu Darya and Syr Darya are also important for irrigation in Pakistan, Kyrgyzstan, Tajikistan and Turkmenistan. In Iran and, to a lesser extent, Pakistan groundwater also plays an important role. The contribution of groundwater is most important in the Arabian Peninsula, where there are no big rivers. Non-conventional water for irrigation is used in 10 countries (Table 18) and plays a fairly important role in the drier regions of the Maghreb, the Arabian Peninsula and on the island of Malta. It consists mostly of treated wastewater (in Morocco untreated wastewater is also used in very specific conditions). In Kuwait, 1 180 ha are irrigated by treated wastewater and about 680 ha in greenhouses are irrigated by desalinated water. In general, desalinated water is still too saline for irrigation purposes. In Kuwait, 39% of the irrigated area is irrigated by non-conventional water, in Malta 37% and in Bahrain 14%. In all the other countries less than 6% is irrigated in this way (Table 18).

Indicators of irrigation intensity

Two indicators are frequently used to assess irrigation intensity: *the rate of use of land equipped for irrigation*, which is that part of the equipped area actually used for production at least once a year, and *the cropping intensity*, which is the ratio between irrigated crops areas (where double or triple cropping areas are counted twice or three times respectively) and the physical areas equipped for irrigation.

The survey showed that in general these figures were either not available or not reliable at country level. Moreover, in most cases in the literature it was not possible to distinguish between these two indicators. It is possible, for example, that part of the equipped area was not used for several reasons (abandoned, water shortage, etc.), while the remaining part was cultivated in double or triple cropping. Moreover, figures vary significantly from one year to another, particularly in the areas where irrigation schemes are facing water availability problems.

Table 17 shows the part of the full and partial control irrigation equipped area which is actually irrigated for the 19 countries for which the information was available. In all countries the rate of use of the equipped area was greater than 50%: in three countries it was less than 60%, in six countries it was reported to be 100%. Overall, the equipped area of these 19 countries represents 23.8 million hectares (52% of the equipped area of the whole region). Of these 23.8 million hectares about 3.8 million hectares, or 16%, are not irrigated.

It was also difficult to get reliable information on cropping intensity for the reasons explained above. Figures available for some countries show a cropping intensity of 1.66 for Egypt, 1.19 for Syria, 1.15 for Oman (all referring to 1993) and 1.07 for Jordan (in 1991). In Saudi Arabia, Bahrain and Kuwait the cropping intensity is reported to be 1, probably because no cropping is possible in the hot season. In Malta, 95% of the equipped area was irrigated in 1990, with a cropping intensity on that area of 2.50. In Cyprus it was 0.85 in 1994 and in Qatar only 0.66 in 1993 because of water shortages.

Irrigated crops

The figures on irrigated crops are very incomplete and do not allow the establishment of statistical tables by country showing the distribution of the major crops under irrigation in the Near East. However, by using all the data available, information can be obtained for about 35.8 million hectares. The results are summarized in Table 10, in which the different crops have been grouped into seven major categories.

TABLE 10
Sub-regional distribution of the main irrigated crops (based on partial information)

Region (unit: '000 ha)	wheat	other cereals *	fodder crops	vege-tables **	cotton	other annual ***	perma-nent crops	total
Maghreb	158 8%	507 25%	125 6%	573 28%	16 1%	179 9%	464 23%	2 022 100%
N-eastern Africa	1 077 16%	2 024 31%	1 098 17%	409 6%	704 11%	858 13%	386 6%	6 556 100%
Arabian Peninsula	916 45%	243 12%	323 16%	178 9%	13 1%	23 1%	326 16%	2 022 100%
Middle East	784 29%	576 21%	38 1%	307 11%	408 15%	396 15%	197 7%	2 706 100%
Central Asia	10 331 46%	3 769 17%	1 187 5%	896 4%	3 846 17%	846 4%	1 615 7%	22 490 100%
Total Near East	13 226 37%	7 119 20%	2 771 8%	2 363 7%	4 987 14%	2 302 6%	2 988 8%	35 796 100%

* *total includes 3 449 000 ha rice, 1 143 000 ha maize, 945 000 ha barley and 568 000 ha sorghum*
** *total includes 225 000 ha tubers (mainly potatoes) and 374 000 ha pulses*
*** *total includes 386 000 ha sugar cane, 286 000 ha oil crops (exc. olives) and 168 000 ha sugar beet*

The most widespread crop is wheat, which represents 37% of the cropped area. It is the most important irrigated crop in Central Asia, the Arabian Peninsula and the Middle East. The largest rice area is in Central Asia (mainly in Pakistan and Iran) and in Egypt (the Nile Delta area). The whole category of irrigated cereals is predominant in each sub-region and represents 57% of the irrigated crops in the whole region, varying between 33% in the Maghreb and 63% in Central Asia. Irrigated cereals are present in each country, except in Bahrain and Malta.

Irrigated fodder crops are important in Egypt, where berseem represents over 20% of the cropped area, and are present in each single country of the Arabian Peninsula, occupying between

12% (Saudi Arabia) and 32% (Qatar) of the cropped area. In Kyrgyzstan fodder crops represent 37% of the irrigated cropped area.

Vegetables are present in all regions and almost every single country. They are the most important irrigated crops in the Maghreb, representing 28% of the total cropped area under irrigation.

In absolute and relative terms, cotton is an important irrigated crop in Central Asia, covering over 3.8 million hectares. It is also a relatively important crop in Egypt, Sudan, Syria and Turkey. Sugar cane is mainly found in North-eastern Africa, Morocco and Iran.

Permanent crops under irrigation (dates, citrus, other fruits, olives) are present in almost all countries of the Maghreb, North-eastern Africa, the Arabian Peninsula and the Middle East. In relative terms, there is the largest concentration in the Maghreb, occupying 23% of the irrigated area, followed by the Arabian Peninsula with 19%. In Bahrain, Oman and the United Arab Emirates they represent over 50% of the total irrigated cropped area. In Cyprus, Jordan and Lebanon they represent around 40% of the irrigated area. In absolute terms the largest area of permanent crops under irrigation is in Iran, occupying almost 1.6 million hectares, which represents 22% of the irrigated cropped area of the country and 53% of the irrigated permanent crop area in the entire Near East.

Salinization, drainage and environment

Arid areas are sensitive to salinization problems because the volume of rainwater dissolving the salts generated by the soil is low. By extracting water from the soil, evaporation and evapotranspiration tend to increase salt concentrations. Direct evaporation from the soil surface causes a rapid accumulation of salt in the top layers. When significant amounts of water are provided by irrigation with no adequate provision for leaching of salts, the soils rapidly become salty and unproductive. Water storage in the reservoirs, where evaporation is intense, tends to increase the salt concentration of the stored water. For all these reasons, the Near East is a region subject to salinization and the problem has been known and recognized for a long time. However, assessment of salinization at national level is a difficult enterprise and very little information on the subject could be found during the survey. Furthermore, no commonly agreed methods exist to assess the degree of irrigation-induced salinization. Figures on areas salinized by irrigation were available for only eight of the 29 countries (Table 19). In the near future, more information on salinization will probably become available and strategies to improve the situation should be defined, as this has been recognized as a priority by most of the Near East countries. On average around 29% of the irrigated areas in the eight countries are reported as having salinization problems, varying from 3.5% in Jordan to over 85% in Kuwait.

One of the measures necessary to prevent irrigation-induced waterlogging and salinization in arid and semi-arid regions is the installation of drainage facilities. Drainage, in combination with adequate irrigation scheduling, allows for the leaching of excess salts from the plant root zone. Figures on drained areas are available for 13 of the 29 countries (Table 16). About 34% of the irrigated areas in these countries have been provided with drainage facilities, varying from 0.6% in Iran to over 90% in Egypt.

Over-extraction of groundwater leads to a lowering of the groundwater table and a deterioration of the groundwater quality due to sea water intrusion and/or the upward diffusion of deeper saline water. Using saline groundwater for irrigation may increase soil salinity. All the countries of the Arabian Peninsula are facing this problem as are the islands of Cyprus and Malta and the coastal zones of countries like Libya and Egypt. Also in Tunisia and Djibouti the groundwater used for irrigation is reported to be rather saline.

The Aral Sea basin

The Aral Sea is not included in the Near East Region. However, since most of the countries of the Aral Sea basin are part of the Near East Region, special attention has been paid to the Aral Sea basin, in view of its particular environmental problems related to water resources and irrigation. Due to the increasing need for water for irrigation in the Aral Sea basin, mainly from the 1950s onwards, the sea level continually fell as more water was diverted from its two major affluents, the Amu Darya and the Syr Darya. Between 1960 and 1965, its surface area shrank by more than half, from 64 500 km^2 to less than 30 000 km^2. At the same time, the level fell by 19 metres and its salinity tripled. Once the world's fourth largest lake, the Aral Sea has now lost so much of its water volume that what remains is contained in three separate highly saline lakes. The dry seabed exposed to weathering has increased soil salinization and desertification around the sea. It is estimated that since the 1970s dust storms yearly spread tens of millions of tons of dust and salt, polluted with pesticides, over the whole region. Traces have been found at a distance of more than 1 000 kilometres. Waterlogging is reported in large areas due to inefficient irrigation combined with insufficient or no drainage. In 1993, all five Aral Sea basin countries (Kazakhstan, Kyrgyzstan, Tajikistan, Turkmenistan and Uzbekistan) signed an agreement to improve the situation in the basin, but so far progress has been slow due to the region's political and economic situation.

Summary tables

TABLE 11
Land use and irrigation potential in the Near East Region

COUNTRY	Area of the country	Cultivable area	Year	Cultivated area			Irrigation potential		COUNTRY
				total	annual	permanent	**	as % of cultivable	
	ha	ha		ha	ha	ha	ha		
	(1)	(2)		(3) = (4) + (5)	(4)	(5)	(6)	100*(6)/(2)	
AFGHANISTAN	65 209 000	8 000 000	1990	3 344 000	3 200 000	144 000	-	-	AFGHANISTAN
ALGERIA	238 174 000	10 742 000	1993	3 372 000	2 841 600	530 400	730 000	6.8	ALGERIA
BAHRAIN	69 526	4 230	1994	3 165	1 483	1 682	4 230	100.0	BAHRAIN
CYPRUS *	925 000	165 433	1994	111 649	69 509	42 140	36 807	22.2	CYPRUS
DJIBOUTI	2 320 000	6 000	1989	407	357	50	1 000	16.7	DJIBOUTI
EGYPT	100 145 000	-	1993	3 246 000	2 862 960	383 040	4 435 000	-	EGYPT
IRAN	164 800 000	51 000 000	1993	14 382 418	12 732 418	1 650 000	15 000 000	29.4	IRAN
IRAQ	43 832 000	11 480 000	1993	3 731 300	3 462 800	268 500	5 554 000	48.4	IRAQ
JORDAN	8 921 000	381 740	1991	214 767	120 077	94 690	85 000	22.3	JORDAN
KUWAIT	1 782 000	25 000	1994	4 770	4 320	450	25 000	100.0	KUWAIT
KYRGYZSTAN	19 850 000	-	1990	1 323 400	1 282 800	40 600	2 247 300	-	KYRGYZSTAN
LEBANON	1 040 000	360 000	1994	189 206	104 120	85 086	177 500	49.3	LEBANON
LIBYA	175 954 000	3 800 000	1987	2 278 648	1 933 648	345 000	750 000	19.7	LIBYA
MALTA	31 600	10 721	1991	10 721	10 091	630	2 000	18.7	MALTA
MAURITANIA	102 552 000	20 129 689	1989	204 000	199 000	5 000	170 000	0.8	MAURITANIA
MOROCCO	44 650 000	8 000 000	1993	7 232 700	6 568 300	664 400	1 653 000	20.7	MOROCCO
OMAN	31 250 000	2 200 000	1993	61 550	18 550	43 000	-	-	OMAN
PAKISTAN	79 610 000	29 900 000	1990	16 556 000	16 100 000	456 000	-	-	PAKISTAN
QATAR	1 100 000	28 000	1994	8 312	5 987	2 325	-	-	QATAR
SAUDI ARABIA	214 969 000	52 684 000	1992	1 608 000	1 512 500	95 500	-	-	SAUDI ARABIA
SOMALIA	63 766 000	8 150 000	1984	998 000	980 000	18 000	240 000	2.9	SOMALIA
SUDAN	250 581 000	105 000 000	1995	7 600 000	7 350 000	250 000	2 784 000	2.7	SUDAN
SYRIA	18 518 000	5 939 047	1993	4 938 670	4 266 256	672 414	1 250 000	21.0	SYRIA
TAJIKISTAN	14 310 000	-	1992	812 000	756 000	56 000	1 570 000	-	TAJIKISTAN
TUNISIA	16 361 000	8 700 000	1993	4 254 600	2 180 800	2 073 800	563 000	6.5	TUNISIA
TURKEY	77 945 000	28 054 000	1991	20 480 000	17 450 000	3 030 000	8 500 000	30.3	TURKEY
TURKMENISTAN	48 810 000	-	1993	1 480 000	1 400 000	80 000	-	-	TURKMENISTAN
UNITED ARAB EM.	8 360 000	66 682	1993	54 512	21 683	32 829	66 682	100.0	UNITED ARAB EM.
YEMEN	52 797 000	3 617 753	1994	1 052 786	852 786	200 000	-	-	YEMEN
NEAR EAST REGION	**1 848 632 126**			**99 553 581**	**88 288 045**	**11 265 536**			**NEAR EAST REGION**
for 25 countries		**358 444 295**							

* ***Cyprus:*** *The figure for area of the country refers to the whole island; other figures refer to the government controlled area.*

** ***Irrigation potential:*** *Figures on irrigation potential cannot be totalled due to possible double counting of shared water resoruces.*

TABLE 12
Population in the Near East Region

COUNTRY	POPULATION 1995							COUNTRY
	TOTAL			URBAN		RURAL		
	inhabitants (1)	% of total	inh/km² (2)	inhabitants (3)	% 100*(3)/(1)	inhabitants (4)	% 100*(4)/(1)	
AFGHANISTAN	20 141 000	3.59	31	4 026 000	20.0	16 115 000	80.0	AFGHANISTAN
ALGERIA	27 939 000	4.98	12	15 591 000	55.8	12 348 000	44.2	ALGERIA
BAHRAIN	564 000	0.10	811	509 000	90.2	55 000	9.8	BAHRAIN
CYPRUS *	742 000	0.13	80	401 000	54.0	341 000	46.0	CYPRUS
DJIBOUTI	577 000	0.10	25	478 000	82.8	99 000	17.2	DJIBOUTI
EGYPT	62 931 000	11.21	63	28 170 000	44.8	34 761 000	55.2	EGYPT
IRAN	67 283 000	11.98	41	39 716 000	59.0	27 567 000	41.0	IRAN
IRAQ	20 449 000	3.64	47	15 258 000	74.6	5 191 000	25.4	IRAQ
JORDAN	5 463 000	0.97	61	3 904 000	71.5	1 559 000	28.5	JORDAN
KUWAIT	1 547 000	0.28	87	1 501 000	97.0	46 000	3.0	KUWAIT
KYRGYZSTAN	4 745 000	0.85	24	1 848 000	38.9	2 897 000	61.1	KYRGYZSTAN
LEBANON	3 009 000	0.54	289	2 622 000	87.1	387 000	12.9	LEBANON
LIBYA	5 407 000	0.96	3	4 649 000	86.0	758 000	14.0	LIBYA
MALTA	366 000	0.07	1 158	327 000	89.3	39 000	10.7	MALTA
MAURITANIA	2 274 000	0.41	2	1 224 000	53.8	1 050 000	46.2	MAURITANIA
MOROCCO	27 028 000	4.81	61	13 071 000	48.4	13 957 000	51.6	MOROCCO
OMAN	2 163 000	0.39	7	285 000	13.2	1 878 000	86.8	OMAN
PAKISTAN	140 947 000	25.10	177	48 742 000	34.6	92 205 000	65.4	PAKISTAN
QATAR	551 000	0.10	50	503 000	91.3	48 000	8.7	QATAR
SAUDI ARABIA	17 880 000	3.18	8	14 340 000	80.2	3 540 000	19.8	SAUDI ARABIA
SOMALIA	9 250 000	1.65	15	2 382 000	25.8	6 868 000	74.2	SOMALIA
SUDAN	28 098 000	5.00	11	6 915 000	24.6	21 183 000	75.4	SUDAN
SYRIA	14 661 000	2.61	79	7 676 000	52.4	6 985 000	47.6	SYRIA
TAJIKISTAN	6 101 000	1.09	43	1 964 000	32.2	4 137 000	67.8	TAJIKISTAN
TUNISIA	8 896 000	1.58	54	5 093 000	57.3	3 803 000	42.7	TUNISIA
TURKEY	61 945 000	11.03	79	42 598 000	68.8	19 347 000	31.2	TURKEY
TURKMENISTAN	4 099 000	0.73	8	1 839 000	44.9	2 260 000	55.1	TURKMENISTAN
UNITED ARAB EM.	1 904 000	0.34	23	1 600 000	84.0	304 000	16.0	UNITED ARAB EM.
YEMEN	14 501 000	2.58	27	4 877 000	33.6	9 624 000	66.4	YEMEN
NEAR EAST REGION	**561 461 000**	**100.00**	**30**	**272 109 000**	**48.5**	**289 352 000**	**51.5**	**NEAR EAST REGION**

* ***Cyprus**: Figures refer to the whole island*

TABLE 13
Renewable water resources in the Near East Region

COUNTRY	ANNUAL AVERAGE PRECIPITATION		ANNUAL RENEWABLE WATER RESOURCES						COUNTRY
			internal		total natural	total actual		Dependency ratio %	
	mm	million m^3	million m^3	m^3 per inhabitant '95	million m^3	million m^3	m^3 per inhabitant '95		
	(1)	(2)	(3)	(4)	(5)	(6)	(7)	(8)	
AFGHANISTAN	-	-	55 000	2 731	65 000	65 000	3 227	15.4	AFGHANISTAN
ALGERIA	68	162 900	13 900	498	14 300	14 300	512	2.8	ALGERIA
BAHRAIN	70	49	4	7	116	116	206	96.6	BAHRAIN
CYPRUS *	503	4 650	900	1 213	900	900	1 213	0.0	CYPRUS
DJIBOUTI	147	3 400	300	520	300	300	520	0.0	DJIBOUTI
EGYPT	18	18 100	1 800	29	86 800	58 300	926	96.9	EGYPT
IRAN	252	415 300	128 500	1 910	137 510	137 510	2 044	6.6	IRAN
IRAQ	154	67 700	35 200	1 721	96 420	75 420	3 688	53.3	IRAQ
JORDAN	94	8 430	680	124	880	880	161	22.7	JORDAN
KUWAIT	176	3 140	0	0	20	20	13	100.0	KUWAIT
KYRGYZSTAN	316	62 730	47 230	9 954	47 230	11 600	2 445	0.0	KYRGYZSTAN
LEBANON	827	8 600	4 800	1 595	4 837	4 407	1 465	0.8	LEBANON
LIBYA	26	46 100	600	111	600	600	111	0.0	LIBYA
MALTA	524	166	16	42	16	16	42	0.0	MALTA
MAURITANIA	99	101 900	400	176	11 400	11 400	5 013	96.5	MAURITANIA
MOROCCO	336	150 000	30 000	1 110	30 000	30 000	1 110	0.0	MOROCCO
OMAN	54	17 000	985	455	985	985	455	0.0	OMAN
PAKISTAN	494	393 300	248 000	1 760	429 370	418 270	2 968	40.7	PAKISTAN
QATAR	75	825	51	93	53	53	96	3.8	QATAR
SAUDI ARABIA	59	126 800	2 400	134	2 400	2 400	134	0.0	SAUDI ARABIA
SOMALIA	253	161 300	6 000	649	15 740	15 740	1 702	61.9	SOMALIA
SUDAN	436	1 092 600	35 000	1 246	154 000	88 500	3 150	77.3	SUDAN
SYRIA	252	46 600	7 000	477	46 080	26 260	1 791	80.3	SYRIA
TAJIKISTAN	-	-	61 800	10 129	81 800	39 500	6 474	24.5	TAJIKISTAN
TUNISIA	207	33 900	3 520	396	4 120	4 120	463	14.6	TUNISIA
TURKEY	643	501 200	196 000	3 164	200 700	183 762	2 967	1.8	TURKEY
TURKMENISTAN	-	-	1 000	244	71 000	71 000	17 321	98.6	TURKMENISTAN
UNITED ARAB EM.	100	8 360	150	79	150	150	79	0.0	UNITED ARAB EM.
YEMEN	170	89 800	4 100	283	4 100	4 100	283	0.0	YEMEN
NEAR EAST REGION			**885 336**	**1 577**					**NEAR EAST REGION**
for 26 countries	**205**	**3 524 850**							

* ***Cyprus**: Figures refer to the whole island*

TABLE 14
Water withdrawal in the Near East Region

COUNTRY	Year	ANNUAL WATER WITHDRAWAL: agricultural million m³ (1)	agricultural % 100*(1)/(4)	domestic million m³ (2)	domestic % 100*(2)/(4)	industrial million m³ (3)	industrial % 100*(3)/(4)	total million m³ (4) = (1) + (2) + (3)	m³ per inhabitant ** (5)	as % of internal renewable water res.	as % of total actual renewable water res.	COUNTRY
AFGHANISTAN	1987	25 849.0	99	261.0	1	0.0	0	26 110.0	1 702	47	40	AFGHANISTAN
ALGERIA	1990	2 700.0	60	1 120.0	25	680.0	15	4 500.0	180	32	31	ALGERIA
BAHRAIN	1991	135.1	56	94.3	39	9.9	4	239.2	465	5980	206	BAHRAIN
CYPRUS *	1993	156.0	74	50.0	24	5.0	2	211.0	331	[23]	[23]	CYPRUS
DJIBOUTI	1985	6.5	87	1.0	13	0.0	0	7.5	20	3	3	DJIBOUTI
EGYPT	1993	47 400.0	86	3 100.0	6	4 600.0	8	55 100.0	913	3061	95	EGYPT
IRAN	1993	64 155.0	92	4 395.0	6	1 484.0	2	70 034.0	1 091	55	51	IRAN
IRAQ	1990	39 380.0	92	1 280.0	3	2 140.0	5	42 800.0	2 367	122	57	IRAQ
JORDAN	1993	737.0	75	214.0	22	33.0	3	984.0	246	145	112	JORDAN
KUWAIT	1994	324.0	60	201.0	37	13.0	2	538.0	348	IRWR negligible	2690	KUWAIT
KYRGYZSTAN	1990	10 465.0	95	259.0	2	312.0	3	11 036.0	2 527	23	95	KYRGYZSTAN
LEBANON	1994	875.0	68	368.0	28	50.0	4	1 293.0	444	27	29	LEBANON
LIBYA	1994	4 000.0	87	500.0	11	100.0	2	4 600.0	880	767	767	LIBYA
MALTA	1995	6.6	12	48.6	87	0.5	1	55.7	152	348	359	MALTA
MAURITANIA	1985	1 499.6	92	101.1	6	29.3	2	1 630.0	923	408	14	MAURITANIA
MOROCCO	1991	10 180.0	92	543.0	5	322.0	3	11 045.0	436	37	37	MOROCCO
OMAN	1991	1 148.0	94	56.0	5	19.0	2	1 223.0	728	124	124	OMAN
PAKISTAN	1991	150 600.0	97	2 500.0	2	2 500.0	2	155 600.0	1 277	63	37	PAKISTAN
QATAR	1994	210.6	74	65.9	23	8.4	3	284.9	528	559	538	QATAR
SAUDI ARABIA	1992	15 308.0	90	1 517.0	9	193.0	1	17 018.0	1 040	709	709	SAUDI ARABIA
SOMALIA	1987	785.7	97	24.3	3	0.0	0	810.0	99	14	5	SOMALIA
SUDAN	1995	16 800.0	94	800.0	4	200.0	1	17 800.0	633	51	20	SUDAN
SYRIA	1993	13 600.0	94	530.0	4	280.0	2	14 410.0	1 017	206	55	SYRIA
TAJIKISTAN	1989	11 088.0	88	630.0	5	882.0	7	12 600.0	2 065	20	32	TAJIKISTAN
TUNISIA	1990	2 727.5	89	261.4	9	86.1	3	3 075.0	382	87	75	TUNISIA
TURKEY	1992	22 900.0	72	5 200.0	16	3 500.0	11	31 600.0	541	16	17	TURKEY
TURKMENISTAN	1989	20 748.0	91	228.0	1	1 824.0	8	22 800.0	6 346	2280	32	TURKMENISTAN
UNITED ARAB EM	1995	1 408.0	67	500.0	24	200.0	9	2 108.0	1 107	1405	1405	UNITED ARAB EM.
YEMEN	1990	2 700.0	92	201.0	7	31.0	1	2 932.0	251	72	72	YEMEN
NEAR EAST REGION		467 892.6	91	25 049.6	5	19 502.1	4	512 444.3	964	58		NEAR EAST REGION

* ***Cyprus:*** *Figures for water withdrawal refer to the government controlled area; figures for water resources refer to the whole island.*

** ***Annual water withdrawal per inhabitant.*** *The figure for the whole region refers to the population in 1993*

TABLE 15
Use of non-conventional sources of water in the Near East Region

COUNTRY	Total water withdrawal		Desalinated water		Reused treated wastewater		Use of desalinated water and treated wastewater		COUNTRY
	Year	million m^3 per year	Year	million m^3 per year	Year	million m^3 per year	million m^3 per year	as % of total withdrawal	
		(1)		(2)		(3)	(4) = (2) + (3)	100*(4)/(1)	
AFGHANISTAN	1987	26 110.0		-		-	-	-	AFGHANISTAN
ALGERIA	1990	4 500.0	1990	64.00		-	64.00	1.422	ALGERIA
BAHRAIN	1991	239.2	1991	44.10	1991	8.03	52.13	21.793	BAHRAIN
CYPRUS *	1993	211.0		-	1995	11.00	11.00	5.213	CYPRUS
DJIBOUTI	1985	7.5	1990	0.10		-	0.10	1.333	DJIBOUTI
EGYPT	1993	55 100.0	1990	25.00	1993	200.00	225.00	0.408	EGYPT
IRAN	1993	70 034.0	1991	2.90		-	2.90	0.004	IRAN
IRAQ	1990	42 800.0		-		-	-	-	IRAQ
JORDAN	1993	984.0	1993	2.00	1991	50.30	52.30	5.315	JORDAN
KUWAIT	1994	538.0	1993	231.00	1994	52.00	283.00	52.602	KUWAIT
KYRGYZSTAN	1990	11 036.0		-	1990	9.00	9.00	0.082	KYRGYZSTAN
LEBANON	1994	1 293.0		-	1991	2.00	2.00	0.155	LEBANON
LIBYA	1994	4 600.0	1994	70.00	1990	100.00	170.00	3.696	LIBYA
MALTA	1995	55.7	1995	31.40	1993	1.56	32.96	59.174	MALTA
MAURITANIA	1985	1 630.0	1990	1.70		-	1.70	0.104	MAURITANIA
MOROCCO	1991	11 045.0	1992	3.40		-	3.40	0.031	MOROCCO
OMAN	1991	1 223.0	1995	34.00	1991	26.00	60.00	4.906	OMAN
PAKISTAN	1991	155 600.0		-		-	-	-	PAKISTAN
QATAR	1994	284.9	1995	98.60	1994	25.20	123.80	43.454	QATAR
SAUDI ARABIA	1992	17 018.0	1995	714.00	1992	217.00	931.00	5.471	SAUDI ARABIA
SOMALIA	1987	810.0	1990	0.10		-	0.10	0.012	SOMALIA
SUDAN	1995	17 800.0	1990	0.40		-	0.40	0.002	SUDAN
SYRIA	1993	14 410.0		-	1993	370.00	370.00	2.568	SYRIA
TAJIKISTAN	1989	12 600.0		-		-	-	-	TAJIKISTAN
TUNISIA	1990	3 075.0	1990	8.30	1993	20.00	28.30	0.920	TUNISIA
TURKEY	1992	31 600.0	1990	0.50		-	0.50	0.002	TURKEY
TURKMENISTAN	1989	22 800.0		-		-	-	-	TURKMENISTAN
UNITED ARAB EM.	1995	2 108.0	1995	385.00	1995	108.00	493.00	23.387	UNITED ARAB EM.
YEMEN	1990	2 932.0	1989	10.00	1992	-	10.00	0.341	YEMEN
NEAR EAST REGION		**512 444.3**		**1 726.50**		**1 200.09**	**2 926.59**	**0.571**	**NEAR EAST REGION**

* ***Cyprus**: The figure for total water withdrawal refers to the government controlled area; other figures refer to the whole island.*

TABLE 16
Types of water control in the Near East Region

COUNTRY	Year	IRRIGATION full or partial control	spate	equipped wetl./ivb	total	as % of cultivated area	as % of irrigation potential	Flood recession cropping	Total water managed	COUNTRY
		ha	ha	ha	ha	area	potential	ha	ha	
		(1)	(2)	(3)	(4) = (1) + (2) + (3)	**	***	(5)	(6) = (4) + (5)	
AFGHANISTAN	1967	2 385 740	-	-	2 385 740	71.3	-	-	2 385 740	AFGHANISTAN
ALGERIA	1992	445 500	110 000	-	555 500	16.5	76.1	-	555 500	ALGERIA
BAHRAIN	1994	3 165	-	-	3 165	100.0	74.8	-	3 165	BAHRAIN
CYPRUS *	1994	39 545	393	-	39 938	35.8	108.5	-	39 938	CYPRUS
DJIBOUTI	1989	674	-	-	674	165.6	67.4	-	674	DJIBOUTI
EGYPT	1993	3 246 000	-	-	3 246 000	100.0	73.2	-	3 246 000	EGYPT
IRAN	1993	7 264 194	-	-	7 264 194	50.5	48.4	10 000	7 274 194	IRAN
IRAQ	1990	3 525 000	-	-	3 525 000	94.5	63.5	-	3 525 000	IRAQ
JORDAN	1991	64 300	-	-	64 300	29.9	75.6	-	64 300	JORDAN
KUWAIT	1994	4 770	-	-	4 770	100.0	19.1	-	4 770	KUWAIT
KYRGYZSTAN	1990	1 072 600	-	-	1 072 600	81.0	47.7	-	1 072 600	KYRGYZSTAN
LEBANON	1993	87 500	-	-	87 500	46.2	49.3	-	87 500	LEBANON
LIBYA	1990	470 000	-	-	470 000	20.6	62.7	-	470 000	LIBYA
MALTA	1990	763	-	-	763	7.1	38.2	-	763	MALTA
MAURITANIA	1994	49 200	-	-	49 200	24.1	28.9	64 000	113 200	MAURITANIA
MOROCCO	1989	1 093 200	165 000	-	1 258 200	17.4	76.1	-	1 258 200	MOROCCO
OMAN	1993	61 550	-	-	61 550	100.0	-	-	61 550	OMAN
PAKISTAN	1990	14 327 000	1 402 448	-	15 729 448	95.0	-	1 230 552	16 960 000	PAKISTAN
QATAR	1993	12 520	-	-	12 520	150.6	-	-	12 520	QATAR
SAUDI ARABIA	1992	1 608 000	-	-	1 608 000	100.0	-	-	1 608 000	SAUDI ARABIA
SOMALIA	1984	50 000	150 000	-	200 000	20.0	83.3	-	200 000	SOMALIA
SUDAN	1995	1 900 000	46 200	-	1 946 200	25.6	69.9	-	1 946 200	SUDAN
SYRIA	1993	1 013 273	-	-	1 013 273	20.5	81.1	-	1 013 273	SYRIA
TAJIKISTAN	1992	718 000	-	-	718 000	88.4	45.7	-	718 000	TAJIKISTAN
TUNISIA	1991	355 000	30 000	-	385 000	9.0	68.4	-	385 000	TUNISIA
TURKEY	1994	4 070 746	-	115 164	4 185 910	20.4	49.2	-	4 185 910	TURKEY
TURKMENISTAN	1993	1 300 000	-	-	1 300 000	87.8	-	-	1 300 000	TURKMENISTAN
UNITED ARAB EM.	1995	66 682	-	-	66 682	122.3	100.0	-	66 682	UNITED ARAB EM.
YEMEN	1994	383 200	98 320	-	481 520	45.7	-	-	481 520	YEMEN
NEAR EAST REGION		**45 618 122**	**2 002 361**	**115 164**	**47 735 647**	**47.9**		**1 304 552**	**49 040 199**	**NEAR EAST REGION**

* *Cyprus: Figures refer to the government controlled area.*

** *Djibouti, Qatar, United Arab Emirates: Total area equipped for irrigation is larger than total cultivated area. No cultivation is possible without irrigation, but not all the equipped area is actually irrigated.*

*** *Cyprus: Total area equipped for irrigation is larger than irrigation potential. Due to the increasing need for water for other purposes in the future, the area under irrigation (potential) will have to be reduced.*

TABLE 17
Full or partial control irrigation techniques in the Near East Region

COUNTRY	Year	FULL OR PARTIAL CONTROL IRRIGATION : Equipped area				% of equipped area actually irrigated	COUNTRY
		surface	sprinkler	micro	total		
		ha	ha	ha	ha		
		(1)	(2)	(3)	(4) = (1) + (2) + (3)	(5)	
AFGHANISTAN	1967	-	114 000	-	2 385 740	-	AFGHANISTAN
ALGERIA	1992	-	40 000	-	445 500	82.2	ALGERIA
BAHRAIN	1994	2 497	130	538	3 165	100.0	BAHRAIN
CYPRUS *	1994	1 977	1 977	35 591	39 545	83.6	CYPRUS
DJIBOUTI	1989	-	-	-	674	60.4	DJIBOUTI
EGYPT	1993	3 046 000	117 000	83 000	3 246 000	100.0	EGYPT
IRAN	1993	7 173 494	47 200	43 500	7 264 194	100.0	IRAN
IRAQ	1990	-	-	8 000	3 525 000	54.9	IRAQ
JORDAN	1991	20 300	5 700	38 300	64 300	-	JORDAN
KUWAIT	1994	3 020	600	1 150	4 770	100.0	KUWAIT
KYRGYZSTAN	1990	931 588	141 000	12	1 072 600	-	KYRGYZSTAN
LEBANON	1993	53 500	21 000	13 000	87 500	-	LEBANON
LIBYA	1990	0	470 000	0	470 000	51.1	LIBYA
MALTA	1990	113	150	500	763	94.8	MALTA
MAURITANIA	1994	-	-	-	49 200	54.4	MAURITANIA
MOROCCO	1989	986 000	103 200	4 000	1 093 200	-	MOROCCO
OMAN	1993	57 820	1 640	2 090	61 550	100.0	OMAN
PAKISTAN	1990	-	-	-	14 327 000	-	PAKISTAN
QATAR	1993	-	-	-	12 520	66.4	QATAR
SAUDI ARABIA	1992	547 000	1 029 000	32 000	1 608 000	100.0	SAUDI ARABIA
SOMALIA	1984	-	-	-	50 000	-	SOMALIA
SUDAN	1995	-	-	-	1 900 000	63.2	SUDAN
SYRIA	1993	981 273	30 000	2 000	1 013 273	-	SYRIA
TAJIKISTAN	1992	-	-	-	718 000	97.2	TAJIKISTAN
TUNISIA	1991	294 000	55 000	6 000	355 000	90.7	TUNISIA
TURKEY	1994	3 806 511	263 849	386	4 070 746	74.0	TURKEY
TURKMENISTAN	1993	-	-	-	1 300 000	-	TURKMENISTAN
UNITED ARAB EM.	1993	25 382	3 748	37 552	66 682	81.7	UNITED ARAB EM.
YEMEN	1994	382 450	350	400	383 200	-	YEMEN
NEAR EAST REGION					**45 618 122**		**NEAR EAST REGION**
for 18 countries (ha)		**18 312 925**	**2 291 544**	**300 019**	**20 904 488**	**84.2**	**for 19 countries**
for 18 countries (%)		**87.6**	**11.0**	**1.4**	**100.0**		

* ***Cyprus***: *Figures refer to the government controlled area.*

TABLE 18
Origin of irrigation water in the Near East Region

COUNTRY	FULL OR PARTIAL CONTROL IRRIGATION: Equipped area					COUNTRY
	Year	Total area	Origin of irrigation water			
			surface water	groundwater	non-conv. water	
		ha	%	%	%	
		(1)	*(2)*	*(3)*	*(4)*	
AFGHANISTAN	1967	2 385 740	84.6	15.4	0.0	AFGHANISTAN
ALGERIA	1992	445 500	-	-	-	ALGERIA
BAHRAIN	1994	3 165	0.0	86.4	13.6	BAHRAIN
CYPRUS *	1994	39 545	48.2	51.3	0.5	CYPRUS
DJIBOUTI	1989	674	0.0	100.0	0.0	DJIBOUTI
EGYPT	1993	3 246 000	95.4	4.5	0.1	EGYPT
IRAN	1993	7 264 194	49.9	50.1	0.0	IRAN
IRAQ	1990	3 525 000	93.8	6.2	0.0	IRAQ
JORDAN	1991	64 300	39.7	54.6	5.7	JORDAN
KUWAIT	1994	4 770	0.0	61.0	39.0	KUWAIT
KYRGYZSTAN	1990	1 072 600	99.8	0.2	0.0	KYRGYZSTAN
LEBANON	1993	87 500	54.3	45.7	0.0	LEBANON
LIBYA	1990	470 000	-	-	-	LIBYA
MALTA	1990	763	0.0	63.3	36.7	MALTA
MAURITANIA	1994	49 200	90.4	9.6	0.0	MAURITANIA
MOROCCO	1989	1 093 200	68.3	31.1	0.6	MOROCCO
OMAN	1993	61 550	0.0	100.0	0.0	OMAN
PAKISTAN	1990	14 327 000	66.0	34.0	0.0	PAKISTAN
QATAR	1993	12 520	0.0	94.2	5.8	QATAR
SAUDI ARABIA	1992	1 608 000	3.2	95.6	1.2	SAUDI ARABIA
SOMALIA	1984	50 000	-	-	-	SOMALIA
SUDAN	1995	1 900 000	96.0	4.0	0.0	SUDAN
SYRIA	1993	1 013 273	39.8	60.2	0.0	SYRIA
TAJIKISTAN	1992	718 000	91.6	8.4	0.0	TAJIKISTAN
TUNISIA	1991	355 000	37.3	60.7	2.0	TUNISIA
TURKEY	1994	4 070 746	83.5	16.5	0.0	TURKEY
TURKMENISTAN	1993	1 300 000	-	-	-	TURKMENISTAN
UNITED ARAB EM.	1993	66 682	0.0	100.0	0.0	UNITED ARAB EM.
YEMEN	1994	383 200	0.0	100.0	0.0	YEMEN
NEAR EAST REGION		**45 618 122**				**NEAR EAST REGION**
for 25 countries		**43 352 622**	**69.0**	**30.9**	**0.1**	**for 25 countries**

* ***Cyprus**: Figures refer to the government controlled area*

TABLE 19
Salinization and drainage in the Near East Region

COUNTRY	Year	Area under irrigation	Area salinized by irrigation		Drained area		COUNTRY
		ha	ha	%	ha	%	
		(1)	*(2)*	*100*(2)/(1)*	*(3)*	*100*(3)/(1)*	
AFGHANISTAN	1967	2 385 740	-	-	-	-	AFGHANISTAN
ALGERIA	1992	555 500	-	-	-	-	ALGERIA
BAHRAIN	1994	3 165	1 065	33.6	1 300	41.1	BAHRAIN
CYPRUS *	1994	39 938	-	-	-	-	CYPRUS
DJIBOUTI	1989	674	-	-	-	-	DJIBOUTI
EGYPT	1993	3 246 000	1 210 000	37.3	2 931 000	90.3	EGYPT
IRAN	1993	7 264 194	2 100 000	28.9	40 000	0.6	IRAN
IRAQ	1990	3 525 000	-	-	-	-	IRAQ
JORDAN	1991	64 300	2 277	3.5	4 000	6.2	JORDAN
KUWAIT	1994	4 770	4 080	85.5	2	0.0	KUWAIT
KYRGYZSTAN	1990	1 072 600	123 400	11.5	149 200	13.9	KYRGYZSTAN
LEBANON	1993	87 500	-	-	10 800	12.3	LEBANON
LIBYA	1990	470 000	-	-	-	-	LIBYA
MALTA	1990	763	-	-	-	-	MALTA
MAURITANIA	1994	49 200	-	-	12 784	26.0	MAURITANIA
MOROCCO	1989	1 258 200	-	-	-	-	MOROCCO
OMAN	1993	61 550	-	-	-	-	OMAN
PAKISTAN	1990	15 729 448	-	-	5 100 165	32.4	PAKISTAN
QATAR	1993	12 520	-	-	-	-	QATAR
SAUDI ARABIA	1992	1 608 000	-	-	44 000	2.7	SAUDI ARABIA
SOMALIA	1984	200 000	-	-	-	-	SOMALIA
SUDAN	1995	1 946 200	-	-	-	-	SUDAN
SYRIA	1993	1 013 273	60 000	5.9	273 030	26.9	SYRIA
TAJIKISTAN	1992	718 000	-	-	-	-	TAJIKISTAN
TUNISIA	1991	385 000	-	-	162 000	42.1	TUNISIA
TURKEY	1994	4 185 910			3 143 000	75.1	TURKEY
TURKMENISTAN	1993	1 300 000	598 000	46.0	-	-	TURKMENISTAN
UNITED ARAB EM.	1995	66 682	-	-	-	-	UNITED ARAB EM.
YEMEN	1994	481 520	-	-	-	-	YEMEN
NEAR EAST REGION		**47 735 647**					**NEAR EAST REGION**

* ***Cyprus:*** *Figures refer to the government controlled area*

TABLE 20
The Near East Region compared to the world

Variable	Unit	World	Near East Region	Near East as % of World
Total area	km^2	134 223 000	18 486 000	13.8
Total population 1995	inhabitants	5 716 407 000	561 461 000	9.8
Population growth 1985-95	%/year	1.7	3.0	176.5
Population growth 1994-95	%/year	1.5	2.7	180.0
Population density	inhabitants/km^2	43	30	71.3
Precipitation	km^3/year	110 000	3 848	3.5
Renewable water resources:				
Total	km^3/year	40 000	885	2.2
Per inhabitant	m^3/year	6 997	1 577	22.5
Water withdrawal:				
- agricultural	km^3/year	2 236	467	20.9
- % of total	%	69	91	132.3
- domestic	km^3/year	259	25	9.6
- % of total	%	8	5	61.1
- industrial	km^3/year	745	20	2.6
- % of total	%	23	4	16.6
Total water withdrawal	km^3/year	3 240	512	15.8
- in % of renewable water resources	%	8	58	713.5
- per inhabitant *	m^3/year	660	964	146.1
Irrigation	hectares	246 408 529	47 735 647	19.4

* The *figure for the 'World' refers to the population in 1990, the figure for the 'Near East region' refers to the population in 1993.*

Figures

FIGURE 1
The five sub-regions of the Near East Region

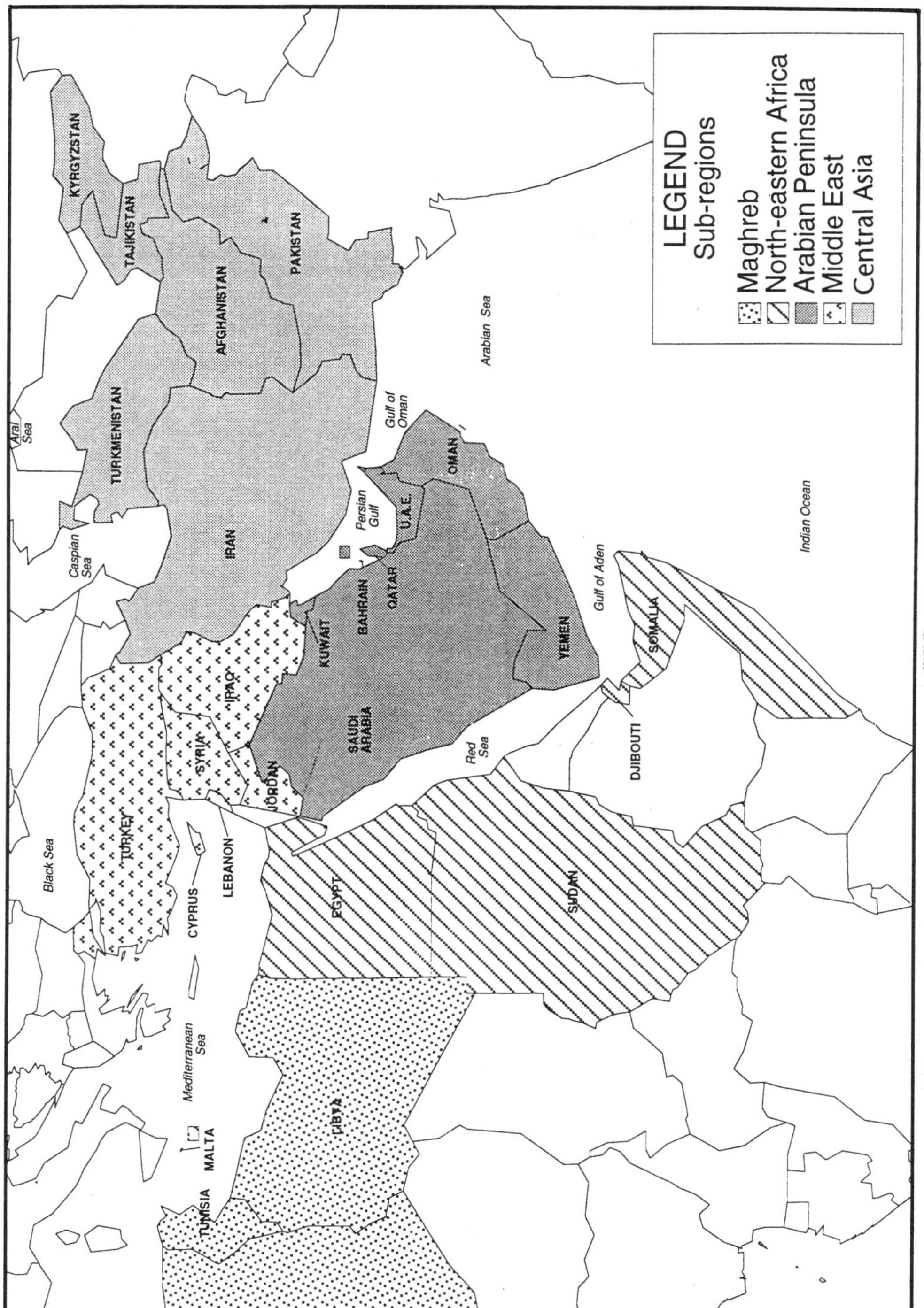

FIGURE 2
Major rivers in the Near East Region

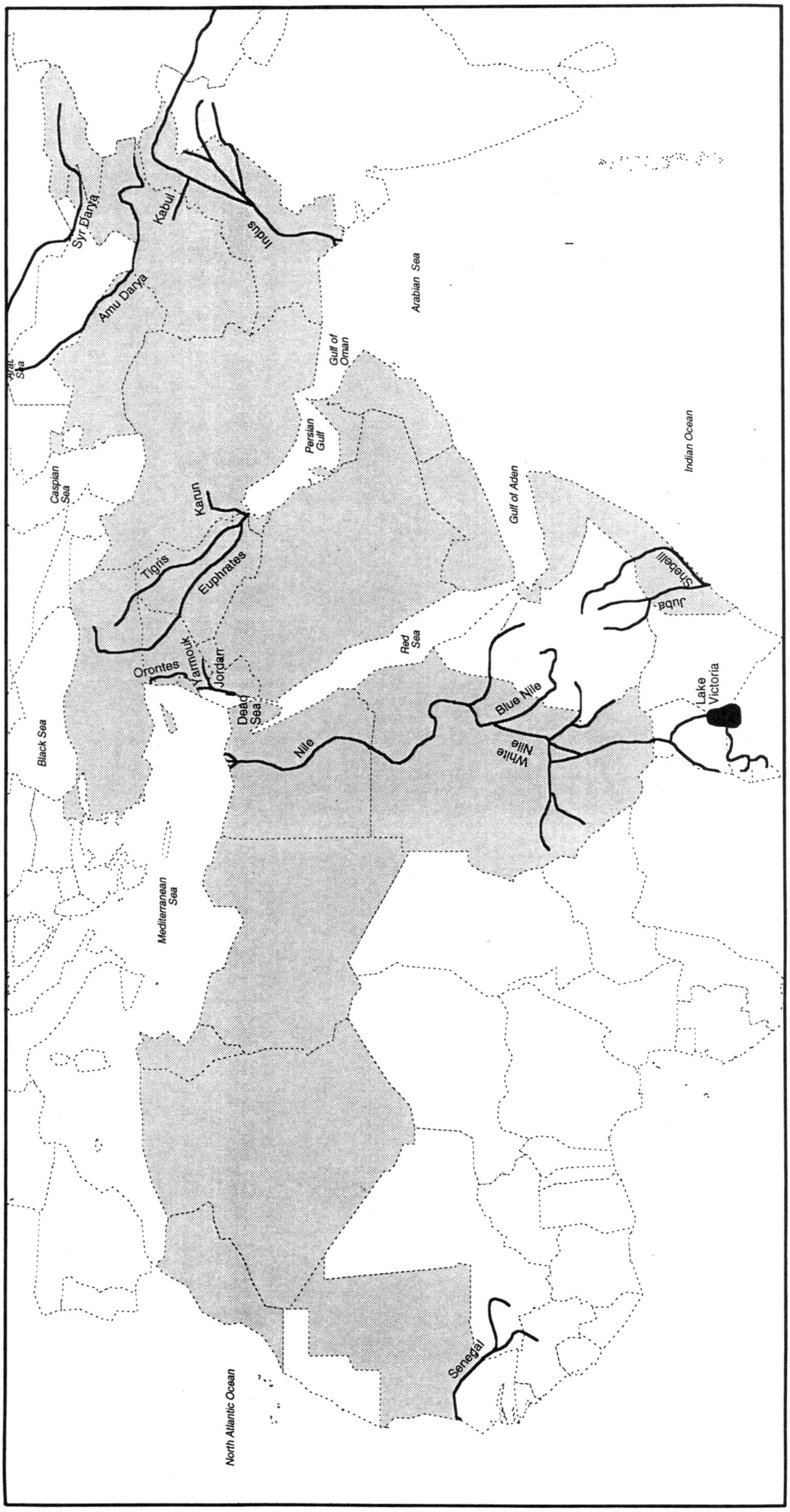

FIGURE 3
Annual internal renewable water resources per inhabitant (1995)

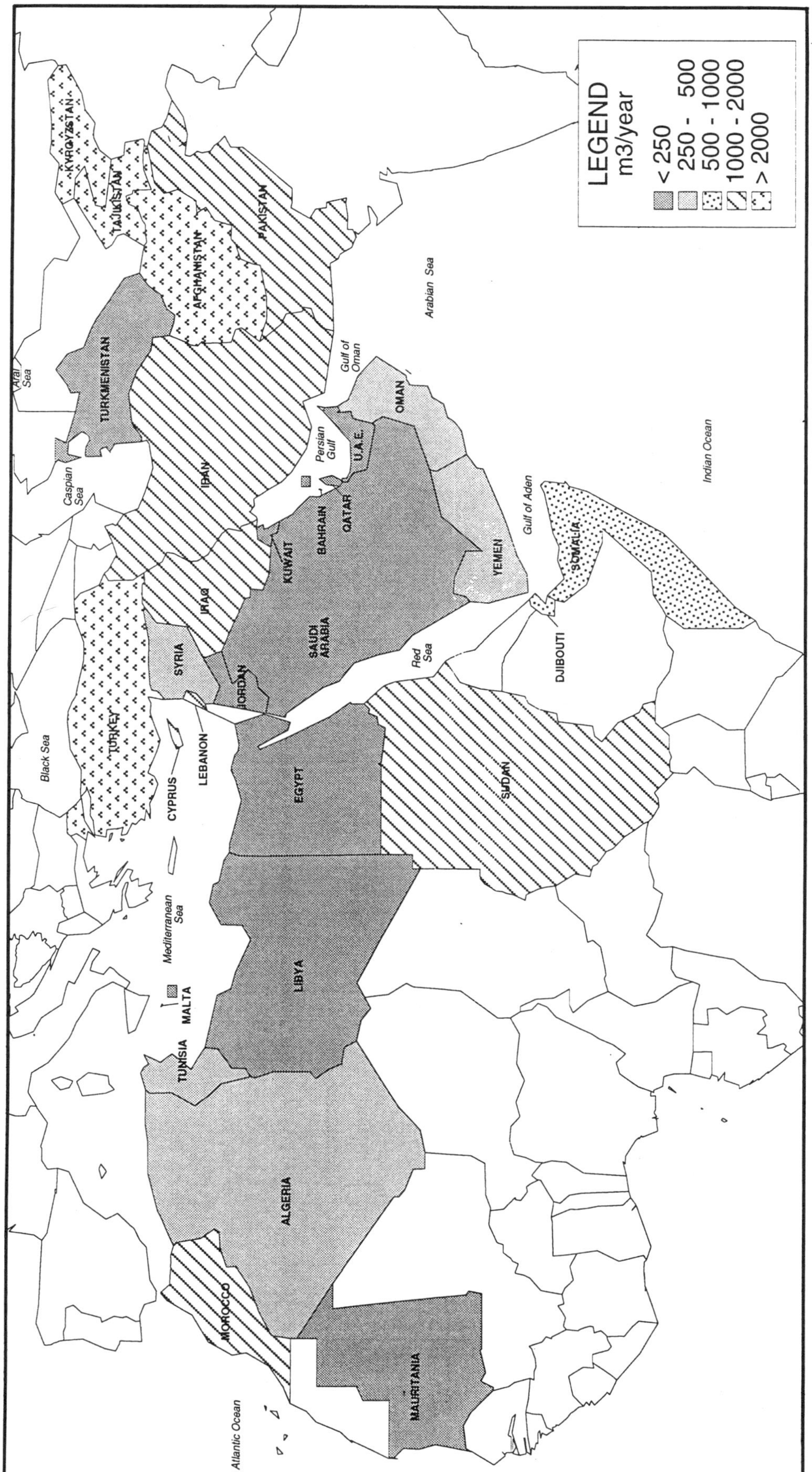

FIGURE 4
Annual actual renewable water resources per inhabitant (1995)

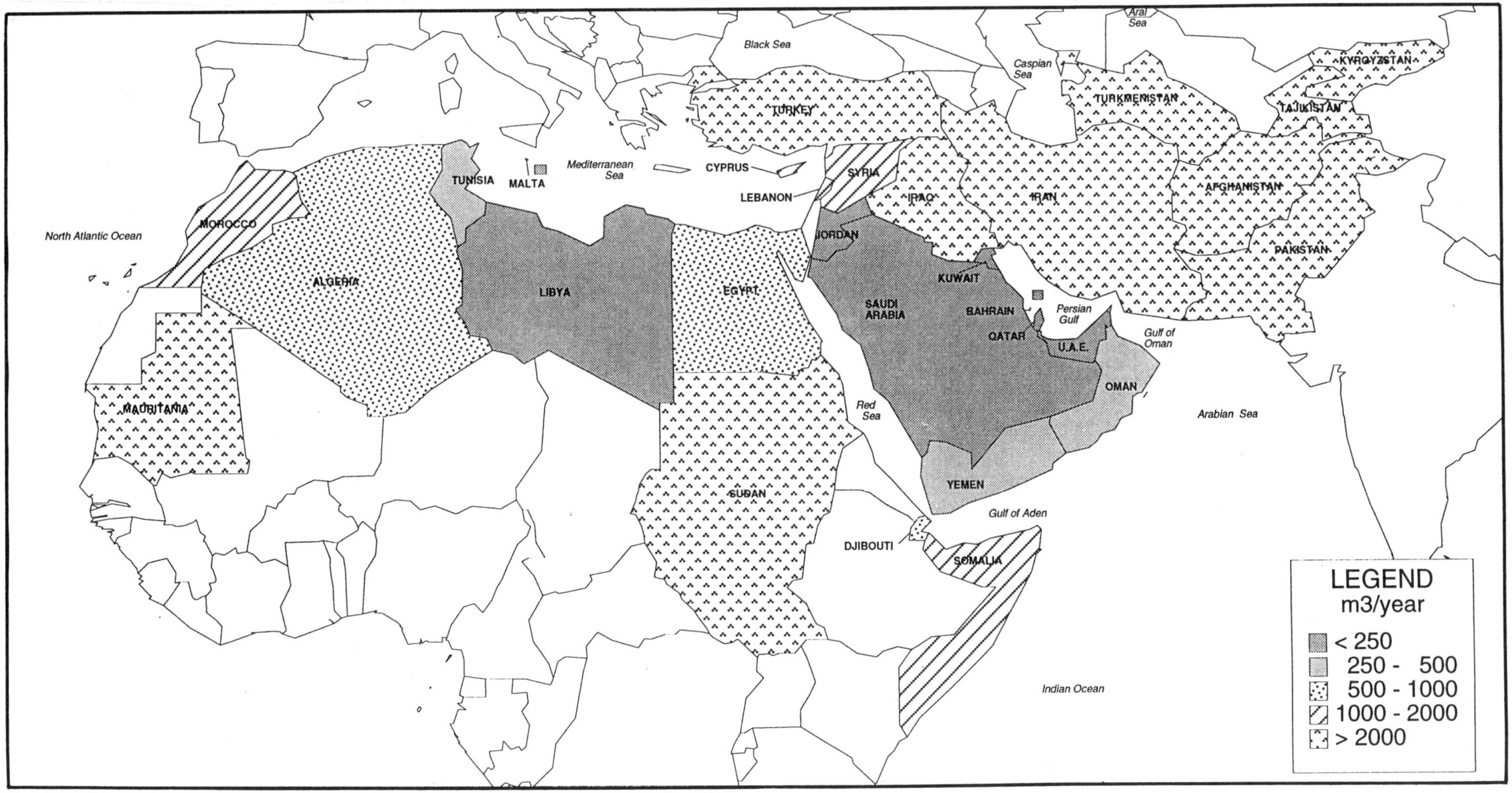

FIGURE 5
Renewable water resources dependency ratio

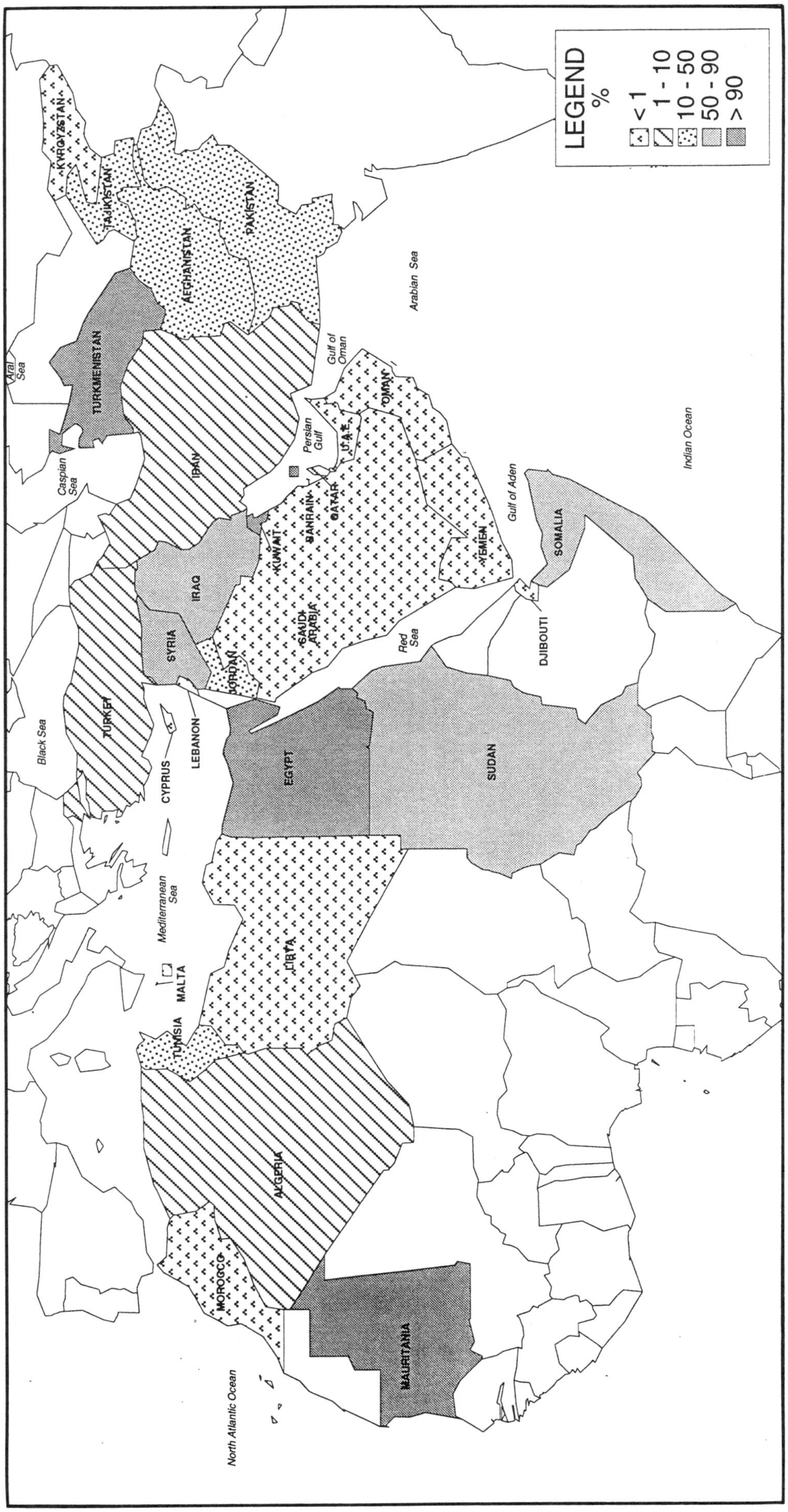

FIGURE 6
Water withdrawal (as a percentage of internal renewable water resources)

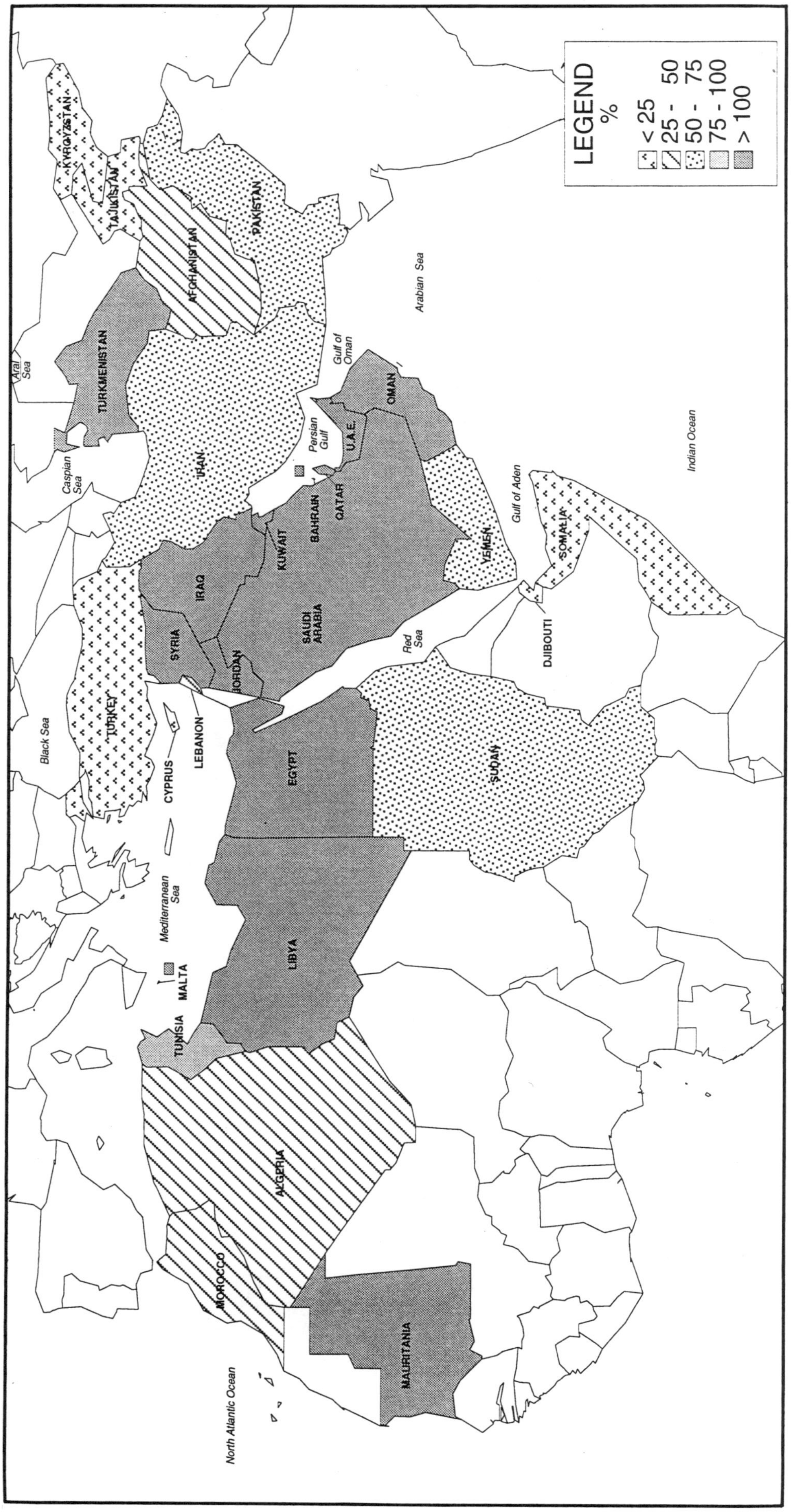

FIGURE 7
Water withdrawal (as a percentage of total actual renewable water resources)

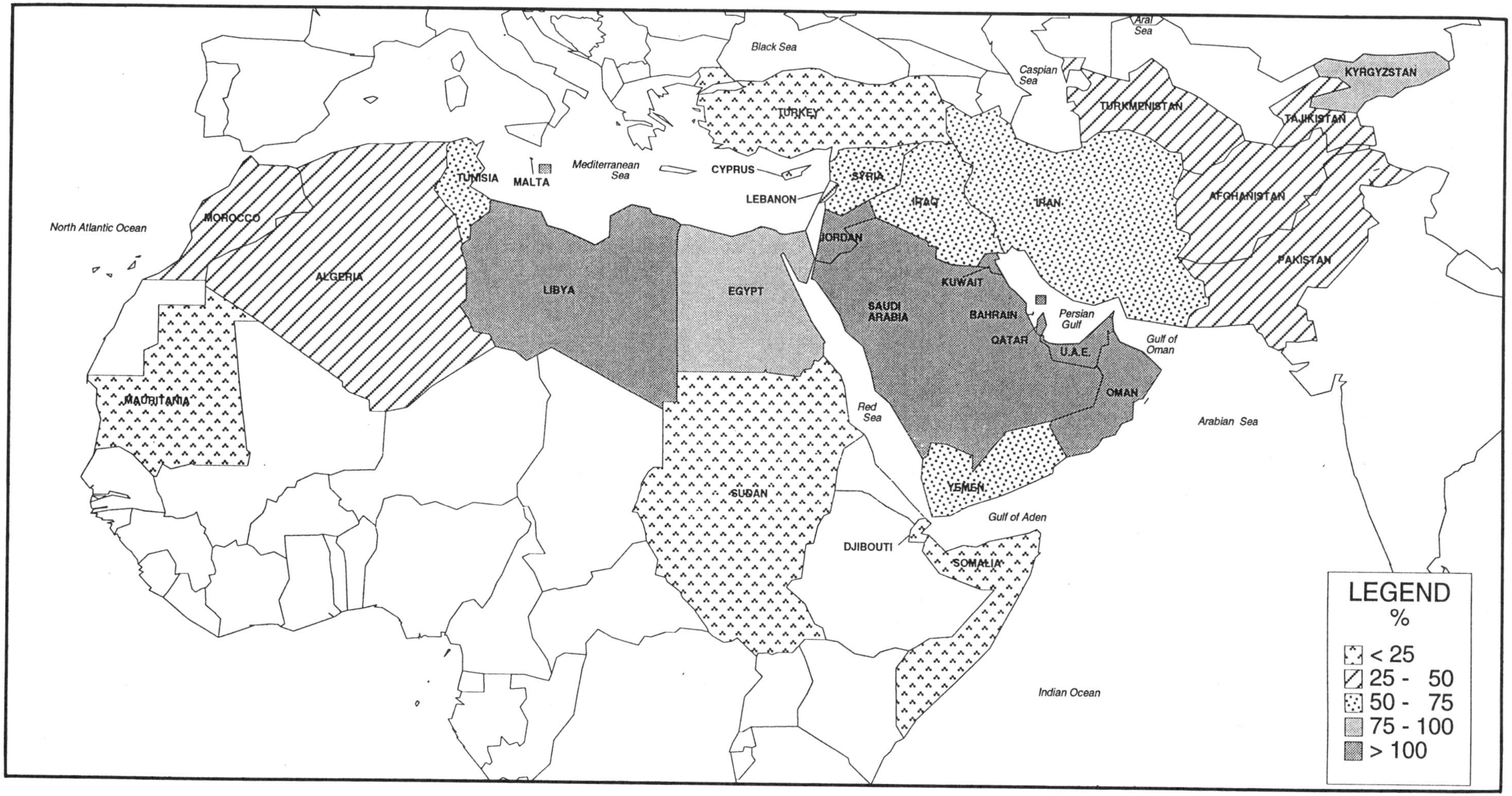

FIGURE 8
Use of non-conventional sources of water (as a percentage of total water withdrawal)

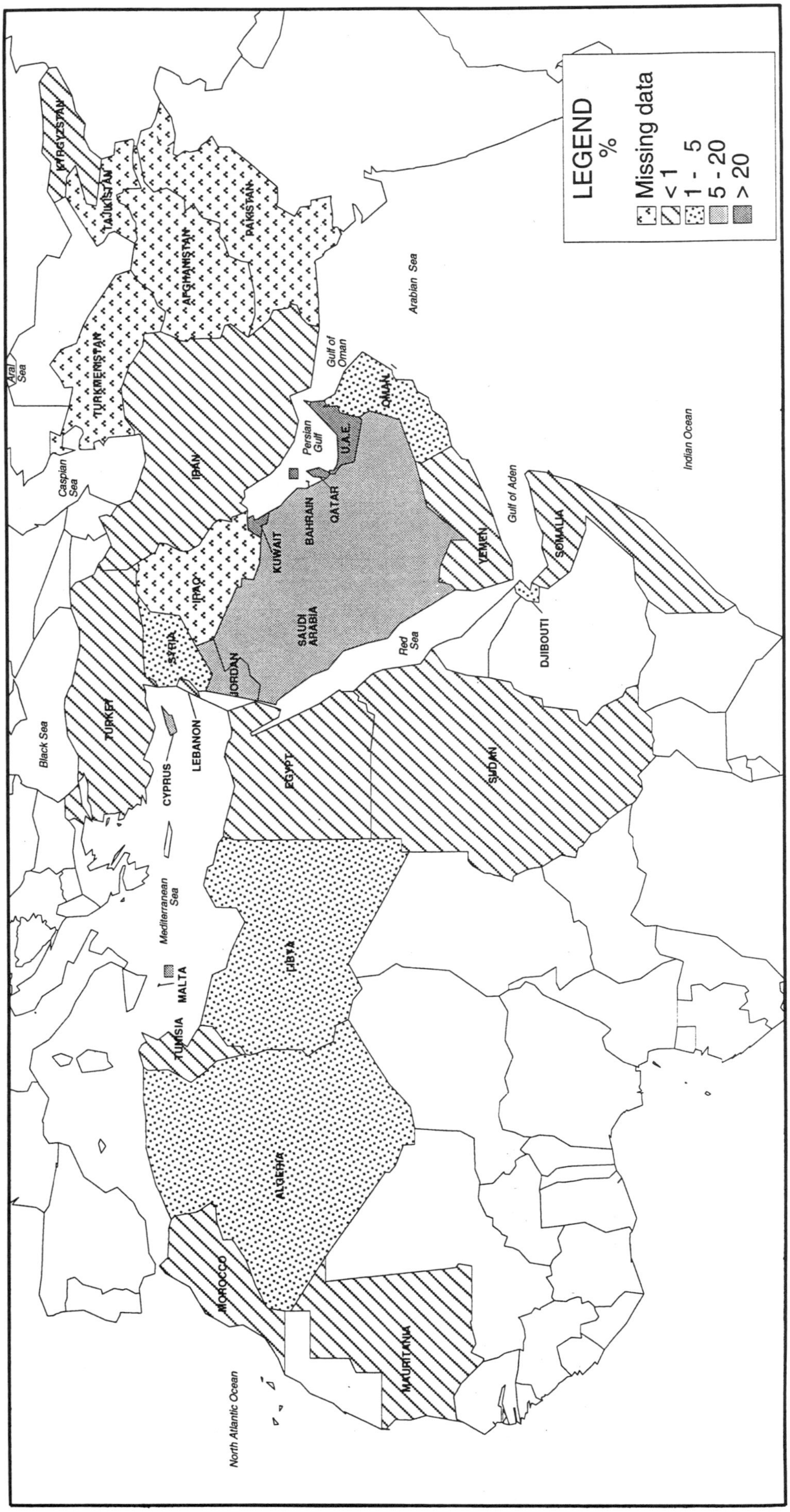

FIGURE 9
Area under irrigation (as a percentage of cultivated area)

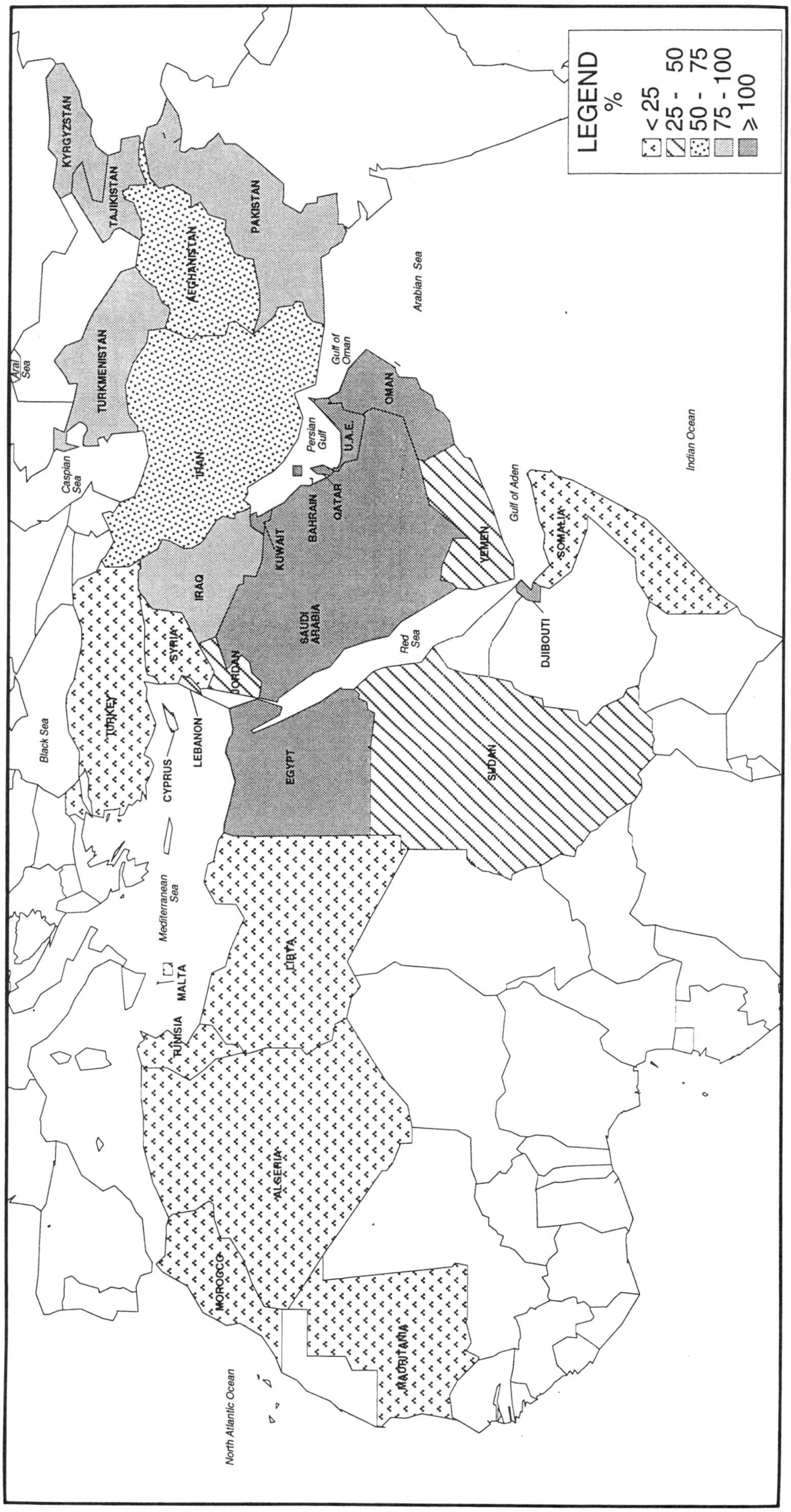

FIGURE 10
Area under irrigation (as a percentage of irrigation potential)

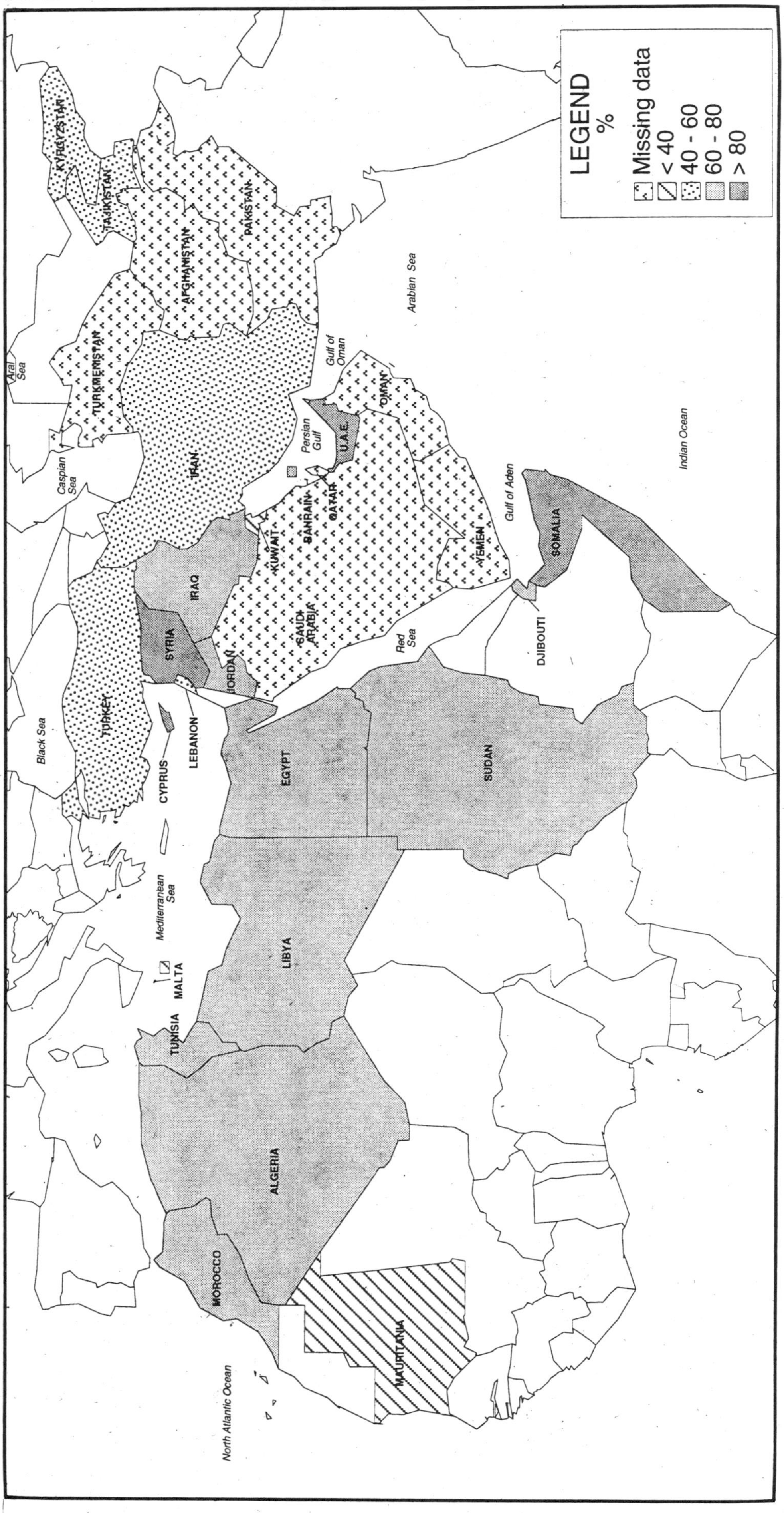

Country profiles

Afghanistan

GEOGRAPHY AND POPULATION

Afghanistan is a landlocked country in Central Asia with a total area of about 652 000 km^2. It is bordered by Turkmenistan, Uzbekistan and Tajikistan to the north, China to the north-east, Pakistan to the east and south and Iran to the west. It is characterized by its rugged terrain and an average elevation of 1 100 metres above sea level, ranging from 150 to 8 000 metres. About three-quarters of the territory consist of mountains and hills, while lowlands include river valleys in the northern part, and desert regions in the southern and south-eastern part.

The political instability prevailing in Afghanistan makes it extremely difficult to obtain reliable information on basic agricultural indicators. Most of the information presented below dates back to years prior to 1978.

TABLE 1
Basic statistics and population

Physical areas:			
Area of the country	1995	65 209 000	ha
Cultivable area	1978	8 000 000	ha
Cultivated area	1990	3 344 000	ha
- annual crops	1990	3 200 000	ha
- permanent crops	1990	144 000	ha
Population:			
Total population	1995	20 141 000	inhabitants
Population density	1995	31	inhab./km^2
Rural population	1995	80	%
Water supply coverage:			
Urban population	1993	38	%
Rural population	1993	17	%

The cultivable area has been estimated at 8 million ha, which is 12% of the total area. In 1991, the area cultivated with annual crops was estimated at 3.2 million hectares, which is only 82% of the area cultivated in 1978. In addition, about 144 000 ha were estimated to consist of permanent crops in 1990.

The total population is 20.1 million (1995), of which 80% is rural. Up to 6 million Afghans, perhaps one third of Afghanistan's 1978 population are estimated to have fled the country because of the war. Another 800 000 to 1 million people are thought to have been displaced by the fighting to the more remote, mountainous regions of the country or to the relative security of the country's few towns and cities. In 1991, 67% of the labour force was employed in the agricultural sector and agriculture accounted for almost 46% of GDP.

CLIMATE AND WATER RESOURCES

Climate

Afghanistan is characterized by a continental climate, though the presence of mountains causes many local variations. Temperatures vary from - 10°C in winter to 34°C in summer. The annual distribution of rainfall shows a picture of an essentially arid country, with more than 50% of the territory receiving less than 300 mm of rain. With the exception of the eastern border regions, which are at the far edges of monsoon influence, about 50% of the precipitation occurs in winter (January to March), much of which falls in the form of snow in

TABLE 2
Water: sources and use

Renewable water resources:			
Average precipitation		-	mm/yr
		-	km^3/yr
Internal renewable water resources		55	km^3/yr
Total (actual) renewable water resources	1995	65	km^3/yr
Dependency ratio	1995	15.4	%
Total (actual) renewable water resources per inhabitant	1995	3 227	m^3/yr
Total dam capacity		-	10^6 m^3
Water withdrawal:			
- agricultural	1987	25 849	10^6 m^3/yr
- domestic	1987	261	10^6 m^3/yr
- industrial	1987	0	10^6 m^3/yr
Total water withdrawal		26 110	10^6 m^3/yr
per inhabitant	1987	1 702	m^3/yr
as % of total (actual) renewable water resources		40.2	%
Other water withdrawal		-	10^6 m^3/yr
Average groundwater depletion		-	10^6 m^3/yr
Wastewater - Non-conventional water sources:			
Wastewater:			
- produced wastewater		-	10^6 m^3/yr
- treated wastewater		-	10^6 m^3/yr
- reused treated wastewater		-	10^6 m^3/yr
Desalinated water		-	10^6 m^3/yr

the central mountainous regions. A further 30% falls in spring (April to June). The runoff from snow melt, in the spring and summer months when day temperatures are high, is the life-blood of Afghan agriculture.

Water resources

Based on the hydrographic systems, the country can be divided into four zones:

- the northern basin (24% of the territory) with the Amu Darya and its tributaries (14%), which drain towards the Aral sea, and the rivers of northern Afghanistan (10%), which disappear within the country before joining the Amu Darya;
- the western region (12%) consisting of the Hari Rud river basin (6%) and the Murgab river basin (6%), both rivers disappearing in Turkmenistan;
- the south-western basin (52%) with the Helmand river flowing towards the Sistan swamps, located on the border of Iran and Afghanistan. In 1972, a document was signed between Afghanistan and Iran to allocate a discharge of 26 m^3/s of Helmand river water to Iran all year round;
- the eastern Kabul basin (12%), which is the only river system having an outlet to the sea, joining the Indus at Attock in Pakistan.

Internal renewable water resources are estimated at 55 km^3/year. The Kunar river, which originates in Pakistan, crosses the border with an average annual flow of 10 km^3 and joins the Kabul river at Jalalabad about 180 km further downstream. The Kabul river flows again into Pakistan 80 km further downstream.

Total water withdrawal was estimated at 26.11 km^3 in 1987, of which 99% for agricultural purposes (Figure 1). Recently, there has been a large development of groundwater use in some provinces.

TABLE 3
Irrigation and drainage

Irrigation potential			-	ha
Irrigation:				
1. Full or partial control irrigation: equipped area		1967	2 385 740	ha
- surface irrigation			-	ha
- sprinkler irrigation		1963	114 000	ha
- micro-irrigation			-	ha
% of area irrigated from groundwater		1967	15.4	%
% of area irrigated from surface water		1967	84.6	%
% of area irrigated from non-conventional sources		1967	0.0	%
% of equipped area actually irrigated			-	%
2. Spate irrigation area			-	ha
3. Equipped wetland and inland valley bottoms (i.v.b.)			-	ha
Total irrigation (1+2+3)		**1967**	**2 385 740**	**ha**
- as % of cultivated area			71	%
4. Flood recession cropping area			-	ha
Total water managed area (1+2+3+4)		**1967**	**2 385 740**	**ha**
- as % of cultivated area			71	%
- increase over last 10 years			-	%
- power irrigated area as % of water managed area			-	%
Full or partial control irrigation schemes:	Criteria			
Large-scale schemes	> - ha		-	ha
Medium-scale schemes			-	ha
Small-scale schemes	< - ha		-	ha
Total number of households in irrigation			-	
Irrigated crops:				
Total irrigated grain production			-	tons
as % of total grain production			-	%
Harvested crops under irrigation (full or partial control)			**-**	**ha**
- permanent crops: total			-	ha
- annual crops: total			-	ha
. wheat		1990	1 030 000	ha
.			-	ha
.			-	ha
.			-	ha
. other annual crops			-	ha
Drainage - Environment:				
Drained area			-	ha
as % of cultivated area			-	%
- drained areas in full or partial control irrigated areas			-	ha
- drained areas in equipped wetland and i.v.b			-	ha
- other drained areas			-	ha
- total drained area with subsurface drains			-	ha
- total drained area with surface drains			-	ha
Flood-protected area			-	ha
Area salinized by irrigation			-	ha
Population affected by water-borne diseases			-	inhabitants

In 1986, there were two dams higher than 15 m. The installed capacity of the hydroelectric plants was 281 MW in 1992, which is about 70% of total installed capacity. There is considerable potential for the generation of hydropower, both by large dams and micro-hydropower stations.

IRRIGATION AND DRAINAGE DEVELOPMENT

By 1978, the surface water potential was more or less fully exploited by existing irrigation systems if no further regulation works were going to be constructed, although the efficiency of exploitation left room for considerable improvement. Irrigated areas could have been expanded by building major dams and other water regulation structures, all of which required large capital investment. There is no estimate, even rough, of irrigation potential.

An estimated 27 to 36% of all irrigation systems were directly affected by war and these figures do not take into account the indirect effects of neglect and abandonment.

FIGURE 1
Water withdrawal (total: 26.11 km^3 in 1987)

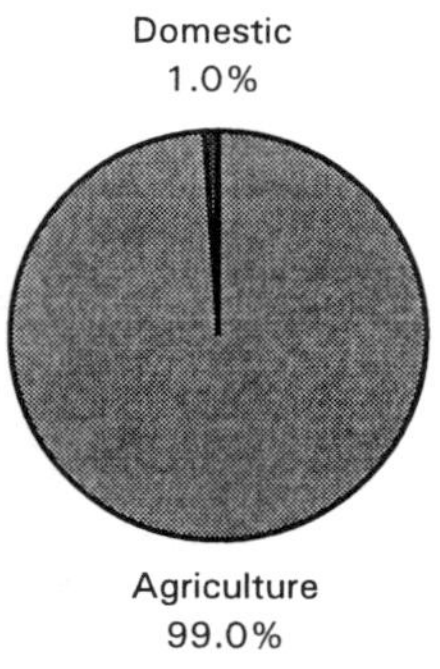

Agriculture was said to be almost entirely dependent on irrigation, although this view tended to ignore the large areas of rainfed wheat grown on the steep rolling hills and steppes in the north. The most reliable figures for irrigated areas date back to 1967, giving a total of 2 385 740 ha. Irrigated land can be divided into 4 classes, according to the origin of the irrigation water. They are presented in Figure 2. In 1963, some 114 000 ha were reported to be equipped for sprinkler irrigation.

FIGURE 2
Origin of irrigation water f/p (total: 2 385 740 ha in 1967)

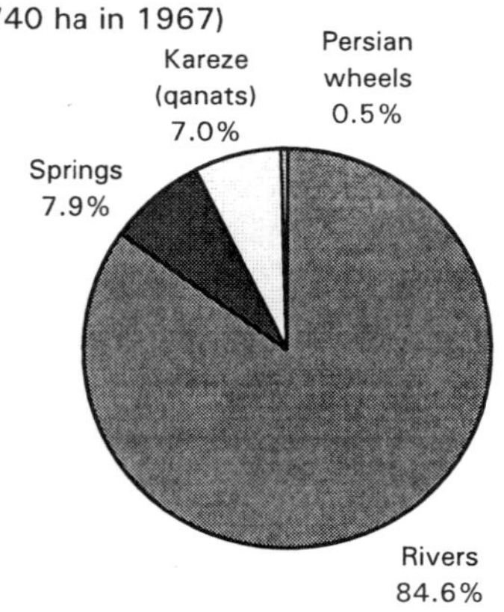

From 1967 up to the present, many different figures have been given for irrigated, rainfed and total cultivated areas. Unfortunately, most of them are unreliable. Some pre-war publications suggest that 2.8 million hectares were cultivated, of which 1.4 million ha had sufficient water to support double-cropping. Figures for rainfed areas are generally cited as 1.4 million hectares, but recent satellite data (1992) indicate that the area might be much larger - a figure of 3.1 million hectares has been suggested.

Irrigation systems can be divided into four main categories:

- Kareze systems. A kareze (qanat) is an unlined tunnel in the hillside, bringing water by free flow from underground aquifers to be used for surface irrigation. Dug by local craftsmen from shafts at close intervals, they are small in size but may be many kilometres in length. It is estimated that 6 470 kareze still supply water to 167 750 ha, as in 1967, the date of the last inventory. It should be noted that kareze are often used for domestic water supply.

- Small-scale informal surface water systems. These are the traditional irrigation systems, many of which have been established for centuries. In the past, maintenance and reconstruction were generally arranged on a traditional informal or communal village basis, and water rights were determined and recognized in a similar manner. Technical knowledge and operational systems were thus dependent on traditional community structures, and were largely retained in the memory of individuals.

- Large-scale informal surface water systems. These are located mainly in the plains and along the main valleys. Although they are called informal, their operation and maintenance was highly structured. Repair and maintenance works can mobilize very large quantities of labour for a long period and farmers in the command area have to contribute in labour, cash or kind. Large parts of these schemes have been abandoned because of the sterilization of the land (waterlogging and salinization), particularly in the Hari Rud, Farah Rud and Helmand valleys.

- Formal irrigation schemes. Formally organized large-scale irrigation systems are a relatively recent innovation. However, by the late 1970s three large-scale modern irrigation systems had been built and were in operation: the Helmand-Arghandab system in the south-west, the Ghaziabad farms near Jalalabad in the east, and the Kunduz-Khanabad system in the northern part of the country. By 1993, only a very small part of these schemes was still operational. Land tenure was different from most traditional systems in that ownership of land was registered. Some schemes were operated under private land ownership agreements, while others were operated as state farms where land ownership was deeded to the State.

In 1993, the average cost of irrigation scheme rehabilitation was estimated at $US 200/ha for small schemes. Rehabilitation costs for large, modern schemes, including main structures, are considerably higher.

Cropping intensity varies widely from system to system according to the relative scarcity of water in relation to land. It might reach 200% in large formal systems with water control (upstream of the river systems, when climatic conditions allow an early wheat crop), while in other systems up to two-thirds of the equipped area are kept fallow each year on a rotation basis.

Per caput wheat consumption in Afghanistan is one of the highest in the world. Pre-war, irrigated land produced 77% of all wheat and 85% of all food and agricultural crops. About 1 million ha of wheat was irrigated in 1990, although the 1978 figure may be as high as 1.3 million ha, and irrigated yields are estimated to be 3 times rainfed yields. Global cereal yields are estimated to have decreased by 30% between 1978 and 1987.

Floods are generally violent and can cause serious damage to agricultural land or inhabited areas. About 50 gabion river protection works and 50 flood protection masonry walls were constructed before the war, mostly in the Nangarhar and Parwan provinces, in the eastern part of the country.

INSTITUTIONAL ENVIRONMENT

The Central Authority for Water and Sewerage is responsible for the urban water supply and sanitation, under the aegis of the Central Authority for Housing and Town Planning, while the Ministry of Rural Development and Rehabilitation is responsible for the rural water supply.

The Ministry of Water and Power was formed in 1988 as the merger of the former separate Ministry of Irrigation and Water Resources and the Ministry of Power. It is in charge of the hydrological network, the development of water resources and of large-scale and modern irrigation systems, especially of diversion structures on the rivers and main canals. The Ministry's Department of Water Management is operational in most provinces.

The Department of Irrigation of the Ministry of Rural Rehabilitation and Development is in charge of small-scale irrigation systems and of irrigation systems downstream of the diversion structures on the main canal.

The Ministry of Agriculture is in charge of production and, therefore, of on-farm infrastructure and water management.

An Irrigation Secretariat within the Ministry of Planning is coordinating irrigation development planning.

In some cases, the Irrigation Development Authorities are still operational. They are in charge of the development, operation and maintenance of large-scale irrigation systems. Government construction and consulting companies involved in irrigation are:

- The Helmand Construction Company (HCC);
- The Spinghar Construction Unit;
- The Water & Power Engineering Company of Afghanistan (WAPECA).

A water law was passed in 1981.

TRENDS IN WATER RESOURCES MANAGEMENT

In 1993, the Ministry of Agriculture's most pessimistic estimates were that only 750 000 ha were still under satisfactory irrigation. A number of programmes have been proposed and some launched in order to rehabilitate the infrastructures, the human and other resources and the institutions and social infrastructure as related to irrigated agriculture.

At the same time, rehabilitation works have to be carried out in a context of scarce human resources and funds, because the people must also address several other urgent needs and repair other basic infrastructures. Therefore, apart from improving water management and reducing water losses to increase irrigated areas, rehabilitation works have to aim at reducing future maintenance requirements through proper design.

Under the last five-year plan, 76% of the budget for the agricultural sector was earmarked for irrigation works.

Due to the increasing use of groundwater in recent years, there is a risk of over-exploitation and depletion in the absence of regulating and licensing authorities, which in some places might lead to the drying out of kareze or qanats, springs and wells, depending on the same water sources.

MAIN SOURCES OF INFORMATION

FAO. 1965. Survey on land and water resources. Afghanistan. Six volumes. Kabul.

UNDP. 1993. Afghanistan rehabilitation strategy: Action plan for immediate rehabilitation. Six volumes. Kabul.

UNORSA. 1993. Afghanistan irrigation subsector profile: Elements for a two-year agriculture rehabilitation action plan. Report prepared by T. Facon, FAO. Islamabad, Pakistan.

Algeria

GEOGRAPHY AND POPULATION

Algeria, with a total area of about 2.4 million km^2, is located in the north-western part of the African continent overlooking the Mediterranean Sea in the north. It is bordered by Tunisia and Libya to the east, Niger, Mali and Mauritania to the south and Morocco to the west.

The country can be divided into three physiographic units:

- The Atlas Mountains. They consist of two principal mountain chains: the El-Atlas El-Talli Mountains, extending along a narrow coastal strip, and the Desert Atlas Mountains with peaks over 2 300 metres above sea level. The two chains are separated by hills that contain a group of depressions known as El-Shatout.

TABLE 1
Basic statistics and population

Physical areas:			
Area of the country	1995	238 174 000	ha
Cultivable area	1995	10 742 000	ha
Cultivated area	1993	3 372 000	ha
- annual crops	1993	2 841 600	ha
- permanent crops	1993	530 400	ha
Population:			
Total population	1995	27 939 000	inhabitants
Population density	1995	12	inhab./km^2
Rural population	1995	44	%
Water supply coverage:			
Urban population	1985	85	%
Rural population	1985	56	%

- The South and South-east Hills. They are a group of isolated hills in the north of the Great Desert called El-AhjarHills with a peak of over 3 000 metres above sea level.

- The Hilly Coastal Depressions. They include the region separating the Atlas Mountains and the El-Ahjar Hills. The elevation is 350 metres in the south-west and about 100 metres in the north-east. The minimum elevation is 21 metres below sea level. Sand dunes are scattered all over this region.

The cultivable area is estimated at about 10.74 million ha, which is 3% of the total area of the country. In 1993, the total area available for agriculture was estimated at about 8.10 million ha, of which 4.73 million ha, or almost 60%, were temporarily fallow. Of the remaining cultivated area of 3.37 million ha, 2.84 million ha consisted of annual crops and 0.53 million ha consisted of permanent crops. Agricultural development is concentrated in the northern part of the country, where the best soils are located and where the climatic conditions are more favourable.

The total population is about 15.6 million (1995), of which 44% is rural. The annual demographic growth rate is estimated at 3.2%. Agriculture employs 25% of the labour force and accounts for 9 to 13% of GDP, depending on the year.

CLIMATE AND WATER RESOURCES

Climate

The climate varies from the desert type in the south to Mediterranean in the north. Average annual rainfall is about 68 mm, but varies from 0 mm in the southern desert up to 1 500 mm

TABLE 2
Water: sources and use

Renewable water resources:			
Average precipitation		68	mm/yr
		162.9	km^3/yr
Internal renewable water resources		13.9	km^3/yr
Total (actual) renewable water resources	1995	14.3	km^3/yr
Dependency ratio	1995	2.8	%
Total (actual) renewable water resources per inhabitant	1995	512	m^3/yr
Total dam capacity	1992	4 300	$10^6\ m^3$
Water withdrawal:			
- agricultural	1990	2 700	$10^6\ m^3/yr$
- domestic	1990	1 120	$10^6\ m^3/yr$
- industrial	1990	680	$10^6\ m^3/yr$
Total water withdrawal		4 500	$10^6\ m^3/yr$
per inhabitant	1990	180	m^3/yr
as % of total (actual) renewable water resources		31.5	%
Other water withdrawal		-	$10^6\ m^3/yr$
Average groundwater depletion		-	$10^6\ m^3/yr$
Wastewater - Non-conventional water sources:			
Wastewater:			
- produced wastewater	1985	660	$10^6\ m^3/yr$
- treated wastewater		-	$10^6\ m^3/yr$
- reused treated wastewater		-	$10^6\ m^3/yr$
Desalinated water	1990	64	$10^6\ m^3/yr$

in the north-eastern coastal area around Skikda. However, even in this region the dry season lasts five months. Precipitation, which mainly occurs in winter and the beginning of spring, is very irregular with considerable variations from year to year.

Water resources

Internal renewable water resources are estimated at 13.9 km^3/year. Incoming surface water has been estimated at 0.4 km^3/year, of which 0.2 km^3 from Morocco and 0.2 km^3 from Tunisia. The water resources, that are potentially available for use in the northern part of the country and the high plateaux have been estimated at 8.1 km^3/year, of which 6.5 km^3 is surface water to be regulated by dams and 1.6 km^3 is groundwater. The safe yield of fossil water in the Sahara varies between 2 and 5 km^3/year according to different hypotheses. However, the extraction of this fossil water is expensive.

FIGURE 1
Water withdrawal (total: 4.5 km^3 in 1990)

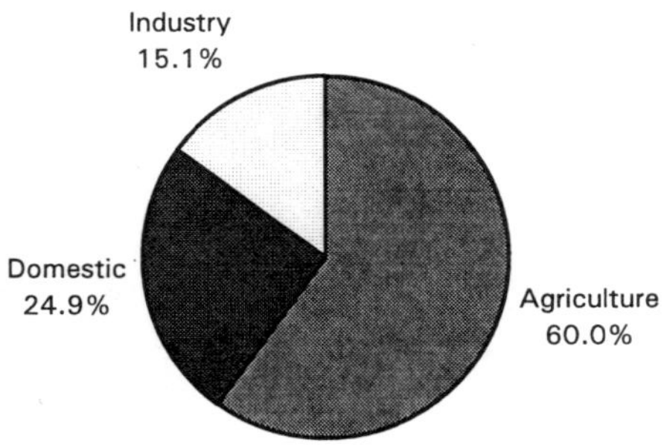

In 1992, 79 dams had been constructed or were under construction with a total dam capacity of 4.3 km^3. The total capacity of dams for irrigation under the schemes of the Regional Offices (see below) was estimated at about 1.2 km^3.

In 1990, total water withdrawal was estimated at 4.5 km^3, of which 60% for agricultural purposes (Figure 1).

TABLE 3
Irrigation and drainage

Irrigation potential	1992	730 000	ha
Irrigation:			
1. Full or partial control irrigation: equipped area	1992	445 500	ha
- surface irrigation		-	ha
- sprinkler irrigation	1992	40 000	ha
- micro-irrigation		-	ha
% of area irrigated from groundwater		-	%
% of area irrigated from surface water		-	%
% of area irrigated from non-conventional sources		-	%
% of equipped area actually irrigated	1992	82.2	%
2. Spate irrigation area	1984	110 000	ha
3. Equipped wetland and inland valley bottoms (i.v.b.)		-	ha
Total irrigation (1+2+3)	**1992**	**555 500**	**ha**
- as % of cultivated area		16	%
4. Flood recession cropping area		-	ha
Total water managed area (1+2+3+4)	**1992**	**555 500**	**ha**
- as % of cultivated area		16	%
- increase over last 10 years		-	%
- power irrigated area as % of water managed area		-	%
Full or partial control irrigation schemes:	Criteria		
Large-scale schemes	> - ha	-	ha
Medium-scale schemes		-	ha
Small-scale schemes	< - ha	-	ha
Total number of households in irrigation		-	
Irrigated crops:			
Total irrigated grain production		-	tons
as % of total grain production		-	%
Harvested crops under irrigation (full or partial control)		**-**	**ha**
- permanent crops: total	1986	145 000	ha
- annual crops: total		-	ha
. vegetables	1986	95 000	ha
. fodder crops	1986	25 000	ha
. cereals	1986	8 000	ha
.		-	ha
. other annual crops		-	ha
Drainage - Environment:			
Drained area		-	ha
as % of cultivated area		-	%
- drained areas in full or partial control irrigated areas		-	ha
- drained areas in equipped wetland and i.v.b		-	ha
- other drained areas		-	ha
- total drained area with subsurface drains		-	ha
- total drained area with surface drains		-	ha
Flood-protected area		-	ha
Area salinized by irrigation		-	ha
Population affected by water-borne diseases		-	inhabitants

IRRIGATION AND DRAINAGE DEVELOPMENT

Irrigation potential has been estimated at 1.24 million ha considering soil resources. If available water resources are also considered, this potential has been estimated at 730 000 ha; if only renewable water resources are considered, it has been estimated at about 510 000 ha. It should be noted that during the very dry period of the years 1980-90, the available water resources were far from sufficient to irrigate the area already equipped for full or partial control irrigation. Irrigation is now of fundamental importance in Algeria, not only for winter crops, but also for spring and summer crops.

In 1992, the total water managed area was estimated at 555 500 ha, of which 110 000 ha benefited from spate irrigation, while 445 500 ha were full or partial control irrigation schemes (Figure 2). The latter can be divided into 45 000 ha of oases located in the south of the country and 400 500 ha of schemes in the north of the country which are usually divided into 'large schemes' and 'small to medium schemes' (Figure 3).

FIGURE 2
Distribution of the water managed areas (total: 555 500 ha in 1992)

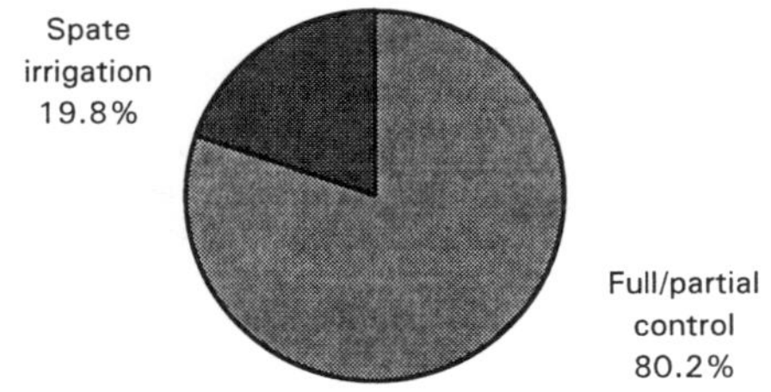

There are 17 large schemes, the size of which varies between 2 200 and 22 500 ha, with a total area of 175 500 ha. They can be regrouped according to which authority is responsible.

FIGURE 3
Distribution of the f/p control irrigation schemes (total 445 500 ha in 1992)

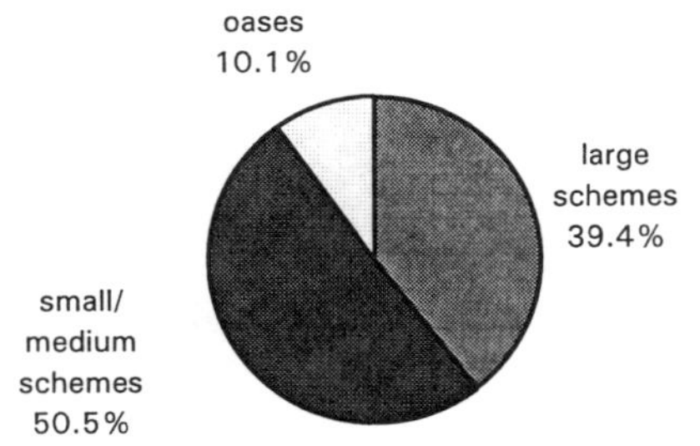

- 10 schemes, from 2 200 to 22 500 ha, regrouped into four Regional Offices (OPIR or Offices de périmètres irrigués régionaux) at El Tarf, Mitjdja, Cheliff and Habre-Sig. The total equipped area is 146 200 ha of which only 80 100 ha are irrigated on a fairly regular basis (Figure 4). The reasons for this are shortage of water, the deterioration of the infrastructures and the abandonment of soils due to salinization. In fact, in 1991 only 32 000 ha were actually irrigated, or 22% of the equipped area, because of a water shortage that year. Most of the schemes are old, 63% having been equipped between 1937 and 1943.

- 7 schemes, from 3 000 to 5 500 ha, regrouped into seven Wilaya Offices (OPIW or Offices de périmètres irrigués de Wilaya). The total equipped area is 29 300 ha of which 16 200 ha are irrigated on a fairly regular basis for the same reasons as explained above. In 1991, only 10 000 ha were actually irrigated, or 34% of the equipped area. Most of these schemes were constructed more recently than the OPIR schemes.

The Saf Saf irrigation scheme, formerly linked to its Wilaya, has been incorporated in the OPIR of El Tarf in 1992, in which the scheme of Guelma, with a total area of 12 000 ha

FIGURE 4
Distribution of the areas equipped by the regional and Wilaya Officer (total: 175 500 ha in 1992)

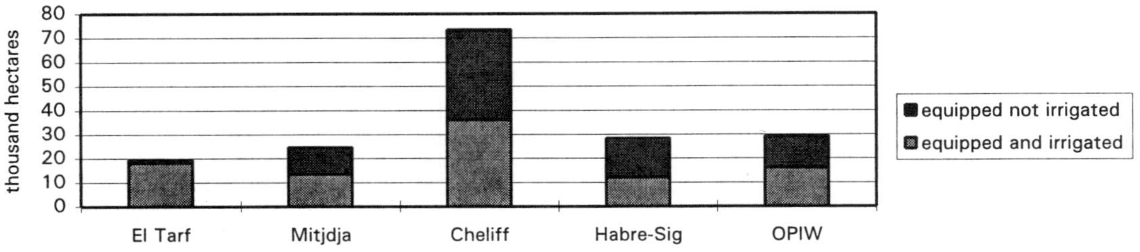

FIlGURE 5
Irrigation techniques in the OPI schemes (total: 96 300 ha in 1992)

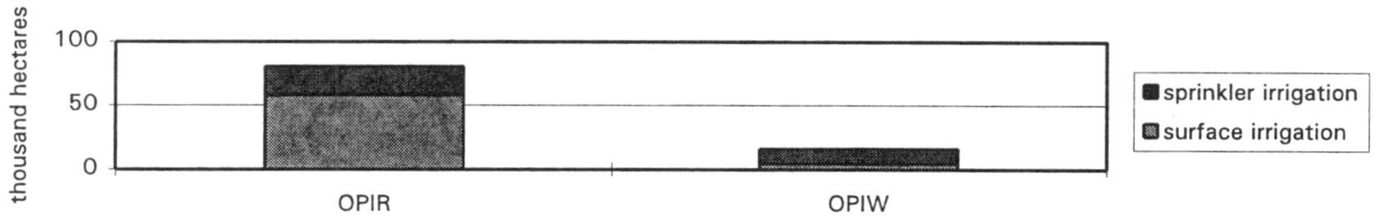

(under construction in 1993), will also be incorporated (this area has not yet been included in the above figures).

FIGURE 6
Origin of irrigation water in the small and medium schemes (total: 225 000 ha in 1992)

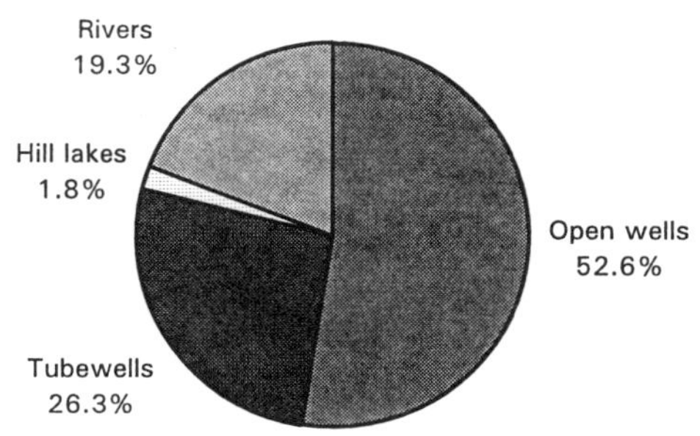

Of the total equipped area of 96 300 ha of the OPIR and OPIW which is irrigated on a fairly regular basis, 62 560 ha are irrigated by surface irrigation techniques, while sprinkler irrigation is practised on 33 740 ha (Figure 5).

In the large schemes (OPIR and OPIW), fruit trees are grown on about 31 000 ha, which represents only 17% of the equipped area but 74% of the area actually irrigated in these schemes in 1991. Priority is given to fruit trees in periods of water shortage. Consequently, in order to avoid frequent restrictions in water distribution to the vegetable crops, many farmers have dug wells on the schemes themselves.

FIGURE 7
Major irrigated crops f/p in 1984

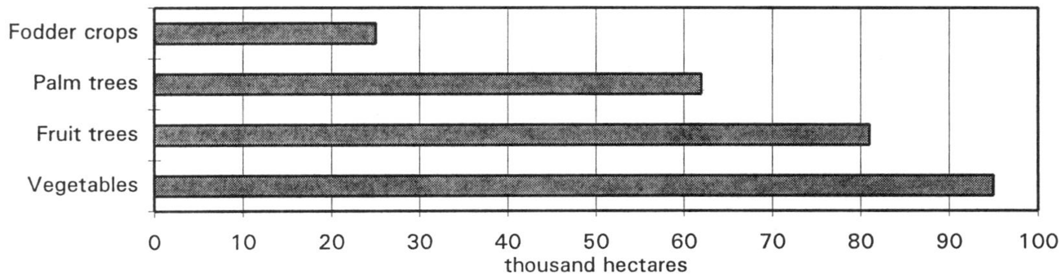

The average cost of irrigation development in the large schemes has been put at $US 11 000/ha for surface irrigation and $US 15 000/ha for sprinkler irrigation. Water charges, which in Algerian Dinar (DA) remained the same between 1985 and 1992, were DA 0.35/m^3 ($US 0.046 in 1989, but only $US 0.015 in 1992). This amount does not cover operation and maintenance costs.

The small and medium schemes covered 225 000 ha in 1992 and were irrigated from open wells, tubewells, small lakes in the hills formed by earth dams, or from small rivers from which water is diverted or pumped (Figure 6).

The major irrigated crops in the full or partial control irrigation schemes are vegetables, fruit trees and palm trees (Figure 7).

INSTITUTIONAL ENVIRONMENT

The following authorities are involved in the water sector:

- The Ministry of Equipment and Housing (MEL or Ministère de l'équipement et du logement). This Ministry covers all the activities of research, exploitation, production, storage and distribution of water for all uses.

At the central level, there are four departments:

◊ Direction de la planification des affaires économiques (DPAE, economic planning);
◊ Direction des grands aménagements et des infrastructures hydrauliques (DGAIH, large-scale water infrastructure);
◊ Direction de l'irrigation de petite et moyenne hydraulique (DIPMH, small- to medium-scale water infrastructure);
◊ Direction de la réglementation, de la protection et de l'usage de l'eau (DRPUE, water regulation, conservation and use).

The agencies, placed under the responsibility of MEL, are:

◊ Agence nationale pour les ressources hydrauliques (ANRH, water resources);
◊ Agence nationale des barrages (ANB, dams);
◊ Agence nationale de réalisation et de gestion des infrastructures hydrauliques pour l'irrigation et le drainage (AGID). This agency is in charge of the design, study and creation of the hydraulic infrastructures that will be operated by the OPI.

- The Ministry of Agriculture (MA or Ministère de l'Agriculture). The following departments are concerned with irrigation:

◊ Direction de la vulgarisation et des institutions rurales (DVIR, extension, rural institutions);
◊ Direction du génie rural (DGR, rural engineering).

- Several institutions at regional level:

◊ the 4 OPIR under the MEL;
◊ the 7 OPIW under the Wali (Wilaya);
◊ a Hydraulic Service (under the MEL) for every Wilaya;
◊ a Direction of the Agricultural Services (under the MA) for every Wilaya.

A water law was passed on 17 July 1992.

TRENDS IN WATER RESOURCES MANAGEMENT

Water resources in Algeria are limited and need to be preserved. Special attention has to be paid to the siltation of the dams, which sometimes reaches worrying levels. The worst case is the Cheurfas dam where, according to some estimates, the siltation has reached 96%.

A close eye also needs to be kept on the quality of the soils. As far as the irrigated areas are concerned, soil salinization has worsened during the dry years of the 1980s and some soils are becoming sterile. On the other hand, the area needing drainage in the OPIR has been estimated at 61 000 ha.

One-third of the cultivated land in Algeria has a slope of more than 12.5%, which poses erosion and soil conservation problems.

At the beginning of the 1990s Algeria drew up an extensive programme of rehabilitation and extension for the existing schemes, which should bring the total area equipped and irrigated in the north of the country up to 500 000 ha by the year 2010.

MAIN SOURCES OF INFORMATION

FAO. 1992. Algérie: projet d'appui à l'irrigation (PAI). Rapport d'identification. FAO Investment Centre. Cooperation programme FAO/World Bank. Report No 4/92 CP-ALG 36. Rome.

Pérennes, J.J. 1993. L'eau et les hommes au Maghreb: Contribution à une politique de l'eau en Méditerranée. Karthala. CNRS.

World Bank. 1988. Democratic and Popular Republic of Algeria: Irrigation Engineering Project. Washington DC.

Bahrain

GEOGRAPHY AND POPULATION

The State of Bahrain is an archipelago of 36 islands located in the Persian Gulf, about midway between Saudi Arabia to the west and Qatar to the east, with a total area of 695 km^2. The largest island is Bahrain where the capital is situated and which represents about 85% of the total area of the country.

The total population is 564 000 (1995), of which 10% is rural. With a population density of 812 inhabitants/km^2, it is one of the world's most densely populated countries. Bahrain has experienced high rates of population growth and urbanization since the early 1960s following the sudden increase in the country's oil revenues leading to a fast increase in its economic base and an improvement in the standard of living. The average annual demographic growth rate was 4% during the period 1980-1991.

TABLE 1
Basic statistics and population

Physical areas:			
Area of the country	1991	69 526	ha
Cultivable area	1994	4 230	ha
Cultivated area	1994	3 165	ha
- annual crops	1994	1 483	ha
- permanent crops	1994	1 682	ha
Population:			
Total population	1995	564 000	inhabitants
Population density	1995	811	inhab./km^2
Rural population	1995	10	%
Water supply coverage:			
Urban population	1995	100	%
Rural population	1995	100	%

Urban development at the expense of agricultural lands has caused significant loss of traditionally agricultural areas. Furthermore, soil salinization resulting from deterioration in the quality of the groundwater used in irrigation has led to a general reduction of the cultivated lands. At present, the total cultivable area is estimated at 4 230 ha, or 6% of the total area of the country. In 1994, the total cultivated area was about 3 165 ha, all irrigated, of which 1 483 ha consisted of annual crops (vegetables and fodder crops) and 1 682 ha consisted of permanent crops (fruit trees and dates).

Agriculture in Bahrain is generally in an unhealthy state with tenancy problems, small farm holdings, labour shortage and lack of financial incentives which restricts investment. In 1992, about 3 200 people were engaged in agricultural activities. This corresponds to about 2% of the labour force. Agriculture accounted for only 1% of GDP in 1991.

CLIMATE AND WATER RESOURCES

Climate

Bahrain has an arid to extremely arid environment. It is characterized by high temperatures, erratic and often scanty rainfall, high evapotranspiration rates (with peaks of over 10 mm/day in July) and high humidity levels due to the surrounding Arabian Gulf waters. Temperature averages vary from 17°C in winter (December-March) to 35°C in summer (June-September). The rainfall season extends from November to April, with an annual average of 70.6 mm,

TABLE 2
Water: sources and use

Renewable water resources:			
Average precipitation		70	mm/yr
		0.049	km^3/yr
Internal renewable water resources		0.004	km^3/yr
Total (actual) renewable water resources	1995	0.116	km^3/yr
Dependency ratio	1995	96.6	%
Total (actual) renewable water resources per inhabitant	1995	206	m^3/yr
Total dam capacity		-	10^6 m^3
Water withdrawal:			
- agricultural	1991	135.1	10^6 m^3/yr
- domestic	1991	94.3	10^6 m^3/yr
- industrial	1991	9.9	10^6 m^3/yr
Total water withdrawal		239.2	10^6 m^3/yr
per inhabitant	1991	465	m^3/yr
as % of total (actual) renewable water resources		206	%
Other water withdrawal	1991	3	10^6 m^3/yr
Average groundwater depletion	1991	96.04	10^6 m^3/yr
Wastewater - Non-conventional water sources:			
Wastewater:			
- produced wastewater	1991	44.9	10^6 m^3/yr
- treated wastewater	1991	44.9	10^6 m^3/yr
- reused treated wastewater	1991	8.03	10^6 m^3/yr
Desalinated water	1991	44.1	10^6 m^3/yr

sufficient only to support the most drought resistant desert vegetation. The mean annual relative humidity is over 67%.

Water resources

Total annual surface runoff is only about 4 million m^3 and there are no perennial streams. Bahrain receives groundwater by lateral under-flow from the Damman aquifer, which forms only a part of the extensive regional aquifer system, called the Eastern Arabian Aquifer. This aquifer extends from central Saudi Arabia, where its main recharge area is located at about 300 metres above sea level, to eastern Saudi Arabia and Bahrain, which are considered the discharge areas. The rate of groundwater inflow has been estimated at about 112 million m^3/year under steady-state conditions (before 1965) and this figure is considered to be the safe groundwater yield in Bahrain. There are no dams in Bahrain.

Desalinated water and treated wastewater

In 1991, the total quantity of desalinated water used was 44.1 million m^3/year. In addition Bahrain treats about 45 million m^3/year of wastewater (secondary treatment). Only 8 million m^3/year receive tertiary treatment and are used for irrigation purposes in government farms and some private farms, while the rest is discharged to the sea. The chemical and hygienic properties of the tertiary treated water are within international limits and are considered good for agricultural purposes. Although government plans for full utilization of the TSE (Treated Sewage Effluent) water through major agricultural projects exist, delay and lack of finances for these projects have caused limitations in the use of these waters.

Water withdrawal

In 1991, total water withdrawal in Bahrain was estimated at more than 239 million m^3, of which 56% or more for irrigation and livestock watering (Figure 1). About 94% of the water

TABLE 3
Irrigation and drainage

Irrigation potential		1994	4 230	ha
Irrigation:				
1. Full or partial control irrigation: equipped area		1994	3 165	ha
- surface irrigation		1994	2 497	ha
- sprinkler irrigation		1994	130	ha
- micro-irrigation		1994	538	ha
% of area irrigated from groundwater		1994	86.4	%
% of area irrigated from surface water		1994	0.0	%
% of area irrigated from non-conventional sources		1994	13.6	%
% of equipped area actually irrigated		1994	100	%
2. Spate irrigation area			-	ha
3. Equipped wetland and inland valley bottoms (i.v.b.)			-	ha
Total irrigation (1+2+3)		**1994**	**3 165**	**ha**
- as % of cultivated area			100	%
4. Flood recession cropping area			-	ha
Total water managed area (1+2+3+4)		**1994**	**3 165**	**ha**
- as % of cultivated area			-	%
- increase over last 10 years		1984-94	9	%
- power irrigated area as % of water managed area		1994	100	%
Full or partial control irrigation schemes:	Criteria			
Large-scale schemes	> 50 ha	1994	280	ha
Medium-scale schemes				ha
Small-scale schemes	< 50 ha	1994	2 885	ha
Total number of households in irrigation		1994	250	
Irrigated crops:				
Total irrigated grain production		1994	0	tons
as % of total grain production		1994	0	%
Harvested crops under irrigation (full or partial control)		**1994**	**3 165**	**ha**
- permanent crops: total		1994	1 682	ha
- annual crops: total		1994	1 483	ha
. vegetables (mainly tomatoes)		1994	1 040	ha
. fodder crops (mainly alfalfa)		1994	443	ha
Drainage - Environment:				
Drained area		1994	1 300	ha
as % of cultivated area			41	%
- drained areas in full or partial control irrigated areas			-	ha
- drained areas in equipped wetland and i.v.b			-	ha
- other drained areas			-	ha
- total drained area with subsurface drains			-	ha
- total drained area with surface drains			-	ha
Flood-protected area		1995	1 300	ha
Area salinized by irrigation		1994	1 065	ha
Population affected by water-borne diseases			-	inhabitants

used in agriculture, including livestock, is groundwater and 6% is treated wastewater, while for domestic and industrial purposes about 60% of the water used is groundwater and the remaining part desalinated water (Figure 2). Non-conventional water sources accounted for almost 22% in the total water withdrawal in 1991.

The excessive pumping of groundwater caused a sharp decrease in groundwater storage and a reduction in potentiometric levels of about 4 metres between 1965 and 1992. As a result, more than half the original

FIGURE 1
Water withdrawal (total: 239.23 million m^3 in 1991)

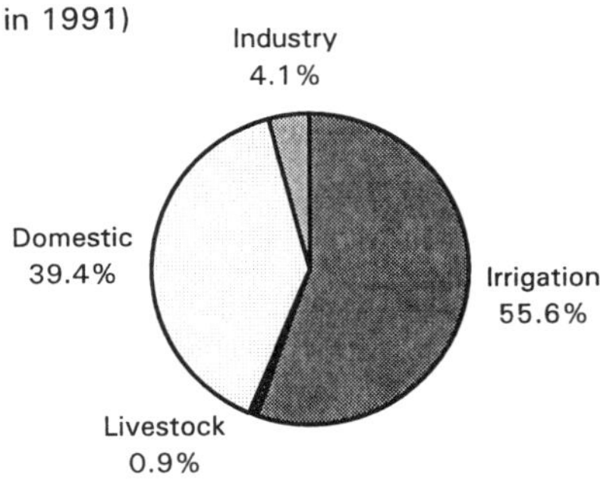

FIGURE 2

Origin of water used by sector (total: 239.23 million m^3 in 1991)

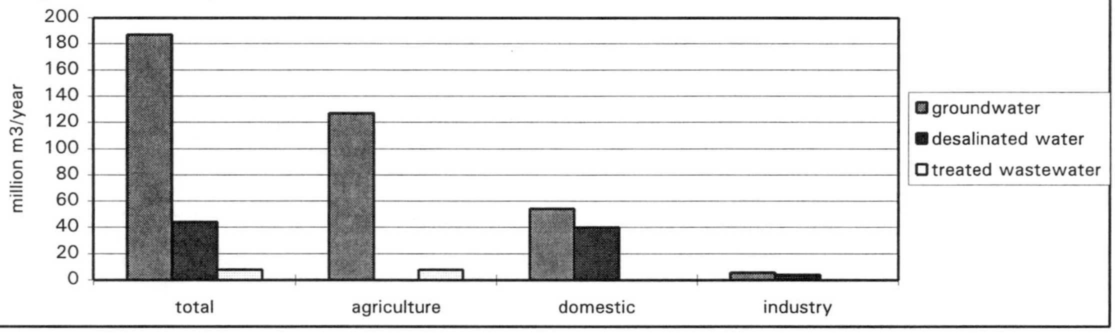

groundwater reservoir has been completely degraded due to seawater intrusion and saline water up-flow from the deeper zones. The table shows that annual extraction is almost twice annual recharge, leading to an ever increasing groundwater deficit. Average annual groundwater depletion over the period 1965-1992 was approximately 40 million m^3, in 1991/92 the groundwater depletion was over 96 million m^3.

Groundwater depletion in Bahrain, 1991/92

Component	Average annual rate (million m^3)
Inflow:	
Recharge by under-flow (aquifer safe yield)	112.00
Recharge by rainfall on outcrop and irrigation return flows	0.28
Total inflow	**112.28**
Outflow:	
Wells abstraction for irrigation, livestock, domestic,industrial and other purposes	190.20
Sabkha natural discharge	12.72
Natural springs discharge	5.40
Total outflow	**208.32**
Total inflow - Total outflow	**- 96.04**

IRRIGATION AND DRAINAGE DEVELOPMENT

The limited availability of good quality soils and water has resulted in the concentration of agricultural development in a relatively narrow strip of land along the north-western coast of Bahrain Island with isolated pockets in the north central areas and along the east coast. Most soils have a sandy texture, traces of organic matter (0.05-1.5%), a deficiency in major nutrients, low water-holding capacity (available moisture 2-6%), and high infiltration rates (>120 mm/hr). In areas along the coastal strip, calcareous impermeable layers are found at varying depths of 1 to 3 metres, causing waterlogging and impeding leaching. Electrical conductivity (EC) in irrigated soils lies within a range of 4-12 mmhos/cm, while in the areas of recently abandoned agriculture (1 065 ha) it could reach 60 mmhos/cm.

At present, out of the total agricultural land of 4 230 ha, drainage works are being carried out on 1 850 ha. The remaining 2 380 ha still suffer from shallow water tables resulting in waterlogging in the crop root zones and an increasing salinization of the top soil. Drainage requirements are exacerbated by the inefficient surface irrigation systems used. In 1994, drainage works had been completed on about 1 300 ha. The average cost of drainage development is estimated at $US 6 600/ha.

In the period from 1956 to 1977, agricultural lands decreased from about 6 460 ha (with 3 230 ha cultivated) to about 4 100 ha (with 1 750 ha cultivated). This decrease was attributed mainly to urban expansion, waterlogging and soil salinization due to deterioration of the quality of the groundwater used in irrigation. In an attempt to reverse the situation, the government initiated a major agricultural development program in the early 1980s represented by:

- the replacement of surface irrigation methods with micro-irrigation (more water efficient) by subsidizing more than 50% of the cost of their implementation;
- the construction of major drainage systems to alleviate waterlogging and salt accumulation;
- the provision of agricultural extension services in terms of educating and advising farmers on types of crops suitable for agriculture under prevailing conditions;
- the introduction of TSE water in irrigation;
- the reclamation of new agricultural lands.

This resulted in a gradual increase and restoration of agricultural lands to about 4 230 ha, with 3 165 ha irrigated at present, all power irrigated. These 4 230 ha can also be considered as the irrigation potential, should there be an increasing future use of non-conventional water sources, in addition to groundwater. The quantity of groundwater available in the future for agriculture is difficult to estimate since groundwater quality, and hence its availability for irrigation, changes with time.

FIGURE 3
Origin of irrigation water f/p (total: 3 165 ha in 1994)

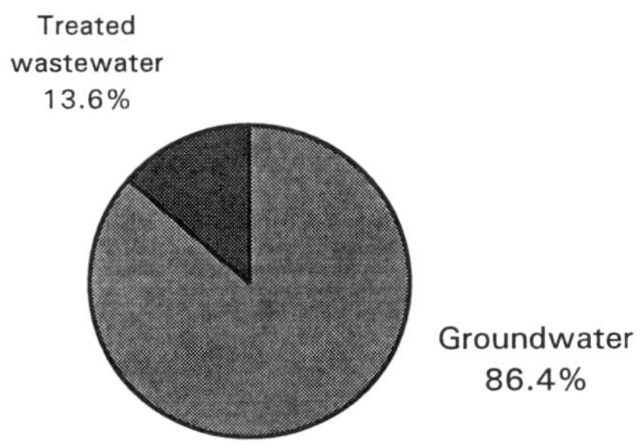

The utilization of 8 million m^3/year of tertiary TSE water in the reclaimed government lands (280 ha) and on some private farms (150 ha) (Figure 3), using modern irrigation techniques (sprinkler and micro-irrigation), had a palpable effect on the increase of agricultural lands and their productivity. While most of the programme is still being carried out at present, government subsidy of the installation of modern irrigation systems has stopped in the 1990s due to lack of funds. Despite efforts to introduce modern irrigation techniques, most farms still use traditional surface irrigation (Figure 4), which causes high water losses, estimated at between 24 and 40%. Sprinkler irrigation is used only in government projects, while micro-irrigation is used in government projects and on a limited number of private farms. Most of the land is worked either directly by the owner, often with hired labour, or by tenant farmers under a 1 to 2 year lease agree-ment. Such short and insecure occupation periods do not encourage tenants to invest in the installation of modern irrigation systems, which cost 40% and even up to 100% more than surface irrigation systems since the government subsidies of the installation of modern irrigation systems has stopped. The small size of agricultural landholdings, ranging between 0.5 and 10 ha with an average of 2.5 ha, and in particular the fragmentation of the agricultural land of farm holdings further restrict investment in the more expensive modern irrigation techniques.

FIGURE 4
Irrigation techniques f/p (total: 3165 ha in 1994)

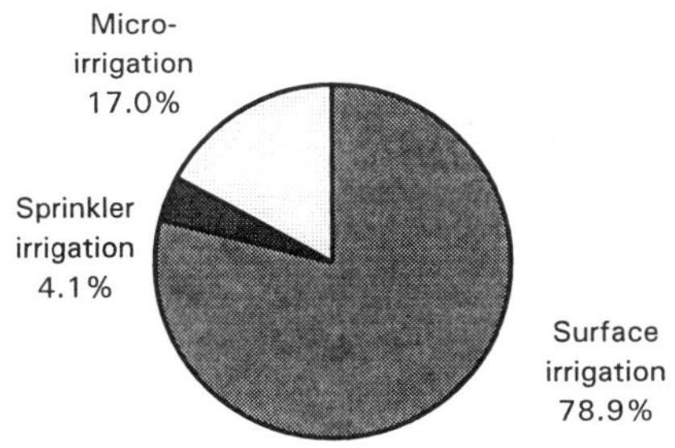

Of the total equipped area of 3 165 ha, 2 885 ha consist of small schemes (< 50 ha) (Figure 5). Most farms in these small-scale schemes are run under the tenancy system and at present there are about 250 households on these schemes. The remaining 280 ha of large schemes (> 50 ha) are owned and completely operated by the government and irrigated by treated wastewater, with a total of 80 government workers, of whom 11 are involved in irrigation.

FIGURE 5
Types of irrigation schemes (total: 3 165 ha in 1994)

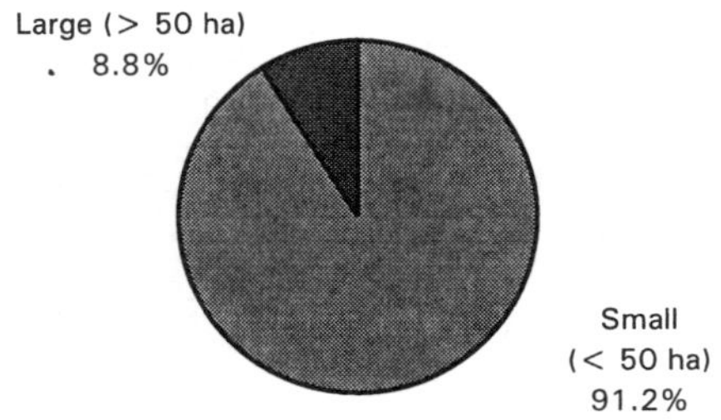

The average cost of irrigation development on small schemes varies between $US 6 600/ha for surface irrigation, $US 9 300/ha for micro-irrigation and $US 13 200/ha for sprinkler irrigation. For large schemes it is $US 16 200/ha for surface irrigation, $US 13 600/ha for micro-irrigation and $US 19 800/ha for sprinkler (central pivot) irrigation. The high costs for large scheme development are attributed to the fact that the major projects are carried out by the government on reclaimed lands. Operation and maintenance costs vary between 10 and 15% of the irrigation development costs on small schemes and between 5 and 15% on large schemes.

The major crops grown are dates and fruit trees with a yield of 7.5 tons/ha, vegetables, mainly tomatoes, with a yield of 11.7 tons/ha, and fodder crops, mainly alfalfa, with a relatively high yield of 74.5 tons/ha (Figure 6). In the last 10 years, there has been an increasing trend in the cultivation of alfalfa for fodder production rather than the cultivation of the traditional date and vegetable crops. Alfalfa tolerates high salinity and is a cash crop grown all year round with high local demand. However, because of the very high irrigation water requirements of alfalfa, it is expected that this trend will have negative implications for the country's groundwater resources.

FIGURE 6
Irrigated crops f/p (total: 3 165 ha in 1994)

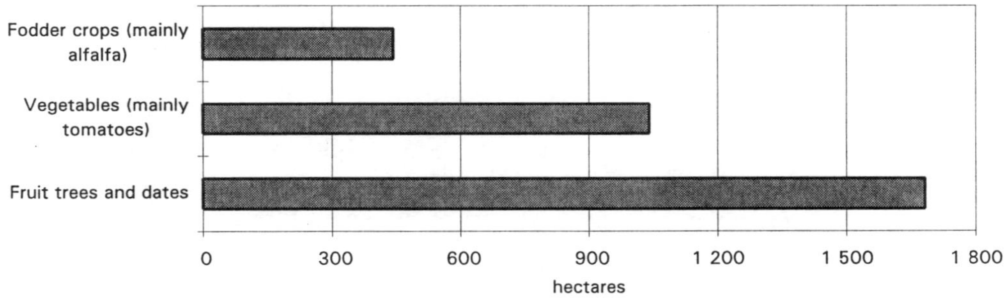

The privately-owned water use rights are the only water rights that exist in Bahrain. The general principle governing these rights is that groundwater is the property of the landowners and, therefore, they have an exclusive right to extract and use water as much as they wish and for any purpose they want without being liable for any damage caused to their neighbours or to the groundwater in general. At present, the agricultural sector's utilization of water is not subjected to any licensing system nor controlled by a pricing system. However, from the mid-1980s onwards, agricultural wells were being metered by the government and the government is in the process of passing a law that would make it compulsory for all well owners to install meters on their wells. The total number of wells metered at present is about 1 670 (86% of total). The final objective of this programme is to observe irrigation water requirements, and subsequently to set up a licensing system for groundwater withdrawal and design an appropriate pricing system for excess water utilization.

The only flood protection works carried out in Bahrain are those against rainfall floods and are developed in one residential, more modern, town located in the west over an area of 1 300 ha, where no agricultural activities are going on.

INSTITUTIONAL ENVIRONMENT

The Bahrain High Council for Water Resources, created in 1982, is the highest authority on water resources, under the chairmanship of the Prime Minister and with the membership of the ministries concerned.

The main duties of the Council are:

- drawing up the country's water policies;
- protecting and developing the country's water resources;
- regulating and coordinating water utilization between competing sectors;
- taking necessary measures to solve any problems that might arise during the implementation of water policies.

The Ministry of Works and Agriculture acts as the advisory body for the Council, and, through its different directorates, is in charge of the country's agricultural development and planning and water resources management.

Furthermore, indicating the government's future reliance on TSE in irrigation, a TSE Utilization Committee, with representatives from the related Ministries, was set up in 1993. The Committee's main responsibilities are the management and planning of TSE utilization in irrigation.

TRENDS IN WATER RESOURCES MANAGEMENT

In the last 15 years, the government has been taking several steps and courses of action to provide solutions to the water crisis in the country and agricultural sector deterioration. These include: water conservation campaigns in all sectors, water pricing in the domestic sector and more reliance on non-conventional water sources (TSE in agriculture and desalinated water for domestic purposes).

Government policy with regard to water use is to reduce groundwater dependency for the domestic water supply, the second main water user, by constructing additional desalination plants with a total capacity of 50 million m^3/year. Groundwater is planned to be exclusively used for irrigation. Additional requirements for future agricultural development would be supplemented by TSE water and the government is planning to increase the TSE utilization volume to about 49 million m^3/year. The additional volume would be used to irrigate an area of about 1 810 ha using modern techniques. This will bring the total area to be irrigated by TSE water to 2 240 ha, leaving the remaining area to be irrigated from groundwater. However, these plans are still awaiting major government funds for the construction of a TSE conveyance system and farmers acceptance. Although the intentions exist, an agricultural licensing system and water pricing are still missing.

Although the government-stated policy indicates that it wishes to develop a modern farming sector on larger production units using mechanization and up-to-date techniques, these aims have not been reflected clearly in the government's capital investment and subsidy programs.

Despite its slower rate at present, the drainage development project, which was initiated in the early 1980s, continues to operate and is expected to cover the remaining 2 380 ha by 2005.

MAIN SOURCES OF INFORMATION

Al-Noaimi, M.A. 1993. Evaluation of available water resources, present utilization, and consuming sectors' future needs, Volume I: Evaluation of available water resources and utilization patterns. Bahrain Center for Studies and Research (in Arabic).

ACE [Associated Consulting Engineers]. 1990. Irrigation master plan: draft report. Ministry of Housing, Municipalities and Environment.

ACE. 1990. Report on treated effluent utilization. Ministry of Works and Agriculture.

Central Statistics Organization. 1993. Statistical abstract 1992. Directorate of Statistics, Bahrain.

GDC [Groundwater Development Consultants]. 1980. Umm Er Radhuma study, Bahrain Assignment. Ministry of Works and Agriculture.

Musayab, R. 1988. Water resources and development in the State of Bahrain. Directorate of Water Supply, Ministry of Electricity and Water.

Ministry of Works and Agriculture. 1994. Annual agricultural statistical report-1994. Directorate of Agricultural Research, Agricultural Economics Section, Ministry of Works and Agriculture.

Ministry of Works and Agriculture/Bahrain Center for Studies and Research. 1992. Country report on water and agriculture in Bahrain. Symposium on the Greening of the GCC Countries, Tokyo, Japan, October, 1992.

UNEP and UNESCWA. 1991. The national plan of action to combat desertification in Bahrain. UNEP/ROWA (92-0725), Bahrain.

Zubari, W.K., and E.J. Lori. 1991. Report on water resources studies: evaluation, utilization, and legislation in Bahrain in the period from 1924 -1991. Bahrain Centre for Studies and Research (in Arabic).

Zubari, W.K., Mubarak, A.M, and I.M. Madany. 1993. Development Impacts on Groundwater Resources in Bahrain. Water Resources Development, Vol. 9 (3), pp. 263-279.

Cyprus

GEOGRAPHY AND POPULATION

Cyprus is the largest island in the Eastern Mediterranean with an area of 9 250 km^2. The main features of the island's topography include two mountain ranges, the Kyrenia in the north, which rises to 1 024 metres, and the Troodos in the centre, rising to 1 951 metres. Between these two main mountain ranges lies a wide plain, the Mesaoria plain, which rises from sea level to about 325 metres. Between the two mountain ranges and the sea there are narrow fertile coastal plains.

The area of the country currently under government control, to which this country profile and the accompanying tables will refer unless otherwise stated, is about 5 807 km^2.

The cultivable area is estimated at 165 433 ha, which is 28% of the total area under government control. In 1994, cultivated land amounted to 111 649 ha, of which 69 509 ha consisted of annual crops and 42 140 ha consisted of permanent crops. Most land is private. There are 51 300 farm holdings and according to the 1994 Census of Agriculture, the average holding size is 3.5 ha divided into 4.5 parcels of land.

TABLE 1
Basic statistics and population

Physical areas:			
Area of the country *	1995	925 000	ha
Cultivable area	1994	165 433	ha
Cultivated area	1994	111 649	ha
- annual crops	1994	69 509	ha
- permanent crops	1994	42 140	ha
Population: *			
Total population	1995	742 000	inhabitants
Population density	1995	80	inhab./km^2
Rural population	1995	46	%
Water supply coverage:			
Urban population	1992	99.6	%
Rural population	1992	98.6	%

* *Figures refer to the whole island*

The total population for the whole island is 742 000 according to United Nations estimates (1995), of which 46% is rural. According to the National Demographic Report of 1994, the total population in the government-controlled areas amounted to 638 300 (as against 734 000 for the whole island), of which 32% was rural, but the method used for to estimate the rural population differs from the one used by United Nations. The annual population growth is 1%. Agriculture, including crop production, livestock, fisheries and forestry accounted for 5.4 % of the GDP in 1994, down from 10% in 1980. The share of agriculture's contribution to GDP can however vary from year to year depending on the prevailing weather conditions, particularly rainfall. Although the share of agriculture in the GDP has had a declining trend in recent years, owing to more rapid development in the other sectors of the economy, agricultural exports still play a significant role in the country's foreign exchange earnings, accounting for over 21% of total exports in 1994. The percentage of the total economically active population engaged in agriculture has also declined in recent years. In 1994 it was 11.6% compared with 17% in 1980. Nevertheless, agriculture is still considered an important sector and efforts are directed towards increasing production and maintaining people in rural areas through integrated development projects.

TABLE 2
Water: sources and use

Renewable water resources: *			
Average precipitation		503	mm/yr
		4.65	km^3/yr
Internal renewable water resources		0.90	km^3/yr
Total (actual) renewable water resources	1995	0.90	km^3/yr
Dependency ratio	1995	0	%
Total (actual) renewable water resources per inhabitant	1995	1 213	m^3/yr
Total dam capacity	1994	299	$10^6\ m^3$
Water withdrawal:			
- agricultural	1993	156	$10^6\ m^3/yr$
- domestic	1993	50	$10^6\ m^3/yr$
- industrial	1993	5	$10^6\ m^3/yr$
Total water withdrawal		211	$10^6\ m^3/yr$
per inhabitant	1993	331	m^3/yr
as % of total (actual) renewable water resources **		[23.4]	%
Other water withdrawal		-	$10^6\ m^3/yr$
Average groundwater depletion *	1989	40	$10^6\ m^3/yr$
Wastewater - Non-conventional water sources: *			
Wastewater:			
- produced wastewater	1995	40	$10^6\ m^3/yr$
- treated wastewater	1995	16	$10^6\ m^3/yr$
- reused treated wastewater	1995	11	$10^6\ m^3/yr$
Desalinated water		-	$10^6\ m^3/yr$

* *Figures refer to the whole island*

** *Figure for total water withdrawal refers to the government-controlled area, the figure for water resources refers to the whole island*

CLIMATE AND WATER RESOURCES

Climate

The climate of Cyprus is typical of the Eastern Mediterranean, with mild winters and hot, dry summers. Average daily minimum temperature is 9°C in January, while the average daily maximum is 35°C in August. The mean annual precipitation for the whole island is 503 mm, ranging from 290 mm in the east to 1 190 mm in the Troodos mountains. Most of the rain falls in the winter months from December to February, with snow falling frequently in the Troodos mountains. Hail storms are common and often cause considerable crop damage. There are large variations in rainfall from year to year.

Water resources

A water balance cannot be easily calculated for the government-controlled area of Cyprus. A water balance for the whole island, however, indicates 900 million m^3/year of renewable water resources. Surface runoff is estimated to represent about 830 million m^3/year. The natural aquifer recharge is estimated at 300 million m^3, of which about 70 million m^3 flows to the sea and 100 million m^3 emerges from springs.

There are 14 main rivers, none of which provides perennial flow. The source of water for these rivers originates in the Troodos mountains. The main groundwater aquifers are the Western Mesaoria (Morphou), Kokkinochoria (South-eastern and Eastern Mesaoria) and the Akrotiri. Smaller aquifers exist in other parts of the country.

TABLE 3
Irrigation and drainage

Irrigation potential		1994	36 807	ha
Irrigation:				
1. Full or partial control irrigation: equipped area		1994	39 545	ha
- surface irrigation		1994	1 977	ha
- sprinkler irrigation		1994	1 977	ha
- micro-irrigation		1994	35 591	ha
% of area irrigated from groundwater		1994	51.3	%
% of area irrigated from surface water		1994	48.2	%
% of area irrigated from non-conventional sources		1994	0.5	%
% of equipped area actually irrigated		1994	83.6	%
2. Spate irrigation area		1994	393	ha
3. Equipped wetland and inland valley bottoms (i.v.b.)			-	ha
Total irrigation (1+2+3)		**1994**	**39 938**	**ha**
- as % of cultivated area			36	%
4. Flood recession cropping area			-	ha
Total water managed area (1+2+3+4)		**1994**	**39 938**	**ha**
- as % of cultivated area			36	%
- increase over last 10 years		1984-94	28	%
- power irrigated area as % of water managed area		1994	87.5	%
Full or partial control irrigation schemes:	Criteria			
Large-scale schemes	> 500 ha	1994	21 746	ha
Medium-scale schemes		1994	2 091	ha
Small-scale schemes	< 100 ha	1994	15 708	ha
Total number of households in irrigation		1994	34 000	
Irrigated crops:				
Total irrigated grain production		1994	8 960	tons
as % of total grain production		1994	14	%
Harvested crops under irrigation (full or partial control)		**1994**	**33 967**	**ha**
- permanent crops: total		1994	16 600	ha
- annual crops: total		1994	17 367	ha
. roots and tubers (mainly potatoes)		1994	7 610	ha
. vegetables		1994	3 512	ha
. cereals (mainly barley)		1994	2 848	ha
. pulses		1994	1 130	ha
. other annual crops		1994	2 267	ha
Drainage - Environment:				
Drained area			-	ha
as % of cultivated area			-	%
- drained areas in full or partial control irrigated areas			-	ha
- drained areas in equipped wetland and i.v.b			-	ha
- other drained areas			-	ha
- total drained area with subsurface drains			-	ha
- total drained area with surface drains			-	ha
Flood-protected area			-	ha
Area salinized by irrigation			-	ha
Population affected by water-borne diseases			-	inhabitants

Dams

In 1995, total dam capacity reached 299 million m^3 on the whole island, up from 6 million m^3 in 1961 and 64 million m^3 in 1974. New dams for storing water for irrigation are planned, particularly in Paphos province in the south-western part of the island. Additional dams are also planned for Lefkosia province in the centre of the island, but a substantial quantity of this water will be diverted for domestic and industrial use and to compensate for the loss of water recharge downstream of the dams.

Non-conventional water sources

At present no desalination of water takes place, but a desalination plant is expected to become operational in 1997, with a total capacity of 7 million m^3/year.

Currently some 40 million m^3 of wastewater are produced annually on the whole island. Only 16 million m^3 of this amount is being treated, mainly in Lefkosia province where Nicosia City is located. Nicosia has a city-wide sewage processing plant, part of which is not under government control. About 11 million m^3 is being reused for irrigation purposes, mainly in the part of the island that is not under government control around Nicosia. Only 1 million m^3/year is reused for the irrigation of hotel gardens and recreation areas in the government-controlled area. Plans are available to utilise more as soon as the central sewage plants at Limassol, Larnaca, Ayia Napa and Paralimni in the government-controlled areas are completed and the water treatment plants are in place.

Water withdrawal

In 1993, total water withdrawal in the government-controlled area was 211 million m^3 of which 74% for agricultural purposes, including both irrigation and livestock (Figure 1). The trend in recent years, which is likely to continue in the future, is that increasing quantities of water will be used for domestic water supplies at the expense of agriculture. This has been necessary in view of an increasing standard of living, an expansion of tourist services and industrialization.

FIGURE 1

Water withdrawal (total: 211 million m^3 in 1993)

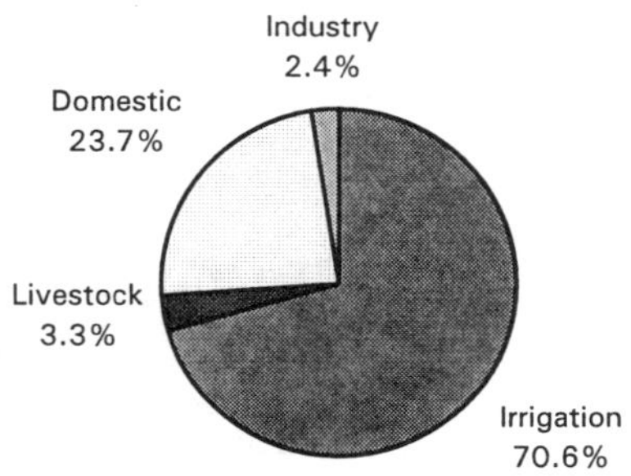

Considering the whole island, 70 million m^3 of groundwater flows to the sea yearly and 270 million m^3 is either pumped out or emerges from springs, leading to a total extraction from the aquifers of 340 million m^3/year. As the annual recharge has been estimated at 300 million m^3, there could be up to 40 million m^3/year of excess pumping over natural recharge. As a result, the total area of Cyprus is experiencing a gradual decline in groundwater yield, a lowering of the water table and, in certain cases, sea water intrusion.

IRRIGATION AND DRAINAGE DEVELOPMENT

In 1994, the water managed area was estimated at 39 938 ha in the government controlled area, of which 39 545 ha were equipped for full or partial control irrigation (Figure 2). Less than 1% consisted of spate irrigation. It is estimated however, on the basis of current and potentially available water when the latter is fully developed, that this water managed area is already larger than the irrigation potential. In 1994 only 32 864 ha, or only 82% of the water managed area, were irrigated because of water shortages. Due to continuous water withdrawal and to increasing future water needs for domestic and industrial purposes, this area is unlikely to increase considerably. For this reason, estimates of an irrigation potential of 36 807 ha have been given, including the possible use of tertiary treated wastewater in the future and a greater water storage capacity in the new dams to be constructed. This irrigation potential area is slightly more than the actually irrigated area in 1994 but less than the total water managed area in 1994.

FIGURE 2
Distribution of the water managed areas (total: 39 938 ha in 1994)

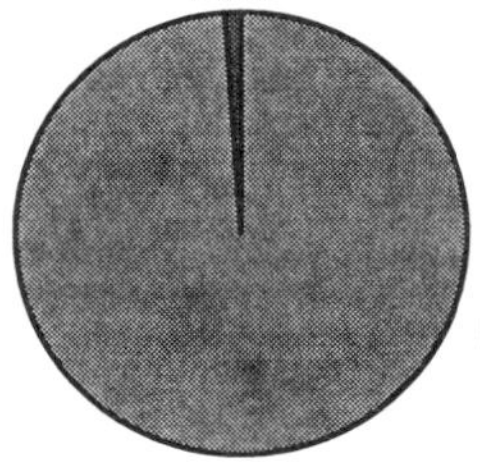

Spring water and groundwater were the first target of water resources development. Traditionally this water was cheap and easy to develop by individual farmers or farmers' Irrigation Divisions or Associations. In the former case the farmers develop the springs or tubewells on an individual basis, whilst in the latter case water resources are developed by a group of farmers who are then eligible for a government subsidy for the capital expenditure. In both cases a government permit is required prior to initiation of any water work. Normally these schemes are small and they cover 1 to 3 ha, although larger schemes have also been developed in recent years.

FIGURE 3
Origin of irrigation water f/p (total: 39 545 ha in 1994)

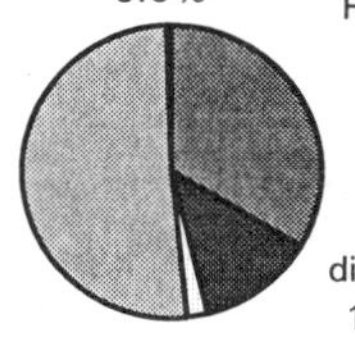

After independence in 1961 and following the full utilization of groundwater resources, emphasis was placed on collecting and storing surface water during the winter and utilizing it throughout the year. In 1994, a little less than half the area was irrigated from surface water (Figure 3).

Public schemes, often based on the joint use of groundwater and surface water have been constructed since the late 1960s. They include:

- The Paphos Irrigation Project to provide annually 36 million m^3 of groundwater and surface water to irrigate 4 600 ha;
- The Vasilikos Pendashinos Project to provide annually 17 million m^3 of surface water to irrigate 1 525 ha and for the domestic water supply;
- The Khrysochou Irrigation Project to provide annually 18 million m^3 of surface water to irrigate 2 790 ha;
- The Pitsilia Integrated Rural Development Project to supply water for irrigation of 850 ha and for domestic use;
- The Southern Conveyor Project intended to provide annually 65 million m^3 of surface water to irrigate 11 244 ha and for the domestic water supply.

All public schemes are operated by the Water Development Department (WDD) of the Ministry of Agriculture, Natural Resources and Environment and the farmers are charged for the water on a volumetric basis. This is possible through the use of water meters which are monitored regularly by the WDD. Irrigation Divisions or Associations usually operate and maintain their own schemes and charge the users either on a volumetric or an hourly basis. Those irrigation works belonging to Irrigation Divisions, that are considered to be major or to involve safety factors or include small dams, are maintained by the government and one third of the operational and maintenance expenses is recovered from the Irrigation Divisions. Individual farmers who possess their own source of water, manage the water resources themselves.

The irrigation network in Cyprus is highly efficient. It generally consists of closed systems with an overall conveyance efficiency averaging 90-95%. Field application efficiency averages 80-90%. In parallel with the government's efforts to increase the water available for agriculture, emphasis was placed on the optimum utilization of water through improved irrigation methods. To encourage farmers to use improved irrigation methods the government offered incentives to participating farmers in the form of subsidies and long-term low interest loans for the purchase and installation of improved irrigation systems. In addition, through extensive demonstrations, the government convinced the farmers that improved irrigation methods, initially sprinklers for vegetables and the hose/basin method for tree crops, to be followed by micro-irrigation systems, not only saved water but also led to increased yields. As a result of these efforts, the area irrigated by surface irrigation methods has declined from about 13 400 ha in 1974 to less than 2 000 in 1995 while the area equipped for micro-irrigation has increased over the same period from about 2 700 ha to almost 35 600 ha (Figure 4). There are few margins for further improvements in water application efficiency. The areas irrigated by surface irrigation methods are mostly cropped with deciduous trees and are found in the hilly areas of the country. They are usually irrigated from small springs which do not lend themselves easily to the adoption of improved irrigation techniques.

FIGURE 4
Evolution of the irrigation techniques

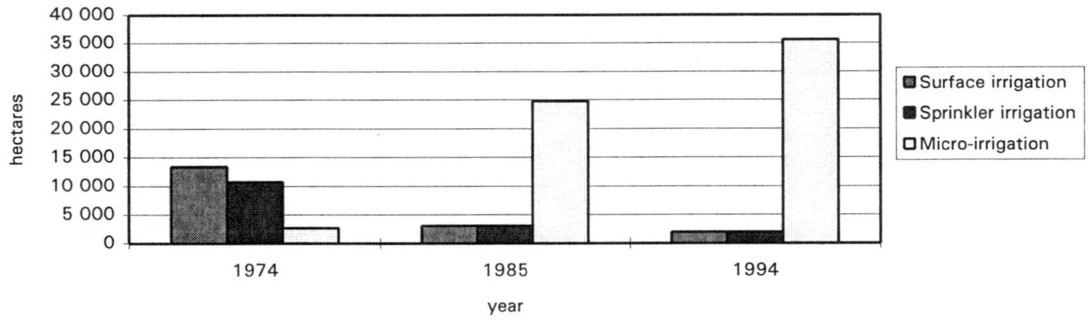

In 1994, 21 746 ha consisted of large schemes (> 500 ha), 2 091 ha of medium schemes and 15 708 ha of small schemes (< 100 ha) (Figure 5).

The cost of irrigation development varies and depends on a number of factors. The average cost of irrigation development using tubewells varies from about $US 3 930/ha for up to one hectare, $US 2 260/ha for two hectares to $US 1 700/ha for three hectares. This includes the cost of on-farm micro-irrigation systems. Excluding the cost of the dam, the development of surface water varies from $US 1 560/ha to $US 2 610/ha including on-farm micro irrigation system. The average annual cost of maintenance varies from $US 300-350/ha for private schemes (tubewells) to $US 50-120/ha for public schemes.

FIGURE 5
Typology of f/p control irrigation schemes (total: 39 545 ha in 1994)

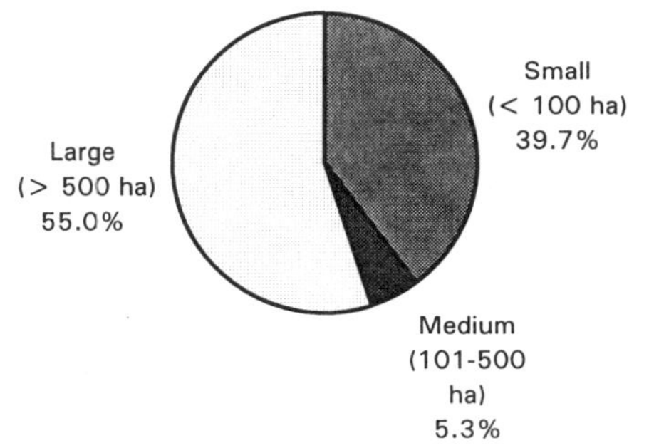

The major irrigated crops are fruit trees and potatoes (Figure 6). For most crops the cost of irrigation water varies from about $US 90 to 270/year per ha. Public schemes currently charge the consumer $US 0.03/m^3, whereas the Irrigation Divisions usually charge their members the full operating costs in addition to a basic charge for repayment of loans (when applicable). These costs are high and discourage the farmers from using irrigation for low value crops such as cereals, pulses, olives, almonds, carrots etc. The above-mentioned cost amounts, for example, to about 23 % of variable costs and 17% of total production costs for oranges or 17% and 11% respectively for spring potatoes. These two crops are the major export crops of Cyprus and cover an area of about 2 200 and 4 600 ha/year respectively. When irrigation is used for other crops, for example wheat or barley, the yields also substantially increase. While the national average yields of rainfed wheat and barley were 0.55 and 0.92 tons/ha respectively in 1994, irrigated crop yields were 4 and 3 tons/ha respectively. Although irrigated crops cover only about 30% of the cultivated land, they account for 60-70% of the production earnings.

FIGURE 6
Irrigated crops (total: 33 967 ha in 1994)

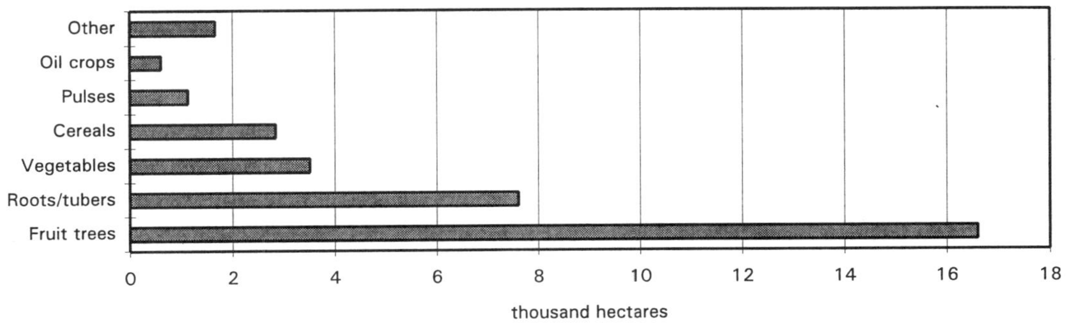

INSTITUTIONAL ENVIRONMENT

The Ministry of Agriculture, Natural Resources and Environment is responsible, through a number of departments, for water resources assessment and development. The Water Development Department (WDD) assesses the surface water resources (groundwater resources are assessed by the Geological Survey Department), plans the water development projects, develops the engineering studies (including civil works needed) and operates and maintains these projects. Project construction is usually subcontracted to the private sector, following bids. The WDD also has the responsibility of recommending plans to government for the allocation of water resources as well as water rates. For irrigation purposes the water rates cannot legally exceed 40% (65% in exceptional cases) of the average total cost of water provision, including capital expenditure as well as operation and maintenance expenditures (for domestic and industrial water supply the full costs are recovered). The Department of Agriculture has the overall responsibility for improved irrigation practices at farm level and the Agricultural Research Institute is involved in all aspects of applied research with regard to irrigation methods, application rates etc.

The Ministry of the Interior also plays a key role. It is responsible for the implementation and enforcement of all water related laws, as well as the issuing of groundwater permits and the registration of water rights. Officers of the Ministry of the Interior also act as Chairpersons of District Water Boards (for domestic water supply) and for the Irrigation Divisions. The Ministry is also involved in the Sewage Boards that manage the sewage systems of the largest towns. The Ministries of Agriculture and of the Interior work closely and coordinate all aspects of water development and utilization.

The main laws concerned with irrigation date back to before to independence in 1961, and include the Government Water Works Law (Cap 341 of 1928) which provides for the control of water and the construction of water works by the government; the Wells Law (Cap 351 of 1946) which covers the installation of wells and their related water rights, and the Irrigation Divisions (Cap 342 of 1938) which regulates the formation of Irrigation Divisions and their operation. The Laws, in general, function effectively and, in the case of the Water Laws, cover all aspects of water development as well as interactions between government and users. All land in Cyprus is registered and owners have deeds or certificates of ownership.

TRENDS IN WATER RESOURCES MANAGEMENT

At present, almost all the renewable water resources in Cyprus are utilized and, in a number of areas, groundwater is rapidly depleting with sea water intrusion occurring in the main coastal aquifers of Morphou (Western Mesaoria), Famagusta and Kokkinochoria (Eastern Mesaoria) and Akrotiri. There is no accurate estimate of the quantity of water extracted in excess of natural recharge, but it could be as high as 40 million m^3/year. Even so, in years of drought or below average rainfall, it is necessary to divert water from agriculture to the domestic and industrial sector. In such cases it is necessary to restrict the amount of water made available for both annual and perennial crops. For instance, in the 1989/90 and 1990/91 seasons, annual crops were limited on average to 70% of the normally irrigated land, whereas perennial crops received only 80% of normal supply. A similar situation occured in 1993.

This precarious situation is unlikely to change in the future since almost all the conventional water resources are already used. This includes water stored in a series of dams on all rivers rising in the southern slopes of the Troodos mountains. Other potential but smaller water storage schemes are planned for rivers arising from the northern slopes of the Troodos mountains. These new schemes, which will not be completed before 2005 - 2010, are unlikely to alter the quantity of water available for irrigation significantly (with the possible exception of the Paphos District), since water from most of these rivers currently recharges the aquifer downstream which is already substantially utilized. Furthermore, water demand for domestic and industrial purposes will undoubtedly continue to increase and will receive priority over water demand for agriculture. This leaves the use of treated wastewater as one of the main sources for increasing water supply for agriculture in the foreseeable future. In view of this restrictive situation, the government is also considering alternative ways of increasing the water supply of the country. In this respect in 1996 it awarded a contract for the construction of a desalination plant with a minimum capacity of 7 million m^3/year.

Other possible steps and options will also have to be considered and/or evaluated. These include:

- the further minimizing of water losses in the domestic water distribution system which now average about 23%, although this figure is already quite low (compared to a current average of 40% for developing countries and 20% for developed countries);

- the shifting of water from marginally economical agricultural activities to other uses especially to domestic use, thus eliminating the water subsidy;

- the inter-regional transfers of water from the better-endowed western part (i.e. Paphos), to the eastern districts, albeit at a high cost.

Additional, but integral components of the government's policy in water resources management will be the improvement of the water delivery system in the hilly areas, and further overall water savings through increasing the price of irrigation water (at present covering 34% of the average cost of water provision) to the maximum allowed by the existing legislation. In both cases the resulting water savings will however be minimal.

Waterlogging, soil salinization, and vector-borne diseases are not present in Cyprus. Contamination of groundwater, especially with fertilizers (particularly nitrates) in certain areas of the island where agriculture is intensively practised, is however occurring and is a cause of concern. There is also the problem seawater intrusion in the main coastal aquifers. This situation overall requires close monitoring.

MAIN SOURCES OF INFORMATION

Department of Statistics and Research. 1995. Agricultural Statistics 1994.

Department of Statistics and Research. 1996. Census of Agriculture.

Petrides, C. 1994. Water Balance and Water Policy in Cyprus. Paper presented at the University of Salonika, Greece.

Water Development Department. 1990. Water Development Department: 50 years, 1939-89.

Water Development Department. 1994. Dam construction (Drawing).

World Bank. 1995. Cyprus Water Planning and Management Strategies. Washington DC.

Djibouti

GEOGRAPHY AND POPULATION

Djibouti, situated in the Horn of Africa, is one of the smallest countries of the continent with a total area of 23 200 km^2. It is bordered in the east by the Gulf of Aden, in the south-east by Somalia, in the south and west by Ethiopia and in the north by Eritrea.

The cultivable area has been estimated at 6 000 ha, which is only 0.26% of the total area. The cultivated area was 407 ha in 1989, which is 7% of the cultivable area. Agriculture is concentrated in the southern part of the country, mainly in the Djibouti and Dikhil districts. These two districts cover about 34% of the country while they contain 75% of the cultivated areas (Figure 1).

TABLE 1
Basic statistics and population

Physical areas:			
Area of the country	1995	2 320 000	ha
Cultivable area	1989	6 000	ha
Cultivated area	1989	407	ha
- annual crops	1989	357	ha
- permanent crops	1989	50	ha
Population:			
Total population	1995	577 000	inhabitants
Population density	1995	25	inhab./km^2
Rural population	1995	17	%
Water supply coverage:			
Urban population	1990	50	%
Rural population	1990	21	%

The total population is 577 000 (1995), of which only 17% is rural. The average population density is 25 inhabitants/km^2. It varies from about 3 in the north to over 400 in the Djibouti district where the capital is located. The annual demographic growth rate was estimated at 3.5% over the period 1980-1991. The economy of the country is mainly based on services, which accounted for 63% of GDP in 1982, while agriculture accounted for 6%. In 1991, agriculture accounted for only 3% of GDP.

FIGURE 1
Distribution of the cultivated area over the districts (total: 407 ha in 1990)

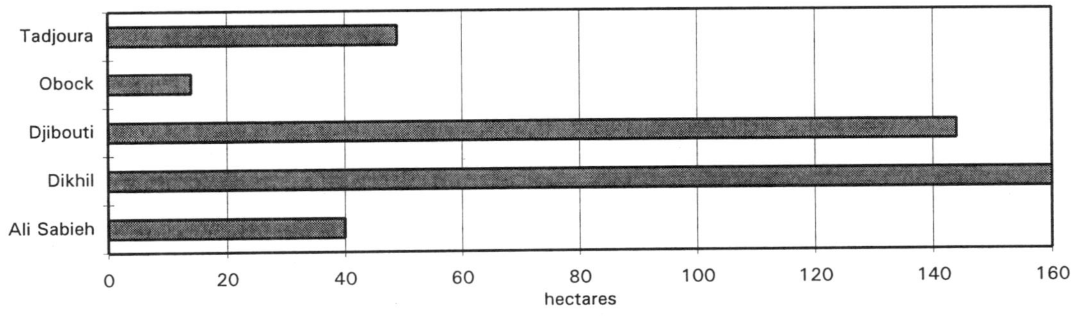

TABLE 2
Water: sources and use

Renewable water resources:			
Average precipitation		147	mm/yr
		3.4	km^3/yr
Internal renewable water resources		0.3	km^3/yr
Total (actual) renewable water resources	1995	0.3	km^3/yr
Dependency ratio	1995	0	%
Total (actual) renewable water resources per inhabitant	1995	520	m^3/yr
Total dam capacity		-	10^6 m^3
Water withdrawal:			
- agricultural	1985	6.5	10^6 m^3/yr
- domestic	1985	1.0	10^6 m^3/yr
- industrial	1985	0.0	10^6 m^3/yr
Total water withdrawal		7.5	10^6 m^3/yr
per inhabitant	1985	20	m^3/yr
as % of total (actual) renewable water resources		2.5	%
Other water withdrawal		-	10^6 m^3/yr
Average groundwater depletion		-	10^6 m^3/yr
Wastewater - Non-conventional water sources:			
Wastewater:			
- produced wastewater		-	10^6 m^3/yr
- treated wastewater		-	10^6 m^3/yr
- reused treated wastewater		-	10^6 m^3/yr
Desalinated water	1990	0.1	10^6 m^3/yr

CLIMATE AND WATER RESOURCES

Climate

The climate of Djibouti is classified as arid to extremely arid. The average temperature varies between 25°C in December and 35°C in July. Average annual rainfall is 147 mm and varies from 50 mm in the north-east to 300 mm in the region west of Tadjoura. Occasions of extreme rainfall recorded since 1901 were 9.5 mm in 1980 and 693 mm in 1989. Because of the low annual rainfall and its irregular distribution there are no perennial rivers and surface runoff takes place during one or two days only after relatively heavy rainfall.

Water resources

The hydrographic system can be divided into two zones, one draining to the sea (45%) and one draining to the plains in the west of the country (55%). Internal renewable water resources are estimated at 300 million m^3/year. The only external surface water resources (about 2 km^3/year) come from Ethiopia, through the border lake Abbe in the west of the country. However, since it is a salt lake, this resource cannot be put into beneficial use.

Groundwater recharge is limited to the infiltration of water in the wadis and is closely linked to the frequency of floods, the infiltration conditions in the wadis, the hydraulic contact with the substratum as well as to the permeability of the substratum. The groundwater resources are rather difficult to prospect and extract. The sediment formations represent only 10% of the total area of the country, the remaining parts consisting of volcanic rock formations. In general, groundwater used for irrigation causes problems because of excessive salinity, even in the alluvial layers. Only in the north-western part of the country is the ionic content of the water below the acceptable limit for irrigation purposes. High boron contents occur most

TABLE 3
Irrigation and drainage

Irrigation potential		1995	1 000	ha
Irrigation:				
1. Full or partial control irrigation: equipped area		1989	674	ha
- surface irrigation			-	ha
- sprinkler irrigation			-	ha
- micro-irrigation			-	ha
% of area irrigated from groundwater		1989	100	%
% of area irrigated from surface water		1989	0	%
% of area irrigated from non-conventional sources		1989	0	%
% of equipped area actually irrigated		1989	60.4	%
2. Spate irrigation area			-	ha
3. Equipped wetland and inland valley bottoms (i.v.b.)			-	ha
Total irrigation (1+2+3)		**1989**	**674**	**ha**
- as % of cultivated area			166	%
4. Flood recession cropping area			-	ha
Total water managed area (1+2+3+4)		**1989**	**674**	**ha**
- as % of cultivated area			166	%
- increase over last 10 years		1980-90	270	%
- power irrigated area as % of water managed area			-	%
Full or partial control irrigation schemes:	Criteria			
Large-scale schemes	> - ha		-	ha
Medium-scale schemes			-	ha
Small-scale schemes	< - ha		-	ha
Total number of households in irrigation		1989	1 158	
Irrigated crops:				
Total irrigated grain production		1989	4	tons
as % of total grain production		1989	100	%
Harvested crops under irrigation (full or partial control)		**1989**	**407**	**ha**
- permanent crops: total		1989	50	ha
- annual crops: total		1989	357	ha
. vegetables		1989	355	ha
. cereals (maize and sorghum)		1989	2	ha
Drainage - Environment:				
Drained area			-	ha
as % of cultivated area			-	%
- drained areas in full or partial control irrigated areas			-	ha
- drained areas in equipped wetland and i.v.b			-	ha
- other drained areas			-	ha
- total drained area with subsurface drains			-	ha
- total drained area with surface drains			-	ha
Flood-protected area			-	ha
Area salinized by irrigation			-	ha
Population affected by water-borne diseases			-	inhabitants

frequently. The water present in the alluvial layers, close to the surface, can be used for small-scale irrigation. Total renewable groundwater resources in the Djibouti district are estimated at between 10 and 21 million m^3/year. In 1985, total water withdrawal was 7.5 million m^3, of which about 6.5 million m^3 for agricultural purposes (Figure 2).

IRRIGATION AND DRAINAGE DEVELOPMENT

Crops can only be grown under irrigation, which in general takes place in the cooler season from the middle of November till the middle of May. In summer, only gourds (cucurbicatees) and palm trees are irrigated. They represent a small area compared to the other crops (mainly vegetables) (Figure 3).

FIGURE 2
Water withdrawal (total: 7.5 million m^3 in 1985)

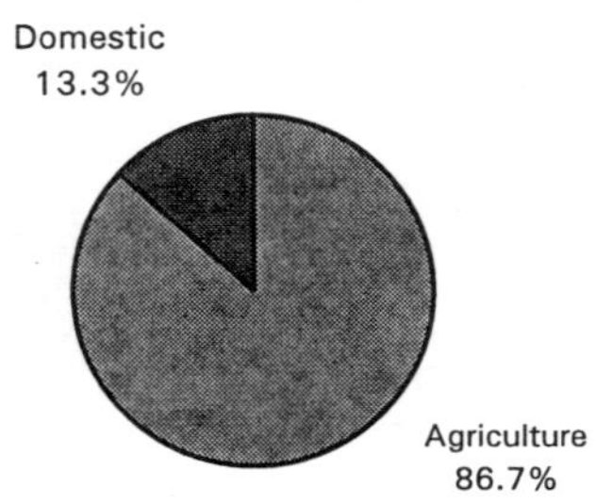

FIGURE 3
Irrigated crops f/p (total: 407 ha in 1989)

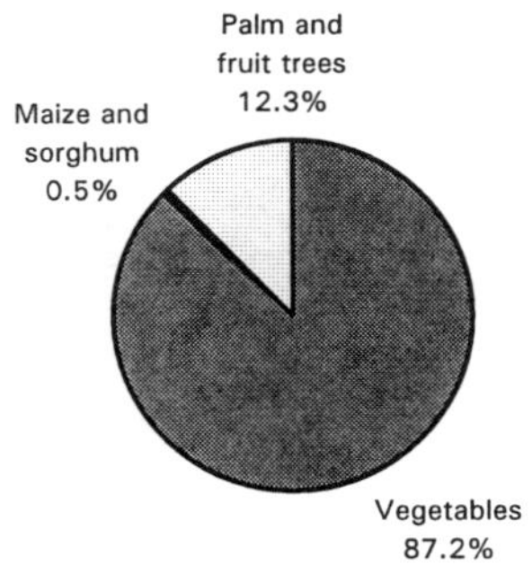

In 1982, there were 450 agricultural households in the whole of the country cultivating an area of 110 ha, or an average of 2 250 m^2 per household. In 1989, 1 158 agricultural households were cultivating 407 ha, or an average of 3 515 m^2 per household (Figure 4). The total area equipped for irrigation was 674 ha, but only 65% of the equipped area was cultivated on average each year.

Agricultural water comes from shallow wells, tubewells or springs. In 1986, the cost of water extraction from tubewells was about \$US 0.28/m^3. For shallow wells it was about \$US 0.11/m^3.

FIGURE 4
Evolution of the irrigated area

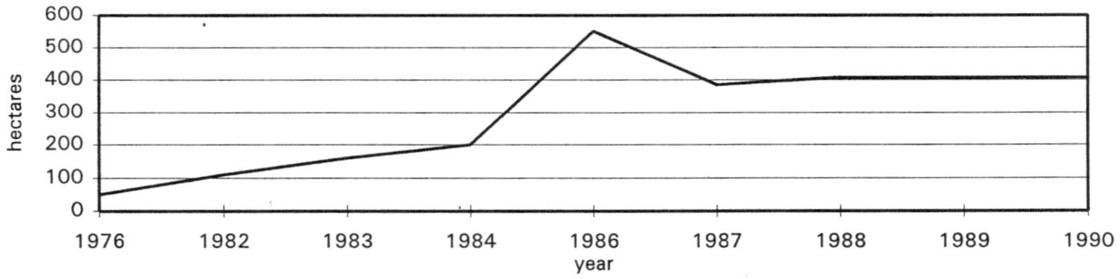

INSTITUTIONAL ENVIRONMENT

Agricultural development is organized in three zones covered by three rural development programmes benefiting from the support of technicians at regional level and from the different services within the Ministry of Agriculture and Rural Development at the centre. The regional support staff is in charge of the execution of the technical programmes fixed by the four central services:

- The Livestock and Fisheries Department;
- The Agriculture and Forestry Service;

- The Rural Engineering Service;
- The Water Analysis Laboratory.

TRENDS IN WATER RESOURCES MANAGEMENT

Agriculture in Djibouti is not possible without irrigation. However, it could be improved in several ways:

- better use of the limited surface water;
- study of the water potential of the alluvial layers and development of irrigation in the wadis;
- demonstration of more economic water use techniques;
- improvement of the canal network (in general earthen canals);
- cropping of the whole equipped area.

MAIN SOURCES OF INFORMATION

Bundesanstalt für Geowissenschaften und Rohstoffe. 1982. Inventaire et mise en valeur des ressources en eau de la République de Djibouti. Report prepared by W. Müller, for project No 78.2233.1, Coopération hydrogéologique allemande. Hannover, Allemagne.

Direction nationale de la statistique, Ministère du commerce, des transports et du tourisme. Annuaire statistique de Djibouti. Résultats de 1990.

FAO. 1986. Compte rendu final du projet FAO TCP/DJI/4505, Renforcement du Service de l'agriculture et des forêts. Rome.

IGADD [Intergovernmental Authority on Drought and Development]. 1990. Forum on environmental protection and development of subregional strategy to combat desertification, 1990.

Egypt

GEOGRAPHY AND POPULATION

Egypt lies in the north-eastern corner of the African continent, with a total area of about 1 million km^2. It is bordered in the north by the Mediterranean Sea, in the east by Israel and the Red Sea, on the south by Sudan and in the west by Libya.

In 1993, the total cultivated land was estimated to be 3.24 million ha, or 3.2% of the total area. About 2.86 million ha, or 88% of the total cultivated area, consisted of annual crops and 0.38 million ha consisted of permanent crops.

TABLE 1
Basic statistics and population

Physical areas:			
Area of the country	1995	100 145 000	ha
Cultivable area	1993	-	ha
Cultivated area	1993	3 246 000	ha
- annual crops	1993	2 862 960	ha
- permanent crops	1993	383 040	ha
Population:			
Total population	1995	62 931 000	inhabitants
Population density	1995	63	inhab./km^2
Rural population	1995	55	%
Water supply coverage:			
Urban population	1993	90	%
Rural population	1993	45	%

Total population is about 62.9 million (1995), of which 55% is rural, with annual demographic growth estimated at 2.1%. Average population density is 63 inhabitants/km^2, but ranges from 2 inhabitants/km^2 over 96% of the total area, to 1 492 inhabitants/km^2 in the Nile Valley and Delta. This area, where population density is among the highest in the world, represents only 4% of the total area. In 1992, agriculture accounted for 17% of Egypt's GDP and provided employment to 38% of the labour force.

CLIMATE AND WATER RESOURCES

Climate

The mean annual rainfall is estimated at 18 mm. It ranges from 0 mm in the desert to 200 mm in the northern coastal region. In many districts rain may fall in large quantity only once in two or three years. During summer, temperatures are extremely high, reaching 38°C to 43°C with extremes of 49°C in the southern and western deserts. The Mediterranean coast has cooler conditions with 32°C as a maximum.

Surface water resources

The Nile river is the main source of water for Egypt. Under the 1959 Nile Waters Agreement between Egypt and Sudan, Egypt's share is 55.5 km^3/year. The 1959 Agreement was based on the average flow of the Nile during the 1900-1959 period, which was 84 km^3/year at Aswan. Average annual evaporation and other losses from the High Dam lake were estimated to be 10 km^3/year, leaving a net usable annual flow of 74 km^3/year, of which 18.5 km^3/year was allocated to Sudan and 55.5 km^3/year to Egypt. Internal surface water resources are estimated at 0.5 km^3/year. This brings the total (actual) surface water resources to 56.0 km^3/year.

TABLE 2
Water: sources and use

Renewable water resources:			
Average precipitation		18	mm/yr
		18.1	km^3/yr
Internal renewable water resources		1.8	km^3/yr
Total (actual) renewable water resources	1995	58.3	km^3/yr
Dependency ratio	1995	96.9	%
Total (actual) renewable water resources per inhabitant	1995	926	m^3/yr
Total dam capacity	1992	169 000	10^6 m^3
Water withdrawal:			
- agricultural	1993	47 400	10^6 m^3/yr
- domestic	1993	3 100	10^6 m^3/yr
- industrial	1993	4 600	10^6 m^3/yr
Total water withdrawal		55 100	10^6 m^3/yr
per inhabitant	1993	913	m^3/yr
as % of total (actual) renewable water resources		94.5	%
Other water withdrawal	1993	1 800	10^6 m^3/yr
Average groundwater depletion	1993	1 400	10^6 m^3/yr
Wastewater - Non-conventional water sources:			
Wastewater:			
- produced wastewater	1993	3 430	10^6 m^3/yr
- treated wastewater	1994	650	10^6 m^3/yr
- reused treated wastewater	1993	200	10^6 m^3/yr
Desalinated water	1990	25	10^6 m^3/yr

Groundwater resources

The volume of groundwater entering the country from Libya is estimated at 1 km^3/year. Internal renewable groundwater resources are estimated at 1.3 km^3/year. This brings the total renewable groundwater resources to 2.3 km^3/year. The main source of internal recharge is percolation from irrigation water, and its quality depends mainly on the quality of the irrigation water. In the northern part of the Delta, groundwater becomes brackish to saline due to sea water intrusion. About half of the Delta contains brackish to saline groundwater. The Nubian Sandstone aquifer, located under the Western Desert and extending to Libya, Sudan and Chad, contains important non-renewable fresh groundwater resources, already developed in the oasis of the new valley. Large irrigation schemes pumping water from the Nubian aquifer are under development in the south-western part of the country (Al Aweinat).

Agricultural drainage water, treated wastewater and desalinated water

In 1994, the quantity of agricultural drainage water flowing back into the Nile river and becoming available again for withdrawal downstream was estimated at 4 km^3/year.

In 1994, the treatment of domestic wastewater was estimated at 650 million m^3/year and in 1993 about 200 million m^3/year of treated wastewater was estimated to have been reused. The quantity of desalinated water was estimated at only 25 million m^3 in 1990.

Water withdrawal

Table 4 shows the actual water availability and water use by the different sectors. Agricultural water withdrawal includes an annual estimated loss of 2 km^3/year due to evaporation from 31 000 km of canals (1 000 km of main canals and 30 000 km of secondary canals). Figure 1 shows the distribution of water withdrawal by sector in 1993.

TABLE 3
Irrigation and drainage

Irrigation potential	1993	4 435 000	ha
Irrigation:			
1. Full or partial control irrigation: equipped area	1993	3 246 000	ha
- surface irrigation	1993	3 046 000	ha
- sprinkler irrigation	1993	117 000	ha
- micro-irrigation	1993	83 000	ha
% of area irrigated from groundwater	1993	4.5	%
% of area irrigated from surface water	1993	95.4	%
% of area irrigated from non-conventional sources	1993	0.1	%
% of equipped area actually irrigated	1993	100	%
2. Spate irrigation area		-	ha
3. Equipped wetland and inland valley bottoms (i.v.b.)		-	ha
Total irrigation (1+2+3)	**1993**	**3 246 000**	**ha**
- as % of cultivated area		100	%
4. Flood recession cropping area		-	ha
Total water managed area (1+2+3+4)	**1993**	**3 246 000**	**ha**
- as % of cultivated area		100	%
- increase over last 10 years		-	%
- power irrigated area as % of water managed area	1993	95	%
Full or partial control irrigation schemes: Criteria			
Large-scale schemes > - ha		-	ha
Medium-scale schemes		-	ha
Small-scale schemes < - ha		-	ha
Total number of households in irrigation		-	
Irrigated crops:			
Total irrigated grain production	1993	14 900 000	tons
as % of total grain production	1993	100	%
Harvested crops under irrigation (full or partial control)	**1993**	**5 378 940**	**ha**
- permanent crops: total	1993	383 040	ha
- annual crops: total	1993	4 995 900	ha
. cereals (wheat, maize, rice, millet, barley)	1993	2 489 340	ha
. berseem (fodder crop)	1993	1 098 300	ha
. vegetables	1993	408 240	ha
. cotton	1993	371 280	ha
. other annual crops	1993	628 740	ha
Drainage - Environment:			
Drained area	1988	2 931 000	ha
as % of cultivated area		90	%
- drained areas in full or partial control irrigated areas	1988	2 931 000	ha
- drained areas in equipped wetland and i.v.b		-	ha
- other drained areas		-	ha
- total drained area with subsurface drains	1988	1 681 000	ha
- total drained area with surface drains	1988	1 250 000	ha
Flood-protected area		-	ha
Area salinized by irrigation	1972	1 210 000	ha
Population affected by water-borne diseases		-	inhabitants

TABLE 4
Water availability and water use in Egypt (1993)

WATER INPUT	km^3/year	WATER USE	km^3/year
Surface water resources	56.0	Agriculture (incl. evaporation)	47.4
Renewable groundwater resources	2.3	Domestic	3.1
Agricultural drainage water	4.0	Industry	4.6
Reused treated wastewater	0.2	Navigation/regulation	1.8
Total water input	62.5	Total water use	56.9

It is estimated that by the year 2000 the total water use will approach 70 km^3/year, which is more than the actual water availability. The additional water is expected to be provided by: the construction of the Jonglei canal in the Sudd swamps in Sudan (2 km^3), non-renewable groundwater (2-2.5 km^3), increasing use of agricultural drainage water (2-2.5 km^3), an increase in treated wastewater (1 km^3), improved water management/irrigation efficiency (1 km^3).

Environmental issues

The likely challenges to the sustainability of water resources in Egypt include salinity, waterlogging, and the decline in fresh water as a result of the continuous discharge of usually untreated domestic and industrial wastewater into the Nile. Agricultural drainage water affects the salinity of the main river downstream and in the delta. The quality of water in the river decreases gradually towards the delta and the coastal plains. Also likely to aggravate pollution is the use of chemical fertilizers, which has increased fourfold in the last two decades, partly in response to the Aswan High Dam's reduction of the flow of silt downstream. The use of herbicides to control submerged weeds in canals and water hyacinths in drains (which, if uncleared, can choke irrigation systems), has caused serious environmental hazards.

FIGURE 1
Water withdrawal (total: 55.1 km^3 in 1993)

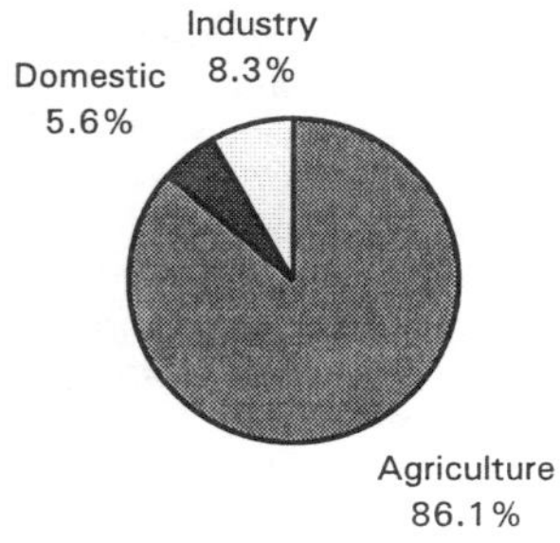

FIGURE 2
Origin of irrigation water f/p
(total: 3 246 000 ha in 1993)

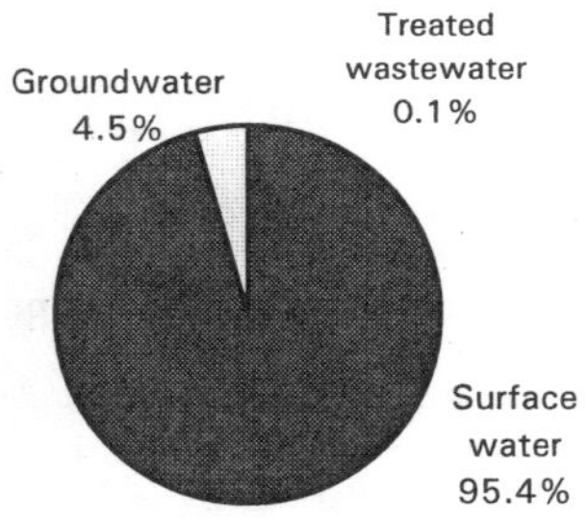

IRRIGATION AND DRAINAGE DEVELOPMENT

Almost all agriculture in Egypt is irrigated. Even the small, more humid area along the Mediterranean coast requires supplementary irrigation to produce reasonable yields. The total water managed area is 3 246 000 ha, of which more than 90% is in the Nile Valley and Delta. Another 920 000 ha are planned to be reclaimed before the year 2000. Irrigation potential has been estimated at 4 435 000 ha. However, all future possibilities for irrigation will depend on the sharing of the Nile waters among the Nile basin countries.

All irrigation is full or partial control irrigation. Over 95% of the area is irrigated from the Nile water. In the provinces of Matrouh, Sinai and New Valley 146 000 ha are irrigated from groundwater. In 1993, an area of 4 200 ha was irrigated from treated wastewater (Figure 2).

The irrigation system in the old lands of the Nile Valley is a combined gravity and water lifting system. The main canal system (first level) takes its water from head regulators, located upstream of the Nile dams. Water is distributed along branches (second level) where the flow is continuous. At the third level, distributaries receive water according to a rotation schedule. Water is pumped from these to irrigate the fields (about 0.5-1.5 m of lift).

The irrigation system in the new lands (reclaimed areas) is based on a cascade of pumping stations from the main canal to the fields, with a total lift of up to 50 m. In the new reclaimed areas, farmers have to use sprinkler or drip irrigation. Surface irrigation techniques are banned by law. The main reason for this ban is that these areas, located at the end of the systems, are more subject to water shortages. In addition, most of the new reclaimed land is sandy soil. Sprinkler and drip irrigation need less water than surface irrigation. Sprinkler irrigation is practised on 117 000 ha, and micro-irrigation on 83 000 ha (Figure 3).

FIGURE 3
Irrigation techniques f/p (total: 3 246 000 ha in 1993)

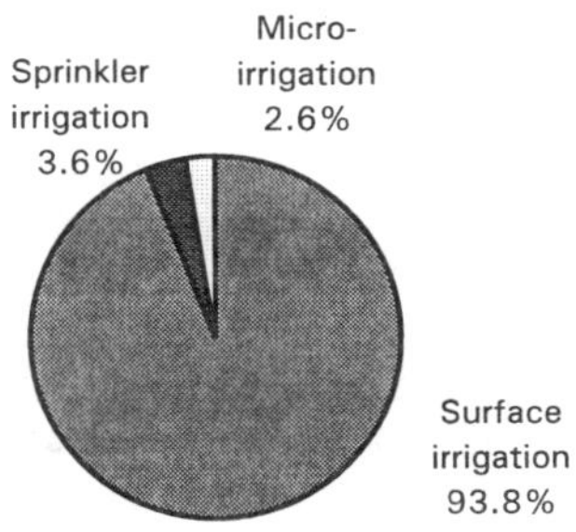

FIGURE 4
Types of drainage methods (total: 2 931 000 ha in 1993)

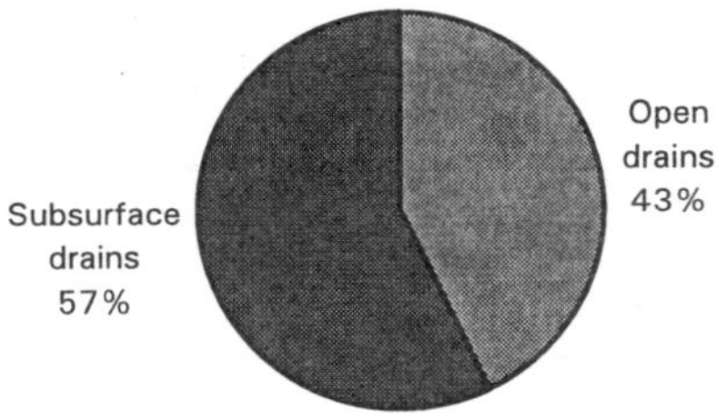

An extensive National Drainage Programme has been carried out over the last 30 years to control waterlogging and salinity. The drainage system consists of open drains, subsurface drains and pumping stations. Of the total irrigated area, 2 931 000 ha (90%) are drained, of which 1 681 000 ha with subsurface drainage (Figure 4). The sub-surface drained area represents nearly 52% of the total cultivated area and more than 74% of the cultivated land in the Valley and the Delta. There are 99 pump stations devoted to the pumping of drainage effluent. The power drained area was estimated at 1 100 000 ha in 1988. Drainage water from agricultural areas on both sides of the Nile Valley is returned to the Nile River in Upper Egypt and in the Southern Delta. Drainage water in the Delta is either pumped back into irrigation canals for reuse or pumped into the northern lakes or the Mediterranean.

In the Fayoum province, irrigation is practised by gravity with no water lifting system. Drainage water is also directed by gravity into Wadi Rayan and the Qaroon lake.

Saline areas in the Nile Valley and Delta were estimated at 1 210 000 ha in 1972, due to sea water intrusion and a high water table. Soil salinity in Upper Egypt developed mainly after the construction of the Aswan High Dam and the conversion of flood irrigation into perennial irrigation. Most of the salt-affected agricultural land in the country has now been provided with drainage systems for leaching the excess salts.

The cropped area is almost 5.4 million ha, leading to an average cropping intensity of 165% (Figure 5). There are three cropping seasons in Egypt: Winter (November to May), Summer (April/May to October), and 'Nili' (July/August to October). On the old lands, cropping intensities can be very high (200%), but on the new lands intensities reach only 150%, mainly because of water shortages and the lack of means of production in those areas. Most crops are grown both in the Delta and the Valley, with the exception of rice (Delta mainly) and sugar cane

FIGURE 5
Irrigated crops f/p (total: 5 378 940 ha in 1993)

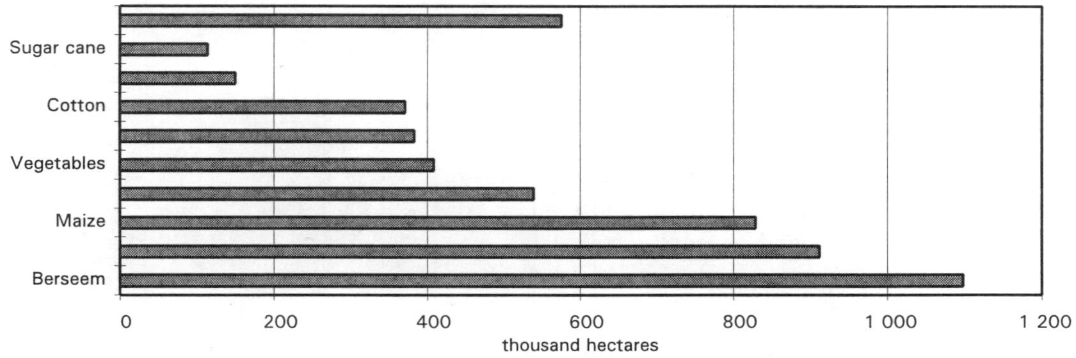

(Valley). The main winter crop is berseem (*Trifolium alexandrinum*), grown either over 3 months with 2 cuts as a soil improver (short berseem), or over 6-7 months, either with 4-5 cuts as a fodder crop or grazed by tethered cattle (long berseem). In addition to berseem, wheat is an important winter crop. Minor winter crops are, amongst others, pulses, barley and sugar beet. The main summer crops are maize, rice and cotton, the latter being the most important Egyptian export crop. Yields have continually increased during past decades, particularly on the old lands. In 1993, yields were 5.4 t/ha for wheat, 7.0 t/ha for maize, 7.7 t/ha for rice, and 1.2 t/ha for cotton.

The average cost of irrigation development is about $US 1 600/ha for micro-irrigation, $US 1 300/ha for mobile sprinkler irrigation, and $US 3 200/ha for stationary sprinkler irrigation.

INSTITUTIONAL ENVIRONMENT

The Ministry of Public Works and Water Resources (MPWWR) is in charge of water resources research, development and distribution, and undertakes the construction, operation and maintenance of the irrigation and drainage networks. At central level, the Planning Sector is responsible for data collection, processing and analysis for planning and monitoring investment projects. Water resources development works are coordinated by the Sector of Public Works and Water Resources. The Nile Water Sector is in charge of cooperation with Sudan and other Nilotic countries. The Irrigation Department provides technical guidance and monitoring of irrigation development, including dams. The Mechanical and Electrical Department is in charge of the construction and maintenance of pumping stations for irrigation and drainage.

Further to these institutions, other public authorities operate in direct relation to the MPWWR. They are the High Aswan Dam Authority, responsible for dam operation; the Drainage Authority, responsible for the construction and maintenance of tile and open drains; and the Water Research Centre. The Water Research Centre comprises 12 institutes and is the scientific body of MPWWR for all aspects related to water resources management.

The Ministry of Agriculture and Land Reclamation (MALR) is in charge of agricultural research and extension, land reclamation and agricultural, fisheries and animal wealth development.

TRENDS IN WATER RESOURCES MANAGEMENT

One important characteristic of Egyptian agriculture is land distribution. More than 95% of the landowners hold less than 2 ha each. Only 3% own 20 ha or more.

It is planned to develop an additional 920 000 ha for agriculture by 2000. This will be achieved through further developing the country's surface water and groundwater resources, increasing drainage water and treated wastewater reuse, and improving irrigation efficiency.

Al Salam canal, now under construction, is planned for the reuse of drainage water from two main drains in the Eastern Delta. This water, added to water extracted from the Damietta branch of the Nile, will be used for the irrigation of a new area of 252 000 ha in the Eastern Delta and North Sinai. The total area planned to be developped for irrigation by different projects in the Sinai is estimated at 630 000 ha.

At present, Egypt and Sudan base their water development plans on the shares stipulated in the 1959 Nile water agreement between the two countries and on future conservation projects in the Sudd area in southern Sudan to increase the yield of the river. However, future developments in upstream countries will have to be taken into account from the point of view of present shares and any future increments that can become available from the conservation projects. In any case, the expansion of irrigation in the Nile basin in the years ahead will require basin-wide cooperation in the management of water resources to meet increasing demands and to face the associated environmental consequences.

MAIN SOURCES OF INFORMATION

Abu Zeid, Mahmoud A., Rady, M.A. 1992. Water resources management and policies in Egypt. In: Country experiences with water resources management: economic, technical and environmental issues. World Bank Technical Paper No. 175. Washington D.C.

ACSAD [Arab Centre for the Study of Arid Zones and Dry Lands]. 1988. Water Resources Assessment in the Arab Region. ACSAD, Damascus.

Attia, F.A.R. 1993. Environmentally sound management of Egyptian groundwater resources. In: [Proceedings of the] Seminar on Techniques of Groundwater Management in the Arab Region. 20-23 December 1993.

CAPMAS [Central Agency for Public Mobilization And Statistics]. 1994. Statistical yearbook, Arab Republic of Egypt.

CAPMAS. 1993. Crop Acreage and cultivated areas in the Arab Republic of Egypt.

Drainage Research Institute. 1989. Land Drainage in Egypt. Amer & de Ridder Editions.

Drainage Research Institute. 1989. Drainage water in the Nile Delta: Re-use monitoring programme. Report No. 32.

FAO. 1993. National action programme - Egypt. Rome.

MPWWR [Ministry of Public Works and Water Resources]. 1994. Irrigation and horizontal expansion sectors.

Nasser Ezzat, M. 1994. Water resources development. Country report: Egypt; prepared for the Nile 2002 Conference.

Othman, Y. 1994. Experiences in integrated land and water management. In: Proceedings of the ISAWIP Final Seminar. Port-Said, Egypt, 18-21 April 1994.

Public Authority for Drainage Projects, Ministry of Public Works and Water Resources. 1994. Nile Delta Drainage V Project. Project completion report.

Rofail, N., & Zahran, M.S. 1994. Available water resources in the Arab world and expected future demand. Paper prepared for the ACSAD preparatory expert meeting for the Arab Ministerial Conference on Sustainable Agriculture and Rural Development, Arab League Centre, Cairo, 25-29 September 1994.

Sarris, A.H. 1991. Structural adjustment and agricultural development in Egypt: Policies, prospectives and options.

Iran

GEOGRAPHY AND POPULATION

The Islamic Republic of Iran covers a total area of about 1.65 million km^2 and is bordered by Azerbaijan, the Caspian Sea and Turkmenistan to the north, Afghanistan and Pakistan to the east, the Gulf of Oman and the Persian Gulf to the south, and Iraq and Turkey to the west. About 52% of the country consists of mountains and deserts and some 16% of the country has an elevation of more than 2 000 m above sea level. The largest mountain massif is that of the Zagros, which runs from north-western Iran first southwards to the shores of the Persian Gulf and then continues eastwards till the most south-eastern province. Other mountain ranges run from the north-west to the east along the southern edge of the Caspian Sea. Finally, along the eastern frontier of Iran several scattered mountain chains exist. The Central or Interior Plateau is located in between these mountain chains and covers over 50% of the country. It is partly covered by a remarkable salt swamp (kavir) and partly by areas of loose sand or stones with stretches of better land near the foothills of the surrounding mountains.

The cultivable area is estimated at about 51 million ha, which is 31% of the total area. In 1993 about 18.5 million ha, or 36% of the cultivable area, were considered usable for agriculture, while 14.4 million ha were actually cultivated. Of this area, 12.8 million ha consisted of annual crops and 1.6 million ha of permanent crops. About 70% of the landholders possess less than 5.5 ha (of which on average 2.13 ha irrigated and 3.25 ha rainfed).

TABLE 1
Basic statistics and population

Physical areas:			
Area of the country	1995	164 800 000	ha
Cultivable area	1993	51 000 000	ha
Cultivated area	1993	14 382 418	ha
- annual crops	1993	12 732 418	ha
- permanent crops	1993	1 650 000	ha
Population:			
Total population	1995	67 283 000	inhabitants
Population density	1995	41	inhab./km^2
Rural population	1995	41	%
Water supply coverage:			
Urban population	1992	91	%
Rural population	1996	86	%

The total population is about 67.3 million (1995), of which 41% is rural. The average population density is 41 inhabitants/km^2, but it ranges from less than 10 in the eastern part of the country up to more than 150 in the Gilan province, located in the Caspian Plain in the north, which is by far the most densely populated region in the country. In the Tehran province, where the capital is located, the population density reaches 400 inhabitants/ km^2. The annual demographic growth rate was estimated at 3.4% over the period 1980-1990 and at 2.6% over the period 1990-1994. In 1991, agriculture employed around 29% of the total labour force and accounted for 21% of GDP, while in 1992 it accounted for 23% of GDP. In 1989, agriculture accounted for 47% of non-oil exports.

TABLE 2
Water: sources and use

Renewable water resources:			
Average precipitation		252	mm/yr
		415.3	km^3/yr
Internal renewable water resources		128.5	km^3/yr
Total (actual) renewable water resources	1995	137.51	km^3/yr
Dependency ratio	1995	6.6	%
Total (actual) renewable water resources per inhabitant	1995	2 044	m^3/yr
Total dam capacity	1993	39 200	10^6 m^3
Water withdrawal:			
- agricultural	1993	64 155	10^6 m^3/yr
- domestic	1993	4 395	10^6 m^3/yr
- industrial	1993	1 484	10^6 m^3/yr
Total water withdrawal		70 034	10^6 m^3/yr
per inhabitant	1993	1 091	m^3/yr
as % of total (actual) renewable water resources		50.9	%
Other water withdrawal	1993	39 228	10^6 m^3/yr
Average groundwater depletion	1993	3 795	10^6 m^3/yr
Wastewater - Non-conventional water sources:			
Wastewater:			
- produced wastewater	1992	4 409	10^6 m^3/yr
- treated wastewater	1992	219	10^6 m^3/yr
- reused treated wastewater		-	10^6 m^3/yr
Desalinated water	1991	2.9	10^6 m^3/yr

CLIMATE AND WATER RESOURCES

Climate

The climate of Iran is one of great extremes due to its geographic location and varied topography. The summer is extremely hot with temperatures in the interior rising possibly higher than anywhere else in the world, certainly over 55°C has been recorded. In winter, however, the great altitude of much of the country and its continental situation result in far lower temperatures than one would expect to find in a country in such low latitudes. Minus 30°C can be recorded in the north-west and minus 20°C is common in many places.

Annual rainfall ranges from less than 50 mm in the deserts to more than 1 600 mm on the Caspian Plain. The average annual rainfall is 252 mm and approximately 90% of the country is arid or semi-arid. Overall, about two-thirds of the country receives less than 250 mm of rainfall per year.

Water resources

Iran can be divided into the following major river basins: the Central Plateau in the middle, the Lake Orumieh basin in the north-west, the Persian Gulf and the Gulf of Oman in the west and south, the Lake Hamoun basin in the east, the Kara-Kum basin in the north-east and the Caspian Sea basin in the north. With an area of 424 240 km^2, the Caspian Sea is the largest landlocked water body in the world and its surface lies about 22 metres below sea level. The rainfall characteristics of the above basins are summarized in Table 4.

TABLE 3
Irrigation and drainage

Irrigation potential	1995	15 000 000	ha
Irrigation:			
1. Full or partial control irrigation: equipped area	1993	7 264 194	ha
- surface irrigation	1993	7 173 494	ha
- sprinkler irrigation	1993	47 200	ha
- micro-irrigation	1993	43 500	ha
% of area irrigated from groundwater	1993	50.1	%
% of area irrigated from surface water	1993	49.9	%
% of area irrigated from non-conventional sources	1993	0.0	%
% of equipped area actually irrigated	1993	100	%
2. Spate irrigation area		-	ha
3. Equipped wetland and inland valley bottoms (i.v.b.)		-	ha
Total irrigation (1+2+3)	**1993**	**7 264 194**	**ha**
- as % of cultivated area		51	%
4. Flood recession cropping area	1993	10 000	ha
Total water managed area (1+2+3+4)	**1993**	**7 274 194**	**ha**
- as % of cultivated area		51	%
- increase over last 10 years	1983-93	30	%
- power irrigated area as % of water managed area	1993	36.0	%
Full or partial control irrigation schemes: Criteria			
Large-scale schemes > 50 ha	1991	708 260	ha
Medium-scale schemes	1991	3 159 924	ha
Small-scale schemes < 10 ha	1991	3 396 010	ha
Total number of households in irrigation	1991	2 620 000	
Irrigated crops:			
Total irrigated grain production	1993	10 000 000	tons
as % of total grain production	1993	61	%
Harvested crops under irrigation (full or partial control)	**1993**	**7 264 194**	**ha**
- permanent crops: total	1993	1 564 884	ha
- annual crops: total	1993	5 699 310	ha
. wheat	1993	2 340 676	ha
. other cereals (barley and rice)	1993	1 256 310	ha
. fodder crops	1993	790 063	ha
. vegetables	1993	425 116	ha
. other annual crops	1993	887 145	ha
Drainage - Environment:			
Drained area	1995	40 000	ha
as % of cultivated area		0.3	%
- drained areas in full or partial control irrigated areas	1995	40 000	ha
- drained areas in equipped wetland and i.v.b		-	ha
- other drained areas		-	ha
- total drained area with subsurface drains	1995	40 000	ha
- total drained area with surface drains		-	ha
Flood-protected area		-	ha
Area salinized by irrigation	1993	2 100 000	ha
Population affected by water-borne diseases		-	inhabitants

TABLE 4
Rainfall in the major basins in Iran

Basin	Total area (km^2)	As % of total area	Rainfall (mm/year)	Rainfall (km^3/year)	As % of total rainfall
Central Plateau	832 000	51	165	138	33
Persian Gulf and Gulf of Oman	431 000	26	366	158	38
Caspian Sea	178 000	11	430	77	19
Lake Hamoun and Kara-Kum	150 000	9	142	21	5
Lake Orumie	57 000	3	370	21	5
Total	1 648 000	100	252	415	100

All these basins, except the Persian Gulf and Gulf of Oman, are interior basins. There are several large rivers, the only navigable one of which is Karun, the others being too steep and irregular. The Karun river, with a total length of 890 km, flows in the south-west of the country to the Shatt El-Arab, which is formed by the Euphrates and the Tigris after their confluence. The few streams that empty into the Central Plateau dissipate into the saline marshes. All streams are seasonable and variable. Spring floods do enormous damage, while there is little water flow in summer when most streams disappear. Water is however stored naturally underground, finding its outlet in subterranean water canals (*qanats*) and springs. It can also be tapped by wells.

Internal renewable water resources are estimated at 128.5 km^3/year. Surface runoff represents a total of 97.3 km^3/year, of which 5.4 km^3/year comes from drainage of the aquifers, and groundwater recharge is estimated at about 49.3 km^3/year, of which 12.7 km^3/year is obtained from infiltration in the river bed. Iran receives 6.7 km^3/year of surface water from Pakistan and some water from Afghanistan through the Helmand river. The flow of the Arax river, at the border with Azerbaijan, is estimated at 4.63 km^3/year. The surface runoff to the sea and to other countries is estimated at 55.9 km^3/year. The total safe yield of groundwater (including non renewable water or unknown groundwater inflow from other countries) has been estimated at 49.3 km^3/year.

Dams

Dams have always played an important role in harnessing Iran's precious water reserves and the long-term objective of Iran's water resources development plan is based on the control and regulation of water resources through dams. In 1994, 27 storage dams were in operation with a total regulation capacity of 39.2 km^3. At the same time, 24 storage dams were under construction with a design regulation capacity of 11.5 km^3. In 1993, the annual electricity production from dams was 25 116 GWH, which is 33% of the total energy production of the country. Dams also play an important role in flood control through routing of floods. Several reservoirs behind the dams seem to offer good sailing and water-skiing facilities, but have not been used for recreation so far.

FIGURE 1
Water withdrawal (total: 70.034 km^3 in 1993)

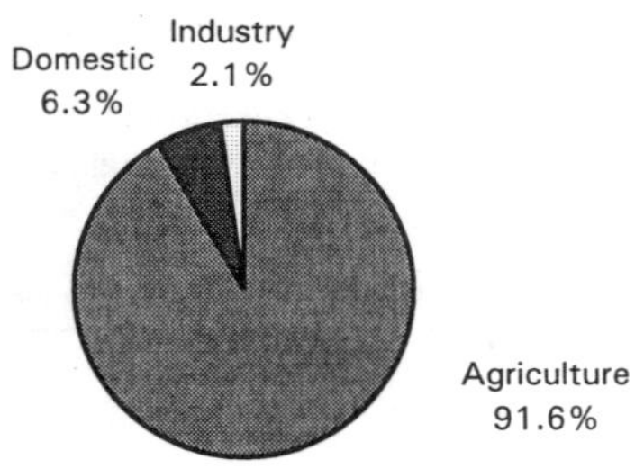

Water use

The total agricultural, domestic and industrial water withdrawal was estimated at about 70 km^3 in 1993 (Figure 1). Although this is equal to 51% of the actual available renewable water resources, current annual abstraction from aquifers (from 49 km^3 in 1990 up to 57 km^3 in 1993) is already more than the estimated safe yield (46 km^3). Further to these 70 km^3, another 39 km^3 of water is used annually, of which about 20 km^3 for electricity production, 11 km^3 for flood control, 2 km^3 for environmental protection (control of downstream parts of rivers), while the remaining part is considered to be surplus water.

IRRIGATION AND DRAINAGE DEVELOPMENT

The problem of water supply has been a constant preoccupation since the beginning of the country's history, thousands of years ago. Its inhabitants learnt to design and implement efficient techniques for harnessing limited water resources and for irrigation. Apart from the qanat, which was a major source of irrigation and domestic water supply for centuries, Iranians have in the past built dams of various types and weirs. Some of these head control structures, built as long as 1000 years ago, are still in good condition.

Agricultural land availability is not a major constraint in the development of Iranian agriculture. The major constraint is the availability of water for the development of these lands. The irrigation potential, based on land and water resources, has been estimated at about 15 million ha, or 29% of the cultivable area. However, this would require optimum storage and water use.

FIGURE 2
Distribution of the water managed area
(total: 7 274 194 ha in 1993)

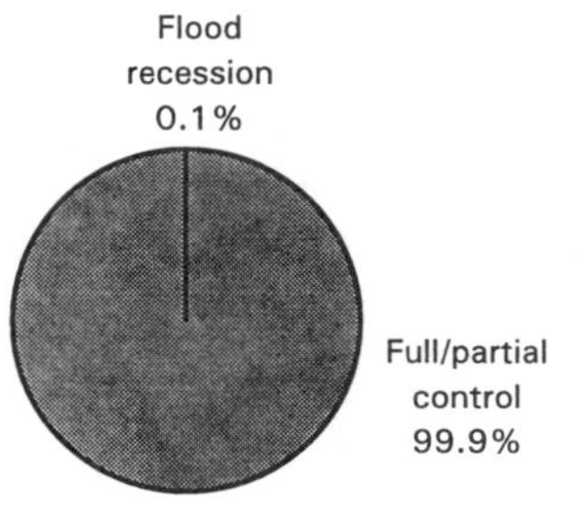

In 1993 out of 14 382 million ha of cultivated land 7 264 million ha, or 51%, were equipped for full or partial control irrigation. Annual crops covered 5 699 million ha and permanent irrigated trees covered 1 565 million ha. In addition, flood recession cropping is practised on an area of about 10 000 ha in the south-west (Figure 2).

FIGURE 3
Irrigation techniques f/p (total: 7 264 194 ha in 1993)

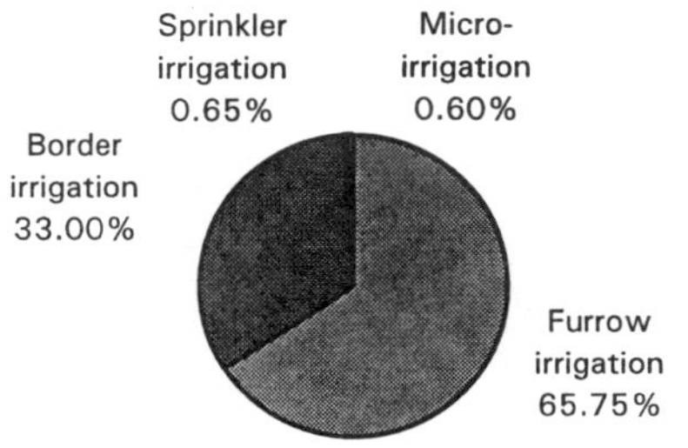

Surface irrigation techniques are used on 98.75% of the area equipped for irrigation and 1.25% benefits from a pressurized irrigation system (Figure 3). About half the area is irrigated from groundwater, including spring water (Figure 4). According to the landholding and technologies which are used, the farming systems are grouped as: small farms (<10 ha) 47%, medium size farms (10-50 ha) 43%, and large farms (>50 ha) 10% (Figure 5).

The cost of surface irrigation development varies from \$US 2 300/ha for large to \$US 2 500/ha for medium and \$US 2 600/ha for small schemes. Average operation and maintenance costs are estimated at \$US 130, 175 and 60 per ha and per year respectively. The cost of micro-irrigation and sprinkler irrigation development is estimated at about \$US 2 200 and 1 200/ha. The average price of water delivered to farmers by government is \$US 0.2 to 0.8/1 000 m^3, while the cost of withdrawal of groundwater by the farmer is \$US 5 to 9/1 000 m^3 and the cost for regulating surface water in existing projects is \$US 3 to 5 per 1 000 m^3. This means that the government heavily subsidizes delivered water, which is probably one of the main reasons for the low irrigation efficiency throughout the country.

On-farm application rates in the country are rather high and in general irrigation has a low efficiency, 32% on average at national level. Major causes of inefficiency include: careless operation, poor maintenance, negligible water prices, fragmentation of responsibilities among different governmental agencies and inadequate training of farmers. Low irrigation efficiency causes waterlogging and salinization in the irrigated areas, which are a major problem in Iran. No comprehensive study has been undertaken regarding the extent of irrigation-induced salinity, but according to an estimate (ICID, 1977) for the year 1974 about 38% of the irrigated area had soils with considerable salinity and drainage problems. Over 2 million ha are estimated to be salt-affected and/or waterlogged at present.

FIGURE 4
Origin of irrigation water f/p (total: 7 264 194 ha in 1993)

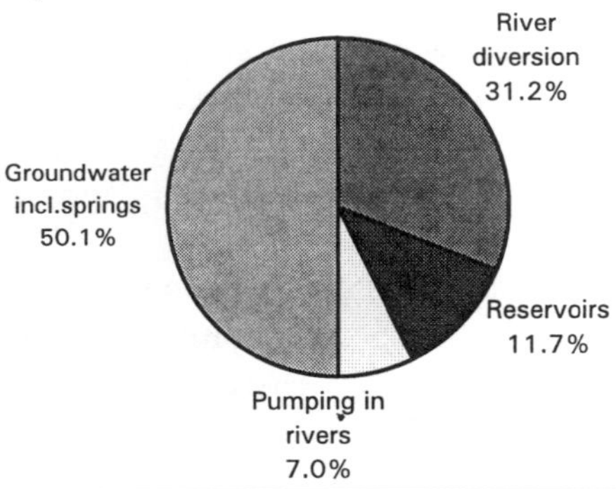

FIGURE 5
Types of f/p control irrigation schemes (total: 7 264 194 ha in 1991)

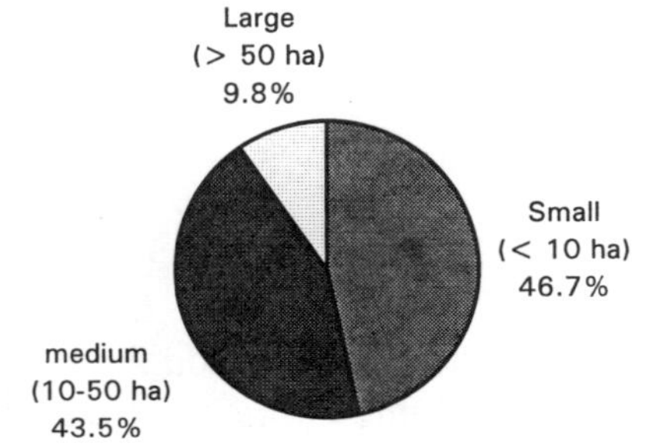

By far the most important irrigated crop is wheat covering almost one-third of the total irrigated area, followed by irrigated fruit trees, covering one fifth of the total area (Figure 6). Other major irrigated crops are barley, rice, vegetables and pulses. Wheat is also by far the most important rainfed crop covering 4.47 million ha, or almost two-thirds of the rainfed area. The yield for irrigated wheat was estimated at 2.78 tons/ha in 1993 against 0.95 tons/ha for rainfed wheat.

FIGURE 6
Irrigated crops f/p (total: 7 264 194 ha in 1993)

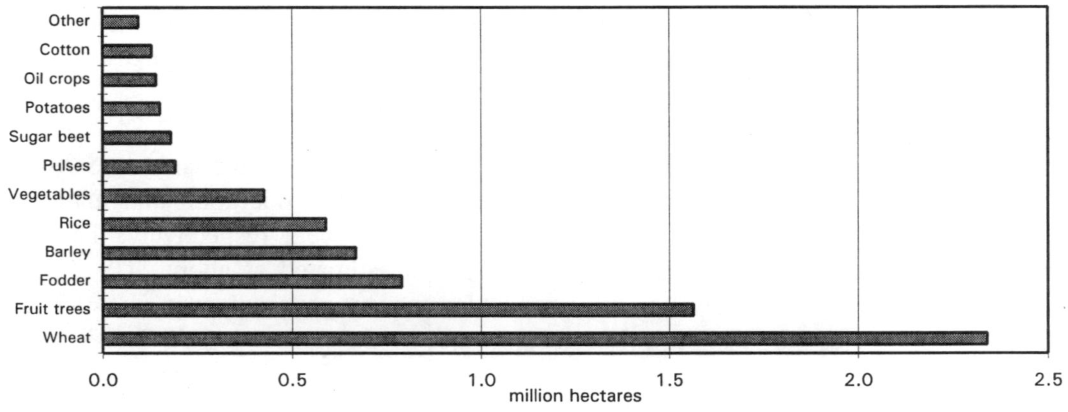

INSTITUTIONAL ENVIRONMENT

According to the water legislation, three ministries are in charge of water resources assessment and development:

- The Ministry of Energy (MOE) has two responsibilities: energy supplies and water resources. In the field of irrigation, it is in charge of the construction of large hydraulic works, including dams and primary and secondary irrigation and drainage canals for the distribution of water. Within MOE, the Water Affairs Department (WAD) is responsible for overseeing and coordinating planning, development, management and conservation of water resources. Fourteen publicly owned Regional Water Authorities (RWA), reporting directly to MOE, are responsible for feasibility studies, project execution and subsequent management. The operation and maintenance of primary and secondary irrigation and drainage canals are operated by operation and maintenance corporations affiliated to MOE.

- The Ministry of Agriculture (MOA) is responsible for supervising rainfed and irrigated crop development. It is in charge of subsurface drains, tertiary and quaternary canals as well as on-farm development and irrigation techniques, planned and operated by the Provincial Agricultural Organizations and the Deputy Ministry for Infrastructure Affairs of the Ministry of Agriculture.

- The Ministry of Jihad-e-Sazandagi (MOJ) deals with watershed management and rural development.

TRENDS IN WATER RESOURCES MANAGEMENT

Agriculture is one of the main priorities in national development plans. The annual increase in irrigated land over 15 years (1978-1993) was 3.8% along with a 4.4% annual increase for agricultural water supply. An increase of 500 000 ha of irrigated land and an increase of 10 km^3/year of agricultural water supply was planned in the second national five-year plan (1995-2000).

At present, a big gap exists between water delivery from the main canals and water application in the field. Compared to the large investments for water resources development, little has been done to improve irrigation water use at farm level. Water is delivered to old traditional irrigation canals and on-farm conveyance and the use of irrigation water is generally rudimentary and wasteful. The use of earth bunds, unlined canals and poor levelling combined with low water charges have resulted in very low levels of water conveyance and use efficiencies (30% as a national average) and caused the emergence of serious drainage problems.

A fundamental review of the organizational chart and institutional changes were made to improve this situation. Since 1992, the Deputy Ministry for Infrastructure Affairs of the Ministry of Agriculture created five departments: farm development, pressurized irrigation systems, water supply, hydraulic constructions and operation and maintenance.

The government policy includes:

- an increase in irrigation efficiency by changing the surface irrigation techniques to pressurized irrigation (2 million ha were planned for the year 1995-2000);
- the establishment of a land Bank to provide loans for on-farm development projects;
- a change in water pricing and delivery methods;
- large-scale privatization.

MAIN SOURCES OF INFORMATION

Bureau of Information and Statistics. 1994. Agricultural Statistics Yearbook 1993. Deputy Ministry of Plan and Project, Ministry of Agriculture.

Bureau of Operation and Maintenance of Dams and Irrigation Networks. 1995. Water utilization in the year 1993. Deputy Ministry of Water Affairs, Ministry of Agriculture.

Deputy Ministry for Infrastructure Affairs. 1991. Summary of the social and agricultural economy of Iran. Ministry of Agriculture.

Keshavarz, M.A. 1993. Improvement of farm level infrastructures. Deputy Ministry for Infrastructure Affairs. Ministry of Agriculture.

Shakiebie. 1994. Seventh Iranian National Seminar on Irrigation and Drainage. IRNCID, Ministry of Energy.

Statistical Centre. 1994. Yearly Statistical Book 1993. Plan and Budget Organization.

Water and Sewage Engineering Co. Situation of water and wastewater in the country in 1992. Ministry of Energy.

World Bank. 1993. Staff appraisal report: Irrigation improvement project. Report No. 11393-IRN.

Yekom Consulting Engineers. 1995. Cost of Irrigation and Drainage Projects - Tender Documents.

Iraq

GEOGRAPHY AND POPULATION

Iraq, with a total area of 438 320 km^2 including 924 km^2 of inland waters, is surrounded by Iran to the east, Turkey to the north, Syria and Jordan to the west, Saudi Arabia and Kuwait to the south, and the Persian Gulf to the south-east. Topographically Iraq is shaped like a basin, consisting of the Great Mesopotamian alluvial plain of the Tigris and the Euphrates rivers. (Mesopotamia means, literally, the land between two rivers). This plain is surrounded by mountains in the north and the east, which can reach altitudes of 3 550 m above sea level, and by desert areas in the south and west, which account for over 40% of the land area. For administrative purposes, the country is divided into 18 governorates, of which three are gathered in an autonomous region.

It is estimated that 11.48 million hectares are cultivable, or 26% of the total area of the country. The total area estimated to be used for agriculture is 8 million ha, which is almost 93% of the cultivable area. However, due to soil salinity, fallow practices and the unstable political situation it is estimated that only 3 to 5 million ha are actually cultivated annually. In 1993, the area actually cultivated was estimated at about 3.73 million ha, of which 3.46 million ha consisted of annual crops and 0.27 million ha consisted of permanent crops.

TABLE 1
Basic statistics and population

Physical areas:			
Area of the country	1995	43 832 000	ha
Cultivable area	1984	11 480 000	ha
Cultivated area	1993	3 731 300	ha
- annual crops	1993	3 462 800	ha
- permanent crops	1993	268 500	ha
Population:			
Total population	1995	20 449 000	inhabitants
Population density	1995	47	inhab./km^2
Rural population	1995	25	%
Water supply coverage:			
Urban population	1990	100	%
Rural population	1990	54	%

The total population is about 20.4 million (1995), of which 25% is rural. Average population density is estimated at 47 inhabitants/km^2, but ranges from 5 inhabitants/km^3 in the Anwar province, in the desert in the western part of the country, to more than 170 inhabitants/km^2 in the Babylon province, in the centre of the country. The average population growth was estimated at 3.6% during 1980-90, but emigration of foreign workers and severe economic hardship have reduced this growth rate since 1990. In 1989, the agriculture sector contributed only 5% to GDP, which was dominated by oil (61%). About 20% of the labour force is engaged in agriculture.

CLIMATE AND WATER RESOURCES

Climate

The climate is mainly of the continental, subtropical semi-arid type, with the north and north-eastern mountainous regions having a Mediterranean climate. Rainfall occurs from December to February (or November to April in the mountains) during cool to cold winters, with a day

TABLE 2
Water: sources and use

Renewable water resources:			
Average precipitation		154	mm/yr
		67.7	km^3/yr
Internal renewable water resources		35.2	km^3/yr
Total (actual) renewable water resources	1995	75.42	km^3/yr
Dependency ratio	1995	53.3	%
Total (actual) renewable water resources per inhabitant	1995	3 688	m^3/yr
Total dam capacity	1990	50 200	10^6 m^3
Water withdrawal:			
- agricultural	1990	39 380	10^6 m^3/yr
- domestic	1990	1 280	10^6 m^3/yr
- industrial	1990	2 140	10^6 m^3/yr
Total water withdrawal		42 800	10^6 m^3/yr
per inhabitant	1990	2 367	m^3/yr
as % of total (actual) renewable water resources		56.7	%
Other water withdrawal	1975	1 790	10^6 m^3/yr
Average groundwater depletion		-	10^6 m^3/yr
Wastewater - Non-conventional water sources:			
Wastewater:			
- produced wastewater		-	10^6 m^3/yr
- treated wastewater		-	10^6 m^3/yr
- reused treated wastewater		-	10^6 m^3/yr
Desalinated water		-	10^6 m^3/yr

temperature of about 16°C dropping at night to 2°C with a possibility of frost. Summers are dry and hot to extremely hot, with a shade temperature of over 43°C during July and August, yet dropping at night to 26°C.

Rainfall is very seasonal and occurs mainly between December and February, except in the north and north-east of the country where the rainy season is longer, from November to April. Average annual rainfall is estimated at 154 mm, but ranges from less than 100 mm over 60% of the country in the south up to 1 200 mm in the north-east.

River basins

There is only one river basin in Iraq, the Shatt Al-Arab basin. The Shatt Al-Arab is the river formed by the confluence downstream of the Euphrates and the Tigris and flows into the Persian Gulf after a course of only 190 km,. Before their confluence, the Euphrates flows for about 1 000 km and the Tigris for about 1 300 km respectively within the Iraqi territory. Nevertheless, due to the importance of the Euphrates and the Tigris, the country is generally divided into three river basins: the Tigris, the Euphrates, and the Shatt Al-Arab (referring to the part downstream of the confluence of the two rivers).

Surface water resources

Both the Tigris and the Euphrates are international rivers originating their source in Turkey. The Tigris river basin in Iraq has a total area of 253 000 km^2, or 54% of the total river basin area.

The average annual flow of the Euphrates as it enters Iraq is estimated at 30 km^3, with a fluctuating annual value ranging from 10 to 40 km^3. Unlike the Tigris, the Euphrates receives no tributaries during its passage in Iraq. About 10 km^3 per year are drained into the Hawr al Hammar (a marsh in the south of the country).

TABLE 3
Irrigation and drainage

Irrigation potential	1990	5 554 000	ha
Irrigation:			
1. Full or partial control irrigation: equipped area	1990	3 525 000	ha
- surface irrigation		-	ha
- sprinkler irrigation		-	ha
- micro-irrigation	1994	8 000	ha
% of area irrigated from groundwater	1990	6.2	%
% of area irrigated from surface water	1990	93.8	%
% of area irrigated from non-conventional sources	1990	0.0	%
% of equipped area actually irrigated	1993	54.9	%
2. Spate irrigation area		-	ha
3. Equipped wetland and inland valley bottoms (i.v.b.)		-	ha
Total irrigation (1+2+3)	**1990**	**3 525 000**	**ha**
- as % of cultivated area		94	%
4. Flood recession cropping area		-	ha
Total water managed area (1+2+3+4)	**1990**	**3 525 000**	**ha**
- as % of cultivated area		94	%
- increase over last 10 years		-	%
- power irrigated area as % of water managed area		-	%
Full or partial control irrigation schemes: Criteria			
Large-scale schemes > - ha		-	ha
Medium-scale schemes		-	ha
Small-scale schemes < - ha		-	ha
Total number of households in irrigation		-	
Irrigated crops:			
Total irrigated grain production		-	tons
as % of total grain production		-	%
Harvested crops under irrigation (full or partial control)		-	**ha**
- permanent crops: total		-	ha
- annual crops: total		-	ha
. wheat	1991	224 490	ha
. barley	1991	200 770	ha
. rice	1991	34 370	ha
.		-	ha
. other annual crops		-	ha
Drainage - Environment:			
Drained area		-	ha
as % of cultivated area		-	%
- drained areas in full or partial control irrigated areas		-	ha
- drained areas in equipped wetland and i.v.b		-	ha
- other drained areas		-	ha
- total drained area with subsurface drains		-	ha
- total drained area with surface drains		-	ha
Flood-protected area		-	ha
Area salinized by irrigation		-	ha
Population affected by water-borne diseases		-	inhabitants

For the Tigris, average annual runoff as it enters Iraq is estimated at 21.2 km^3. All the Tigris tributaries are on its left bank. From upstream to downstream:

- the Greater Zab, which originates in Turkey and is partly regulated by the Bakhma dam. It generates 13.18 km^3 at its confluence with the Tigris; 62% of the 25 810 km^2 of river basin is in Iraq;
- the Lesser Zab, which originates in Iran and is equipped with the Dokan dam (6.8 km^3). The river basin of 21 475 km^2 (of which 74% is in Iraqi territory) generates about 7.17 km^3, of which 5.07 km^3 of annual safe yield after the Dokan construction;
- the Al-Adhaim (or Nahr Al Uzaym), which drains about 13 000 km^2 entirely in Iraq. It generates about 0.79 km^3 at its confluence with the Tigris. It is an intermittent stream subject to flash floods;

- the Diyala, which originates in Iran and drains about 31 896 km^2, of which 75% in Iraqi territory. It is equipped with the Darbandikhan dam and generates about 5.74 km^3 at its confluence with the Tigris;
- the Nahr at Tib, Dewarege (Doveyrich) and Shehabi rivers, draining together more than 8 000 km^2. They originate in Iran, and bring together in the Tigris about 1 km^3 of highly saline waters;
- the Al-Karkha, whose course is mainly in Iran and, from a drainage area of 46 000 km^2, brings about 6.3 km^3 yearly into Iraq, namely into the Hawr Al Hawiza during the flood season, and into the Tigris river during the dry season.

The Karun river, originating in Iran flows with its mean annual flow of 24.7 km^3 into the Shatt Al-Arab. It brings a large amount of fresh water into the Shatt Al-Arab, just before it reaches the sea.

The Euphrates and the Tigris are subject to large and possibly disastrous floods. The level of water in the Tigris can rise at the rate of over 30 cm/hour. In the southern part of the country, immense areas are regularly inundated, levees often collapse, and villages and roads must be built on high embankments. The Tharthar reservoir was planned *inter alia* in the 1950s to protect Baghdad from the ravages of the periodic flooding of the Tigris by storing extra water discharge upstream of the Samarra barrage.

International agreements

In 1980, a Joint Technical Committee on Regional Waters was created by Turkey and Iraq, on the basis of a former protocol (1946) concerning the control and management of the Euphrates and the Tigris. Syria joined the committee afterwards.

The Euphrates represents the most critical issue for Iraq's water strategy as more than 90% of its water comes from outside the country (against only 50% for the Tigris). According to an agreement between Syria and Iraq (1990), Iraq shares the Euphrates' waters with Syria on a 58% (Iraq) and 42% (Syria) basis, based on the flow received by Syria at its border with Turkey. Since Turkey has unilaterally promised to secure a minimum flow of 15.8 km3/year at its border, this agreement would *de facto* represent 9 km^3/year for Iraq. Up to now, there has been no global agreement between the three countries concerning the Euphrates' waters.

Groundwater resources

Good quality subterranean water has been found in the foothills of the mountains in the north-east of the country and in the area along the right bank of the Euphrates:

- the aquifer in the north-east of the country has an estimated sustainable discharge of between 10 and 40 m^3/s, at depths of five to fifty metres. Its salinity increases towards the south-east of the area, where it reaches 1 mg/l;
- the aquifers on the right bank of the Euphrates river are found at depths up to 300 m, and have an estimated discharge of 13 m^3/s. Salinity varies between 0.3 and 0.5 mg/l.

In other areas of the country, groundwater is also found, but always with a salinity level higher than 1 mg/l.

Dams

In 1977, the on-river dams, all of which are located in the Tigris river basin, had a total capacity of 13.7 km^3. There was an important programme of dam construction in Iraq in the 1980s. The programme consisted of the construction of the Saddam dam on the Tigris (11.1 km^3), the Kadisiyya (Qadisia) multipurpose dam on the Euphrates (8.2 km^3), the Bakhma dam on the upper Zab, one of the Tigris tributaries (17.1 km^3), the Badush dam on the Tigris river (0.5 km^3), and several other desert dams totalling about 0.5 km^3. The total on-river storage capacities for the Tigris will thus amount to 42 km^3, and 8.2 km^3 for the Euphrates river, i.e. a total of 50.2 km^3. However, the Bakhma dam was completely destroyed. At present, the Al-Adom dam on the Tigris river, with a capacity of 3.8 km^3, is under construction.

Two off-river storage lakes have been created with the construction of the Tharthar dam (85 km^3) in the Tigris river basin, filled with the Wadi Tharthar waters and, since 1985, with Euphrates waters, and the Habbaniya dam (3.3 km^3), which can be filled from upstream Euphrates waters and which drains into the Euphrates downstream.

New watercourses

In order to increase water transport efficiency, minimise losses and waterlogging, and improve water quality, a number of new watercourses were constructed, especially in the southern part of the country. The Saddam river (or Third river) functions as a main out-fall drain collecting drainage waters of more than 1.5 million hectares of agricultural land from north of Baghdad to the Gulf, between the two main rivers (the Euphrates and the Tigris). The length of the watercourse, completed in December 1992, is 565 km, with a total discharge of 210 m^3/s. Other watercourses were also constructed to reclaim new lands or to reduce waterlogging.

FIGURE 1

Water withdrawal (total: 42.8 km^3 in 1990)

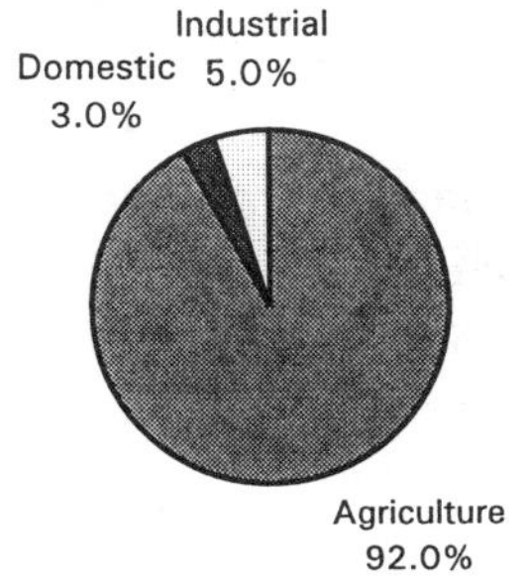

Water withdrawal

Total water withdrawal is estimated at 42.8 km^3 in 1990, of which 92% for agricultural purposes (Figure 1). According to the most recent estimates, 85% of river water withdrawal is used for agricultural purposes.

Water supply

In 1991 safe water supplies reached 100% of urban areas but only 54% of rural areas. The situation has deteriorated as a result of the Gulf war as regards the water supply and sanitation sector, due *inter alia* to shortages in chlorine imports for water treatment.

IRRIGATION AND DRAINAGE DEVELOPMENT

The history of irrigation started 7500 years ago in the land between the Tigris and the Euphrates when the Sumerians built a canal to irrigate wheat and barley.

Irrigation potential was estimated in 1990 at over 5.5 million ha, of which 63% in the Tigris basin, 35% in the Euphrates basin, and 2% in the Shatt Al-Arab basin. Considering the soil resources, it is estimated that about 6 million hectares are classified as excellent, good or moderately suitable for flood irrigation. With the development of water storage facilities, the regulated flow has increased and changed the irrigation potential significantly, since it was estimated at 4.25 million hectares only in 1976. However, irrigation development depends to a large extent on the volume of water released by the upstream countries.

Irrigated areas

The total water managed area was estimated at 3.5 million ha in 1990, all of it being equipped for full or partial control irrigation.

The areas irrigated by surface water are estimated at 3 305 000 ha, of which 105 000 ha (3%) in the Shatt Al-Arab river basin, 2 200 000 ha (67%) in the Tigris river basin, and 1 000 000 ha (30%) in the Euphrates river basin (Figure 2). However, it should be noted that all these areas are not actually irrigated, since a large part has been abandoned due to waterlogging and salinity. Only 1 936 000 ha were estimated to be actually irrigated in 1993.

FIGURE 2
Areas equipped for f/p control irrigation from surface water, by river basin (total 3 305 000 ha in 1990)

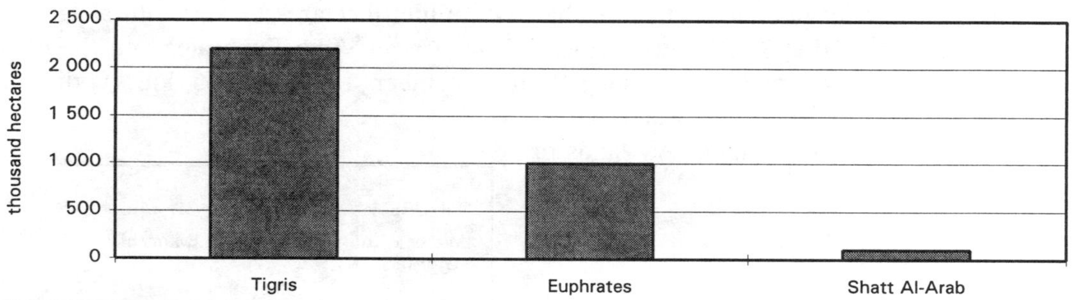

The areas irrigated from groundwater were estimated at 220 000 ha in 1990, with some 18 000 wells (Figure 3). About 8 000 ha were reported equipped for micro-irrigation, but these techniques were not used.

FIGURE 3
Origin of irrigation water f/p
(total: 3 525 000 ha in 1990)

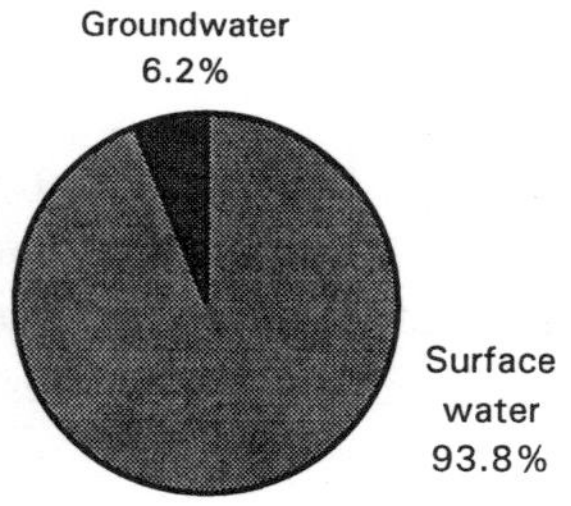

Salinity

Salinity has always been a major issue in this area and it was already recorded as a cause of crop yield reductions some 3800 years ago.

It is estimated that in 1970 half the irrigated areas in central and southern Iraq were degraded due to waterlogging and salinity. The absence of drainage facilities and, to a lesser extent, the irrigation practices used (flooding) were the major causes of these problems. In 1978, a land rehabilitation programme was undertaken, comprising concrete lining for irrigation canals, installation of field drains and collector drains. By 1989, a total of 700 000 ha had been reclaimed at a cost of around $US 2 000/ha.

Recent estimations have nevertheless shown that 4% of the irrigated areas were severely saline, 50% medium saline and 20% slightly saline, i.e. a total of 74% of the irrigated areas suffered from some degree of salinity. The Ministry of Irrigation estimated at 17 million tons the amount of salt transported to the Gulf by the Saddam river in 1995. Irrigation with highly saline waters (more than 1500 ppm) has been practised for date palm trees since 1977. The use of brackish groundwater is also reported for tomato irrigation in the south of the country.

Crops

In 1991, there were 224 490 ha of irrigated wheat, with an average yield of 2.7 tons/ha, while the rainfed wheat area was estimated at 508 620 ha, with an average yield of 1.7 tons/ha. There were 200 770 ha of irrigated barley, with an average yield of 1.8 tons/ha, while the rainfed barley area was estimated at 323 730 ha, with an average yield of 1.3 tons/ha. The other main irrigated crops are rice, maize, vegetables, but also date trees which are important for the economy of the southern part of the country.

INSTITUTIONAL ENVIRONMENT

The Ministry of Agriculture is responsible for organising the ownership of agricultural lands, contracts with farmers, cooperatives and agricultural companies in addition to enhancing agricultural investment activities. In particular, the Ministry is responsible for providing agricultural inputs to all farmers, and for marketing the agricultural commodities.

The Ministry of Irrigation is in charge of water resources development, irrigation and drainage development, as well as its operation and maintenance. Its major functions are to assess water requirements and resources, control running water, reservoirs, wetland and marshes, underground water, the construction of dams, canals and drainage systems, soil conservation, classification, land evaluation and use, and research and studies on land and water. The Ministry is executing most of the water resources development projects with the assistance of a number of State companies.

The State company for Water Wells Drilling (SCWWD) has launched a programme of well-drilling, mainly for irrigation purposes.

TRENDS IN WATER RESOURCES MANAGEMENT

Two main sprinkler irrigation projects were planned in 1990: the Saddam irrigation project, supplied by waters from the Third river, was designed for a total area of 250 000 ha, and the El-Jazira irrigation project which should cover an area of 325 000 ha. The Um Almaarik river in the south west region has recently been excavated in order to irrigate 20 000 ha upon completion.

However, the development of irrigation as it is planned in the upstream countries, particularly the South-east Anatolian (GAP) project in Turkey, and the irrigation projects in Syria will reduce Iraqi irrigation potential unless an agreement is reached on the sharing of waters between the riparian countries. Since water shortages are forecast to occur with the development of irrigation, solutions have to be found for an integrated basin-level planning of water resources development.

Another main issue in water resources management is protection of water quality. The level of salinity in the Euphrates river is high and is expected to increase with the development of irrigation in the basin and, as a consequence, the diminution of the water flow, particularly in the dry season.

Treatment of municipal and industrial wastewater is considered as one way to preserve river water quality. Reuse of treated wastewater for irrigation was also envisaged before 1990. Some industries are already obliged to desalinate the Euphrates waters before using it.

Other measures which could be undertaken would need regional cooperation, for a better management of the flood waters and the dams. But the regulation capacities on the Euphrates river are already greater than the entire average flow, because of the lack of coordination between the riparian countries.

MAIN SOURCES OF INFORMATION

Badry M. M. , Mehdi M. S. and Khawar J. M., 1979. Water resources in Iraq. In: *Irrigation and agricultural development*, a joint publication of ECWA, FAO and the Foundation for Scientific Research of Iraq, edited by S.S. Johl. Pergamon Press.

ECWA. FAO. 1984. Agricultural resources management and desertification control in Iraq. Report E/ECWA/AGR/83/8. Baghdad, Iraq.

ESCWA. 1992. Water resources database in the ECSWA region. Report E/ECSWA/ENR/1992/6.

FAO. 1994. Country information brief, Iraq. FAO-Representation in Iraq.

Fisher W. 1994. Iraq: Physical and social geography in the Middle East and North Africa". Europe publications limited.

UNDTCD. 1990. Progress in the implementation of the Mar del Plata Action plan in the region of Near east and North Africa. Report prepared by Minther J. Haddadin. Amman, Jordan.

Jordan

GEOGRAPHY AND POPULATION

Jordan, with a total area of about 89 210 km^2, lies to the east of the Jordan river and is divided into eightgovernorates: Amman, Zarqa, Irbid, Mafraq, Balqa, Karak, Tafileh and Ma'an. It is bordered in the north by Syria, in the north-east by Iraq, in the south-east and south by Saudi Arabia, in the far south-west by the Gulf of Aqaba (northern shores of the Red Sea) and in the west by Israel.

The country can be divided into four physiographic regions:

- the Ghors (lowlands) in the western part of the country, which consist of 3 zones: the Jordan valley which starts at Lake Tiberias in the north (220 m below sea level), the lowlands along the Dead Sea (405 m below sea level) and the Wadi Araba which extends in a southerly direction to the northern shores of the Red Sea (total area: 5 000 km^2);
- the highlands, which run from north to south at an altitude of between 600 and 1 600 m above sea level (total area: 5 510 km^2);
- the plains, which extend from north to south along the western borders of the desert (Badiah) (total area: 10 000 km^2);
- the desert region (Badiah) in the east, which is an extension of the Arabian desert (total area: 68 700 km^2).

The cultivable land was estimated at 381 740 ha in 1992, or 4.3% of the total area of the country. In 1991, the total cultivated area was estimated at 214 767 ha, which is about 56% of the cultivable area, of which 120 077 ha consisted of annual crops and 94 690 ha consisted of permanent crops. In addition, each year up to 50 000 ha of rainfed land can be left fallow due to fluctuating and unevenly distributed annual rainfall.

TABLE 1
Basic statistics and population

Physical areas:			
Area of the country	1995	8 921 000	ha
Cultivable area	1992	381 740	ha
Cultivated area	1991	214 767	ha
- annual crops	1991	120 077	ha
- permanent crops	1991	94 690	ha
Population:			
Total population	1995	5 463 000	inhabitants
Population density	1995	61	inhab./km^2
Rural population	1995	29	%
Water supply coverage:			
Urban population	1993	100	%
Rural population	1993	97	%

The total population is almost 5.5 million (1995), of which 29% is rural. The population growth is estimated at 3.4% not including fluctuations caused by international political events. Currently, about 90% of the population is concentrated in the north-west quadrant of the country, where rainfall is highest and where most of the water resources are located. Agriculture accounted for 6% of Jordan's GDP in 1992 and for 12% of its exports earnings, mainly fruits and vegetables, in 1991. In 1992, 10% of the labour force were employed in agriculture.

TABLE 2
Water: sources and use

Renewable water resources:			
Average precipitation		94	mm/yr
		8.43	km^3/yr
Internal renewable water resources		0.68	km^3/yr
Total (actual) renewable water resources	1995	0.88	km^3/yr
Dependency ratio	1995	22.7	%
Total (actual) renewable water resources per inhabitant	1995	161	m^3/yr
Total dam capacity	1993	143	10^6 m^3
Water withdrawal:			
- agricultural	1992	737	10^6 m^3/yr
- domestic	1992	214	10^6 m^3/yr
- industrial	1992	33	10^6 m^3/yr
Total water withdrawal		984	10^6 m^3/yr
per inhabitant	1992	246	m^3/yr
as % of total (actual) renewable water resources		111.8	%
Other water withdrawal		-	10^6 m^3/yr
Average groundwater depletion	1993	172	10^6 m^3/yr
Wastewater - Non-conventional water sources:			
Wastewater:			
- produced wastewater	1993	232	10^6 m^3/yr
- treated wastewater	1991	50	10^6 m^3/yr
- reused treated wastewater	1993	50	10^6 m^3/yr
Desalinated water	1993	2	10^6 m^3/yr

CLIMATE AND WATER RESOURCES

Climate

The climate in Jordan is characterized by a long, dry, hot summer, a rainy winter and a drier-than-spring autumn. The temperature increases towards the south, with the exception of some southern highlands. Rainfall varies considerably with location, due mainly to the country's topography. Annual rainfall ranges between 50 mm in the eastern and southern desert regions to 650 mm in the northern highlands. Over 90% of the country receives less than 200 mm of rainfall per year. Average annual rainfall for the country as a whole gives a total volume of 8.43 km^3.

Surface water resources

Surface water resources are unevenly distributed among 15 basins. The largest source of external surface water is the Yarmouk river, at the border with Syria. Originally, the annual flow of the Yarmouk river was estimated at about 400 million m^3 (of which about 100 million m^3 are withdrawn by Israel). Total flow is now much lower than 400 million m^3 as a result of the upstream Syrian development works which have been done in the 1980's. The Yarmouk river accounts for 40% of the surface water resources of Jordan, including water contributed from the Syrian part of the Yarmouk basin. It is the main source of water for the King Abdullah canal and is thus considered to be the backbone of development in the Jordan valley. Other major basins include Zarqa, Jordan river side wadis, Mujib, the Dead Sea, Hasa and Wadi Araba. Internally generated surface water resources are estimated at 400 million m^3/year.

TABLE 3
Irrigation and drainage

Irrigation potential		1992	85 000	ha
Irrigation:				
1. Full or partial control irrigation: equipped area		1991	64 300	ha
- surface irrigation		1991	20 300	ha
- sprinkler irrigation		1991	5 700	ha
- micro-irrigation		1991	38 300	ha
% of area irrigated from groundwater		1991	54.6	%
% of area irrigated from surface water		1991	39.7	%
% of area irrigated from non-conventional sources		199	5.7	%
% of equipped area actually irrigated			-	%
2. Spate irrigation area			-	ha
3. Equipped wetland and inland valley bottoms (i.v.b.)			-	ha
Total irrigation (1+2+3)		**1991**	**64 300**	**ha**
- as % of cultivated area			30	%
4. Flood recession cropping area			-	ha
Total water managed area (1+2+3+4)		**1991**	**64 300**	**ha**
- as % of cultivated area			30	%
- increase over last 10 years			-	%
- power irrigated area as % of water managed area		1992	45	%
Full or partial control irrigation schemes:	Criteria			
Large-scale schemes	> 1 000 ha	1991	6 400	ha
Medium-scale schemes		1991	0	ha
Small-scale schemes	< 10 ha	1991	57 900	ha
Total number of households in irrigation		1991	251 000	
Irrigated crops:				
Total irrigated grain production		1991	28 000	tons
as % of total grain production		1991	30	%
Harvested crops under irrigation (full or partial control)		**1991**	**68 910**	**ha**
- permanent crops: total		1991	27 220	ha
- annual crops: total		1991	41 690	ha
. vegetables		1991	28 960	ha
. cereals (wheat and barley)		1991	8 890	ha
. potatoes		1991	3 840	ha
Drainage - Environment:				
Drained area		1992	4 000	ha
as % of cultivated area			2	%
- drained areas in full or partial control irrigated areas		1992	4 000	ha
- drained areas in equipped wetland and i.v.b			-	ha
- other drained areas			-	ha
- total drained area with subsurface drains			-	ha
- total drained area with surface drains			-	ha
Flood-protected area			-	ha
Area salinized by irrigation		1989	2 277	ha
Population affected by water-borne diseases			-	inhabitants

Groundwater resources

Jordan's groundwater is distributed among 12 major basins. Total internally produced renewable groundwater resources have been estimated at 500 million m^3/year, of which 220 million m^3 constitute the base flow of the rivers. Groundwater resources are concentrated mainly in the Yarmouk, Amman-Zarqa and Dead Sea basins.

The safe yield of renewable groundwater resources is estimated at 275 million m^3/year. Most of it is at present exploited at maximum capacity, in some cases beyond safe yield. Of the 12 groundwater basins, 6 are being overextracted, 4 are balanced with respect to abstraction and

FIGURE 1
Evolution of water withdrawal by sector

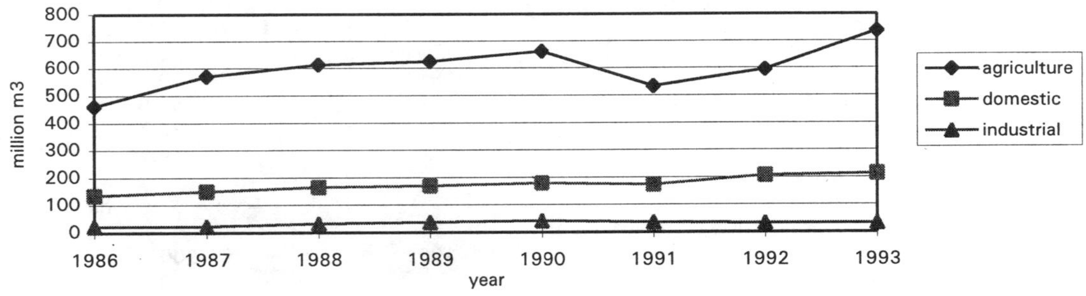

2 are under-exploited. Average groundwater depletion was estimated at 190 million m^3/year in 1990. Over-extraction of groundwater resources has degraded water quality and reduced exploitable quantities, resulting in the abandonment of many municipal and irrigation water well fields, such as in the area of Dhuleil.

The main non-renewable aquifer presently exploited is the Disi aquifer (sandstone fossil), in southern Jordan with a safe yield estimated at 125 million m^3/year for 50 years. Other non-renewable water resources are found in the Jafer basin, for which the annual safe yield is 18 million m^3. In total it is estimated by the Water Authority of Jordan that the safe yield of fossil groundwater is 143 million m^3/year.

Dam capacity

Total dam capacity in Jordan is estimated at 143 million m^3, including the desert dams. The largest dam (the King Talal dam on the Zarqa river) has a total capacity of 80 million m^3. The other main dams are located on the Wadis Araba (20 million m^3), Ziglab (4.3 million m^3), Kafrein (3.8 million m^3) and Shueib (2.3 million m^3). The proposed design of the Wahda (Unity) dam on the Yarmouk river, following a treaty between Jordan and Syria, allows for a dam of 100 m in height with a gross storage capacity of about 230 million m^3. Another proposed dam is the Karameh dam on Wadi Mallaha with a gross storage capacity of 55 million m^3. Following the signature of the Peace Treaty with Israel (1995), investigations have been initiated to assess the need for future storage facilities on the Jordan and Yarmouk rivers.

Wastewater

The produced wastewater was estimated at 232 million m^3/year in 1993 and the quantity of reused treated wastewater reached 50 million m^3, of which 48 million m^3 for irrigation and 2 million m^3 for industrial purposes. The reuse of treated wastewater in Jordan reaches one of the highest levels in the world. The treated wastewater flow in the country is returned to the Search river and the King Tall dam, where it is mixed with the surface flow and used in the pressurized irrigation distribution system in the Jordan valley. The importance of reused wastewater is an essential element of Jordan's water strategy.

Water withdrawal

In 1993, total annual water withdrawal was estimated at 984 million m^3, up from 619 million m^3 in 1986 (Figure 1). Agricultural water withdrawal accounted for 75% of the total water withdrawal, including the use of treated wastewater (Figures 2 and 3). Due to limited and widely scattered sources of water, the construction of important water-conveyance facilities was undertaken between 1962 and 1987 in order to meet the demand of the population which is concentrated in some areas. Some shortages have been observed during recent years, but they are generally limited to less than 10% of the demand. However, during the dry year of 1990, the water shortage affected 17% of the water demand.

FIGURE 2

Water withdrawal (total: 984 million m^3 in 1993)

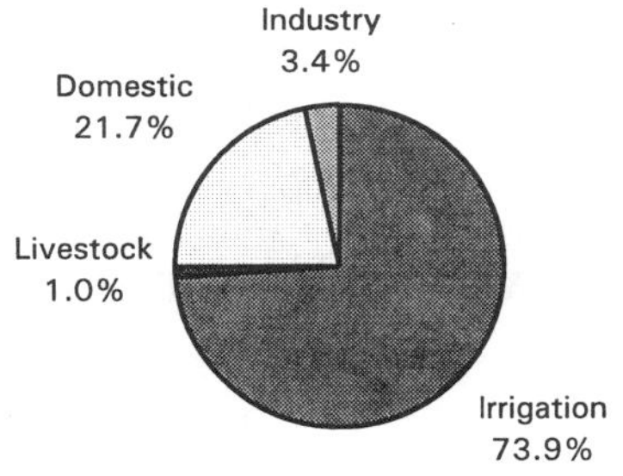

FIGURE 3

Origin of water used by sector (total: 984 million m^3 in 1993)

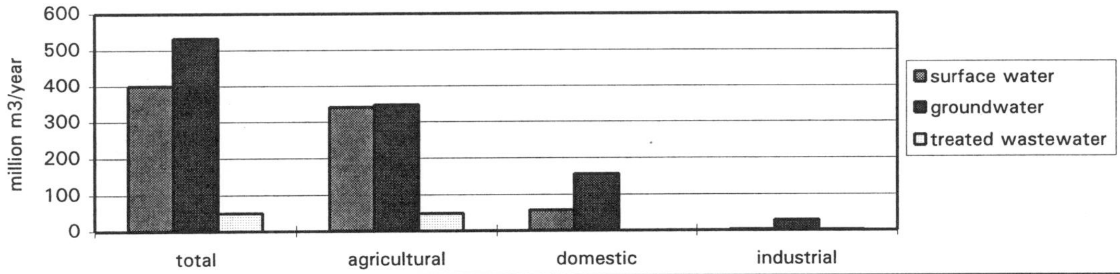

Environment

The level of the Dead Sea falls each year by 85 centimetres due to extensive water use in the Jordan basin. Irrigated soils along the Jordan valley are showing signs of salinization since natural floods are no longer available to flush the irrigated land and leach salts.

International rivers

Most of Jordan's water resources are shared with other countries. The Jordan river is the largest river of the country, where water allocation to riparian countries is one of the most difficult regional issues. Failure so far to develop a unified approach to managing these water resources has encouraged unilateral development by the various riparian countries. Other important shared water resources are the groundwater resources of North Jordan (the Azraq, Yarmouk and Amman-Zarqa basins), where a significant percentage of the natural recharge probably occurs in Syria. Additional Syrian development of groundwater in these basins would endanger current Jordanian development and greatly reduce the safe yield available to Jordan.

The Yarmouk/Jordan river system is the main surface water resource. Given the extreme scarcity of water resources in the region, the importance of developing acceptable water resources allocation strategies was recognized as early as the 1930s. Although no comprehensive agreement exists on sharing the jointly-owned water resources, eleven plans for water use were prepared between 1939 and 1955. The last one was the Johnston Plan of 1955, allocating water between Jordan and Syria.

IRRIGATION AND DRAINAGE DEVELOPMENT

The potential for irrigated cultivation is estimated at around 840 000 ha. However, taking into consideration potentially available water resources, the irrigation potential is about 85 000 ha, including the area currently irrigated.

Although irrigation has been reported in Jordan for a very long time, particularly in the Jordan valley, intensive irrigation projects have been implemented since 1958 when the Government decided to divert part of the Yarmouk river water and constructed the East Ghor canal (later named King Abdullah canal). The canal was 70 km long in 1961 and was extended three times between 1969 and 1987 to reach a total length of 110.5 km. The construction of dams on the side wadis and the diversion of the flows from other wadis has allowed the development of irrigation over a large area. At the same time, wells were drilled in the Jordan valley to abstract groundwater, not only for domestic purposes but also for irrigation.

Apart from in the Jordan valley, irrigation is also reported in the highlands, mainly dependent on groundwater resources. The irrigation system here relies on wells, from 100 to 5 000 m deep, and pumps which deliver water to the agricultural land. There are three types of irrigated farming entities in these areas:

- private holders who have received loans from the ACC (Agriculture Credit Corporation) for drilling, pumps and farm irrigation systems;
- Bedouin-settlement irrigation projects operated and maintained by the Ministry of Agriculture and the Water Authority;

private companies operating large-scale projects in the south-east of the country.

In 1995, the total area equipped for irrigation was estimated at 72 850 ha; in 1991, it was 64 300 ha (38 300 ha under micro-irrigation, 5 700 ha under sprinkler irrigation) (Figure 4). One of the main sources of water is the King Talal dam on the Zarqa river from which water is diverted into the King Abdullah canal. Irrigation water was originally distributed by open canals. This design was changed to reduce losses and all new developed zones have been provided with pipe systems. The techniques used by farmers have gradually changed from surface irrigation (furrow and basin) to micro-irrigation.

FIGURE 4
Irrigation techniques f/p (total: 64 300 ha in 1991)

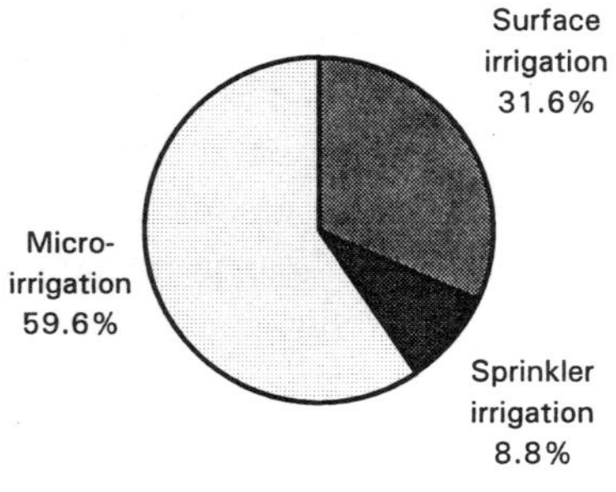

In 1992, it was estimated that about 55% of the area was irrigated from groundwater, 39% from surface water and 6% from treated wastewater (Figure 5). The Disi irrigation project, one of the largest schemes in Jordan covering a total area of 3 000 ha, is supplied with fossil groundwater.

FIGURE 5
Origin of irrigation water f/p
(total: 64 300 ha in 1992)

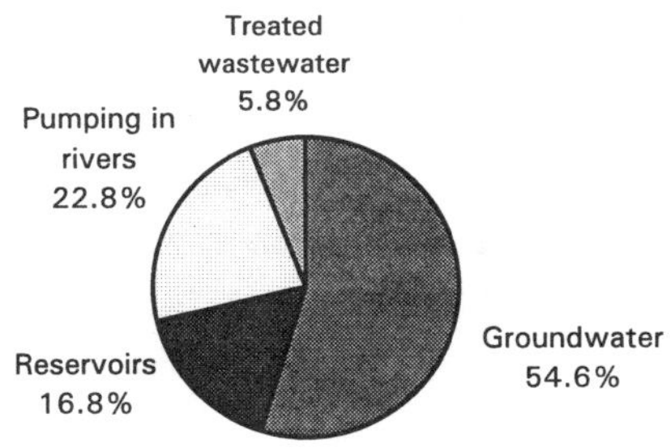

In 1991, two different types of schemes coexisted in Jordan (Figure 6):

FIGURE 6
Types of irrigation schemes f/p
(total: 64 300 ha in 1991)

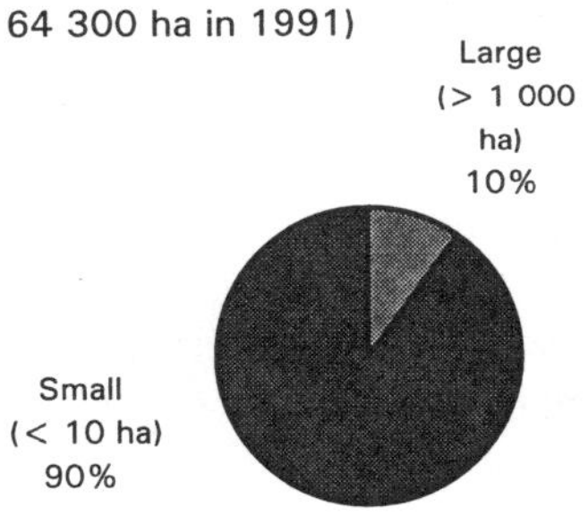

- In the Disi-Mudwara area, the schemes larger than 1 000 ha cover a total area of 6 400 ha. They are owned by large companies. In total, about 1 000 households are working on these schemes. All water is pumped from groundwater and no water charge exists. The average cost of development of such an irrigation project is estimated at $US 30 000 (e.g. the government's investments in the Ghors), while annual operation and maintenance costs can reach $US 4 800/ha.

- The small schemes, generally less than 10 ha each, cover a total area of 57 900 ha. The total number of beneficiaries is 250 000 households considering only the Jordan valley. The price charged for irrigation water has been $US 11.4 every 1 000 m^3 since October 1994 (it was only $US 8.6 previously). The average cost of irrigation development for the production of vegetables varies between $US 6 000 and 9 000/ha and the annual average operation and maintenance costs are about $US 4 800/ha.

The main crops grown in Jordan are wheat and barley (used largely as animal feed), which are generally rainfed. In 1991, irrigated cereals covered an area of 8 890 ha, of which 6 320 ha was wheat and 2 570 ha barley. The average yield for irrigated wheat was 3.5 tons/ha in the case of supplementary irrigation and up to 6 tons/ha under permanent irrigation in south-eastern areas. The yield of rainfed wheat varied between 0.7 and 1.4 tons/ha. The yield of irrigated barley was 2.5 tons/ha as against 0.6 tons/ha for rainfed barley.

The major irrigated crops are vegetables, fruit trees and olives (Figure 7). The cropping intensity of irrigated annual crops is 112%. Vegetables covered a total area of 36 120 ha in 1991 of which 80% was irrigated. They contributed on average 30% to the total gross value of agricultural production during the period 1980-90. They also remain the highest contributor to food exports, providing 58% of total food exports in 1990. In 1991, olives occupied 67 880 ha with about 20% irrigated, and fruit trees occupied 26 810 ha with over 50% irrigated. Irrigation covers 80% of the total agricultural production.

FIGURE 7
Irrigated crops (total: 68 910 ha in 1991)

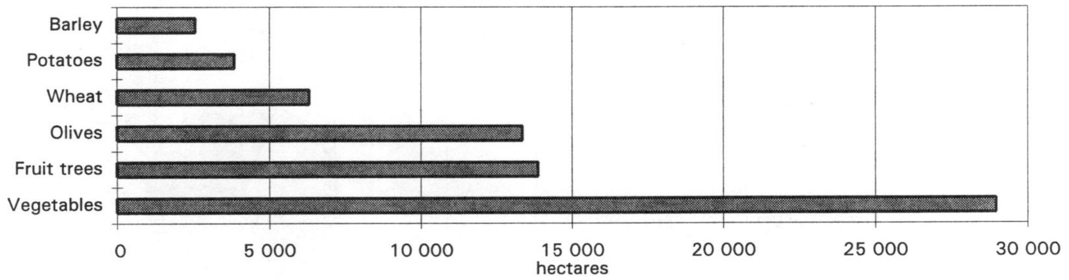

In 1992, drainage existed on about 4 000 ha of the irrigated area, mainly through open drains, and all by gravity. The average cost of drainage development is estimated at $US 2 000/ha.

INSTITUTIONAL ENVIRONMENT

The following ministries/institutions are involved in the water sector in Jordan:

- The Ministry of Water and Irrigation (MOWI), with the Water Authority of Jordan (WAJ) and the Jordan Valley Authority (JVA) as its operational entities;
- The Ministry of Agriculture;
- The National Centre for Agricultural Research and Technology Transfer;
- The Ministry of Municipal and Rural Affairs and the Environment (MMRAE);
- The Water and Environment Research and Study Centre.

The Ministry of Water and Irrigation is the body responsible in Jordan for the formulation and implementation of water and wastewater development programmes. A policy package, covering all aspects of the water sector, was in preparation in 1995.

Municipal water use was made more systematic with the creation of the Water Authority in 1985. Prior to that, many agencies and municipalities were responsible for the production and distribution of municipal water.

A land and water settlement law (No. 40) was enacted to distribute water among farmers. Water is distributed by shares, with each dunum (equivalent to 0.1 hectare) situated on a source having one share. In some cases, water is broken into time units known as fasl. The fasl allows the farmer to use the entire flow of water from a channel for a fixed time, usually 3, 6 or 12 hours.

TRENDS IN WATER RESOURCES MANAGEMENT

Jordan's past economic development plans reveal that surface water resources have been extensively developed by the Government, with priority given to the construction of dams and irrigation projects in the Jordan valley in order to maximize the utilization of this resource before its drainage to the Dead Sea. Limited additional untapped surface water resources could be developed in the Jordan valley side wadis and in the Mujib, Zarqa, Ma'an and Zara basins, subject to specific conditions.

It is no surprise that future trends in irrigation are closely linked to water development possibilities. Aquifers in the Disi-Mudwara, Jafer and Hamad areas are prime fresh water sources, able to supply an additional 80 million m^3/year of water for 100 years. Full utilization of these resources would require adequate management practices to avoid salinization and should be accompanied by reducing extraction from over-utilized aquifers. Plans are under way for a better assessment of the Hamad and Al-Sarhan basins' potential.

Jordan, Israel and the West Bank are presently over-exploiting their water resources by between 10 and 20%. Water levels are dropping, groundwater resources are being mined, salinization and salt water intrusion are observed and the domestic water supply does not reach adequate standards. The following actions are envisaged to remedy this crisis:

- reduction of water demand for irrigation;
- importation of water from water-rich countries like Turkey;
- desalinization of sea water.

As part of the efforts towards joint management of water resources, the Jordanian-Israeli Peace Treaty includes the following arrangements:

- 20 million m^3 of Yarmouk water will be stored by Israel in the winter and released to Jordan in the summer;
- 10 million m^3 will be released from the Tiberias lake outside the summer season for Jordan until the construction of a desalinization plant;
- construction of storage facilities on the Yarmouk and Jordan rivers and groundwater potential in wadi Araba are under investigation;
- 50 million m^3 of drinking water should be further allocated to Jordan through cooperation between both parties.

Although the potential for irrigation development in the highlands is great, a very small increase in irrigated agriculture is anticipated due to the unavailability of water resources. The average water consumption for irrigation in the Jordan valley and southern Ghor is less than 10 000 m^3/ha per year, which is much less than in the highlands where it reaches on average 16 000 m^3/ha per year. This may be due to the fact that in the highlands irrigation water is mostly groundwater pumped individually by farmers with less supervision than in the Jordan valley where the Jordan Valley Authority controls the delivery of water.

Some of the water resources which would be made available for irrigation purposes are recycled water from municipal wastewater treatment plants. The total quantity of reused treated wastewater is expected to grow from 50 million m^3 in 1993 to about 89 million m^3 in 2000 and 237 million m^3 in 2020.

If the Jordan valley plans to improve water use efficiency through conversion from surface to pressurized irrigation networks, it is expected that this will increase the available water resources by 25 million m^3/year. Future development of water resources in order to meet the increasing demand for water for agriculture will require the implementation of expensive projects for the development and conveyance of water to the land. These projects would place a heavy burden on the national budget and would seriously affect the national economy.

Future projects include the construction of storage facilities on major river and side wadis (the Wehdad, Karameh, Walah, Tannur, Mujub and Yabis dams) to alleviate part of the water shortages for agricultural activities. The total yield of these dams for agricultural purposes will be about 200 million m^3/year. The total cost of water development projects for the decade 1995-2005 is estimated at around $US 1 860 million, for an estimated yield of between 385 and 475 million m^3 of water.

Water harvesting techniques have not yet been developed in Jordan. However, small research pilot projects are trying to see whether this technique would suit the country's characteristics.

A revision of the National Water Master Plan (NWMP), first prepared in 1977, began in 1993, with the assistance of GTZ.

MAIN SOURCES OF INFORMATION

Bani Hadi, Mohammed A. 1995. Water policy reform in Jordan. Paper prepared for the 12th FAO Regional Land and Water Commission, Beiruth, December 1996.

Department of Statistics. 1991. Statistical Yearbook 1990.

El Naser, H. and Elias, Z. 1993. Jordan's water sector. Country paper presented at the regional symposium on water use and conservation, organized by ESCWA and CEHA.

ESCWA/FAO/Ministry of Agriculture. 1993. National Farm Data Handbook, Jordan. Amman, Jordan.

FAO. 1992. Agricultural sector programming mission. Rome, Italy.

FAO/ESCWA. 1994. Land and water policies in the Near East Region. Case studies on Egypt, Jordan and Pakistan. Amman, Jordan.

PRIDE and Chemonics International. 1992. A water management study for Jordan. USAID/Ministry of Water and Irrigation.

Qasem, S. and Shatanawi, M. 1993. Policies and strategies for the optimization of development and performance of irrigated agriculture in Jordan. Ministry of Agriculture.

Salameh, H. 1991. Jordan's water resources and their future potential. Proceedings of the Symposium 27-28 October 1991. Published by Friedrich Ebert Stiftung. 1992. Amman, Jordan.

World Bank. 1993. Jordan River Basin Study. The Regional Office for Integrated Development.

Kuwait

GEOGRAPHY AND POPULATION

Kuwait, with a total area of 17 820 km^2, lies at the head of the Persian Gulf. It is bordered in the north and north-west by Iraq, in the south-west and west by Saudi Arabia and it overlooks the Persian Gulf to the east. The land is generally flat with slightly undulating desert plains sloping gently towards the north-east, reaching an altitude of about 300 metres above sea level. Most of the area is desert with a few oases.

About 154 000 ha have been judged potentially cultivable land. However, it is almost completely covered by permanent pasture. Estimates for crop production potential vary between 25 000 and 37 500 ha, mainly located near the southern border (Al-Wafra), near the northern border (Al-Abdally) and in the centre of the country (Al-Sulaibiya). In 1994, the total cultivated area was 4 770 ha, of which 4 320 ha consisted of annual crops, and 450 ha consisted of permanent crops, mainly date palms.

TABLE 1
Basic statistics and population

Physical areas:			
Area of the country	1995	1 782 000	ha
Cultivable area	1994	25 000	ha
Cultivated area	1994	4 770	ha
- annual crops	1994	4 320	ha
- permanent crops	1994	450	ha
Population:			
Total population	1995	1 547 000	inhabitants
Population density	1995	87	inhab./km^2
Rural population	1995	3	%
Water supply coverage:			
Urban population	1993	100	%
Rural population	1993	100	%

The total population is over 1.5 million (1995), of which only 3% is rural. However, exact figures are difficult to give because of the large amount of immigrant labour: in 1994 about 63% of the total population were estimated to be non-Kuwaiti residents. According to the 1995 national census, only 0.7% were considered as rural, living in the areas of Al-Abdally and Al-Wafra, mainly labour and farm advisers. Most farm owners are investors and also have other sources of income. The average population density is 87 inhabitants per km^2, but varies widely from one region to another. The annual population growth, including both Kuwaiti and non-Kuwaiti residents, is estimated at 3.8%. Only 1% of the labour force is employed in agriculture, almost all of them being foreigners (more than 99.7% of the rural population in 1994). Agriculture accounted for less than 1% of GDP during the period 1988-1990.

CLIMATE AND WATER RESOURCES

Climate

Kuwait has a desert climate characterized by a long, dry, hot summer, with temperatures reaching more than 45°C with frequent sandstorms, and a cooler winter, with temperatures sometimes even falling below 4°C. The rainy season extends from October to May. Over an area of about 100 km^2 annual rainfall is less than 100 mm, in the remaining part it varies between 100 and 300 mm. The long-term average annual rainfall for the whole country was

TABLE 2
Water: sources and use

Renewable water resources:			
Average precipitation		176	mm/yr
		3.14	km^3/yr
Internal renewable water resources		0	km^3/yr
Total (actual) renewable water resources	1995	0.02	km^3/yr
Dependency ratio	1995	100	%
Total (actual) renewable water resources per inhabitant	1995	13	m^3/yr
Total dam capacity		-	10^6 m^3
Water withdrawal:			
- agricultural	1994	324	10^6 m^3/yr
- domestic	1993	201	10^6 m^3/yr
- industrial	1993	13	10^6 m^3/yr
Total water withdrawal		538	10^6 m^3/yr
per inhabitant	1994	348	m^3/yr
as % of total (actual) renewable water resources		2 690	%
Other water withdrawal		-	10^6 m^3/yr
Average groundwater depletion		-	10^6 m^3/yr
Wastewater - Non-conventional water sources:			
Wastewater:			
- produced wastewater	1994	119	10^6 m^3/yr
- treated wastewater	1994	103	10^6 m^3/yr
- reused treated wastewater	1994	52	10^6 m^3/yr
Desalinated water	1993	231	10^6 m^3/yr

about 176 mm, giving slightly more than 3.1 km^3. In recent years rainfall has decreased to an average of between 106 and 134 mm/year.

Surface water resources

There are no permanent surface water flows. Rainwater accumulates in the natural depressions where water remains for several weeks. Only a small part of this water percolates into the ground because of the high evaporation and the presence of an impervious layer in some regions.

Groundwater resources

There are two major aquifers: the Kuwait group (upper layer) and the Damman group (lower layer). Groundwater inflow has been estimated at about 20 million m^3/year through lateral under-flow from Saudi Arabia.

There are three classes of groundwater: fresh water with salinity below 1 000 ppm which is used for drinking and domestic purposes, slightly saline water with salinity ranging between 1 000 and 10 000 ppm which is used for irrigation, and highly saline water with salinity exceeding 10 000 ppm which is used in special cases only. In general groundwater quality and quantity are deteriorating due to the continuous pumping of water. In Al-Wafra in the south, 50% of the wells pumped water with a salinity level higher than 7 500 ppm in 1989. This figure is expected to reach 75-80% and 85-90% in the years 1997 and 2002 respectively. In Al-Abdally in the north, 55% of the deep drilled wells pumped water with a salinity level higher than 7 500 ppm in 1989. This is expected to reach 75 and 90% after 5 and 10 years of operation respectively.

TABLE 3
Irrigation and drainage

Irrigation potential	1994	25 000	ha
Irrigation:			
1. Full or partial control irrigation: equipped area	1994	4 770	ha
- surface irrigation	1994	3 020	ha
- sprinkler irrigation	1994	600	ha
- micro-irrigation	1994	1 150	ha
% of area irrigated from groundwater	1994	61.0	%
% of area irrigated from surface water	1994	0.0	%
% of area irrigated from non-conventional sources	1994	39.0	%
% of equipped area actually irrigated	1994	100	%
2. Spate irrigation area		-	ha
3. Equipped wetland and inland valley bottoms (i.v.b.)		-	ha
Total irrigation (1+2+3)	**1994**	**4 770**	**ha**
- as % of cultivated area		100	%
4. Flood recession cropping area		-	ha
Total water managed area (1+2+3+4)	**1994**	**4 770**	**ha**
- as % of cultivated area		100	%
- increase over last 10 years		-	%
- power irrigated area as % of water managed area	1994	100	%
Full or partial control irrigation schemes: Criteria			
Large-scale schemes > - ha		-	ha
Medium-scale schemes		-	ha
Small-scale schemes < - ha		-	ha
Total number of households in irrigation		-	
Irrigated crops:			
Total irrigated grain production	1994	2 000	tons
as % of total grain production	1994	100	%
Harvested crops under irrigation (full or partial control)	**1994**	**4 770**	**ha**
- permanent crops: total	1994	450	ha
- annual crops: total	1994	4 320	ha
. vegetables	1994	1 900	ha
. fodder crops (mainly alfalfa)	1994	1 470	ha
. potatoes	1994	670	ha
. cereals (barley and wheat)	1994	220	ha
. pulses	1994	60	ha
Drainage - Environment:			
Drained area	1994	2	ha
as % of cultivated area		0.04	%
- drained areas in full or partial control irrigated areas		-	ha
- drained areas in equipped wetland and i.v.b		-	ha
- other drained areas		-	ha
- total drained area with subsurface drains	1994	2	ha
- total drained area with surface drains		-	ha
Flood-protected area		-	ha
Area salinized by irrigation	1994	4 080	ha
Population affected by water-borne diseases		-	inhabitants

Desalinated water and treated wastewater

Kuwait relies on water desalination as a primary source of fresh water for drinking and domestic purposes. The first desalination plant was established in 1953 with a total capacity of 4 545 m^3/day. In 1994 there were 6 desalination plants with a maximum capacity of 950 000 m^3/day. The quantity of desalinated water produced in 1993 was 231 million m^3. Fresh water is obtained by mixing distilled water with low salinity groundwater (with a proportion of 8% groundwater) in order to get water suitable for drinking according to the official standards.

The quantity of wastewater produced was 119 million m^3 in 1994. About 103 million m^3 was treated and of this 52 million m^3 has been reused, while the remaining part was directed to the sea.

Water withdrawal

FIGURE 1
Water withdrawal (total: 538 million m^3 in 1994)

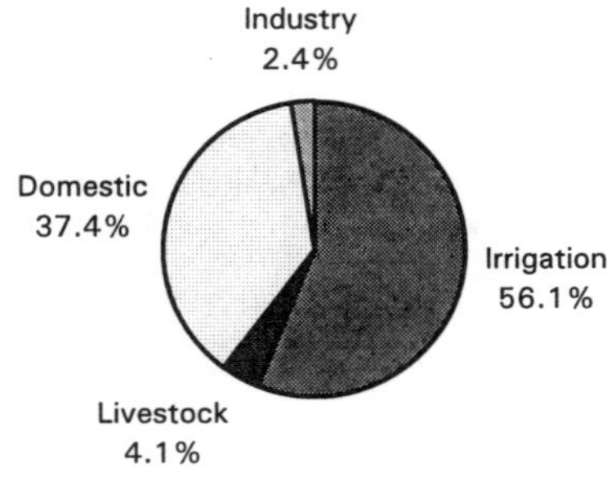

In 1993, total water withdrawal was estimated at 538 million m^3, of which 60% for agriculture, livestock and fish ponds (Figure 1).

Desalinated water (231 million m^3) was used mainly for domestic purposes and a smaller quantity for industry and for greenhouse irrigation (Figure 2). As far as the reuse of treated wastewater is concerned (52 million m^3), a pilot farm was established in 1976, where secondary treated sewage water was initially used. From 1981 onwards tertiary treated sewage water was used. It is mainly used for the irrigation of fodder crops and date palms and for landscaping. The remaining water withdrawal, mainly for agriculture and livestock, consists of groundwater (255 million m^3) leading to an extraction of more than 12 times the annual groundwater inflow. Farmers are only allowed to withdraw water from the Kuwait group aquifer and there were about 1 767 wells in 1994. The water used for livestock purposes is pumped by the Ministry of Electricity and Water (MEW) from the Damman group aquifer through deep artesian wells. It is expected that continued heavy extraction will lead to a decline of the groundwater level of 200 metres by the year 2000.

FIGURE 2
Origin of water used by sector (total: 538 million m^3 in 1994)

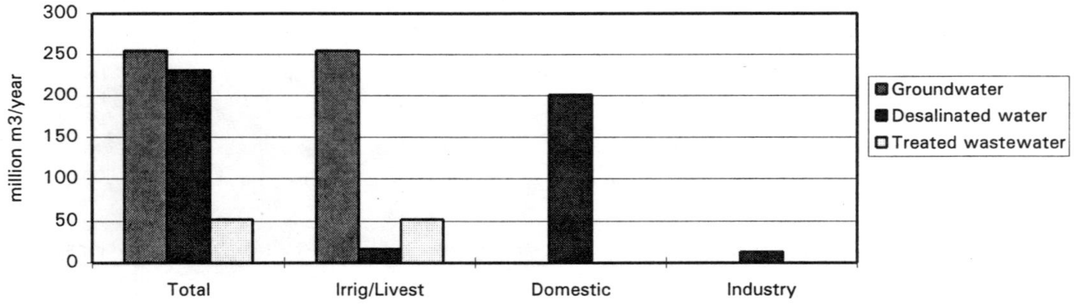

IRRIGATION AND DRAINAGE DEVELOPMENT

Irrigation in Kuwait started in the late 1950s. Initially surface irrigation techniques (furrow and basin irrigation) were used. Sprinkler irrigation was introduced in 1977, using treated wastewater. Micro-irrigation was introduced in 1979, first for agricultural production in greenhouses, but from 1981 onwards also for irrigation in the open field in order to preserve the water resources (Figure 3).

In 1994 the total water managed area, all full or partial control irrigation, was 4 770 ha, which is in fact equal to the cultivated area, as the entire cultivated area benefits from irrigation. Out of this area, almost 61% was irrigated from groundwater (Figure 4).

FIGURE 3
Irrigation techniques f/p
(total: 4 770 ha in 1994)

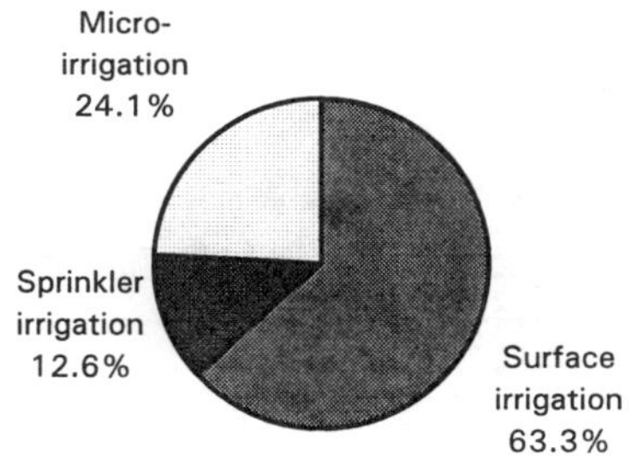

FIGURE 4
Origin of irrigation water f/p
(total: 4 770 ha in 1994)

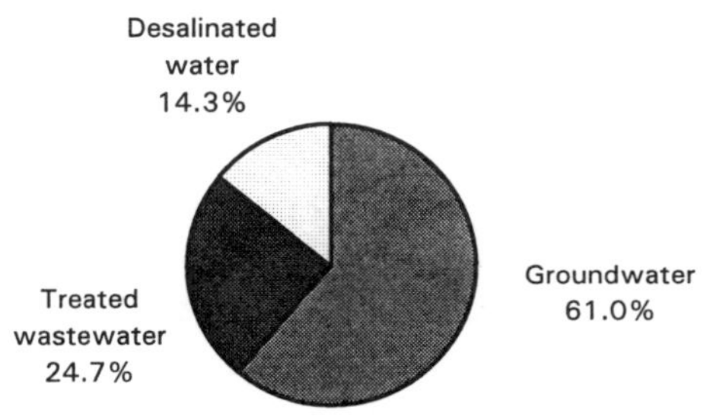

There are three types of farming in the irrigation sector:

- private farms, which are leased by the government to investors (25 years renewable) and operated by labourers. These are the most numerous. The smaller ones are mostly located in Al-Wafra in the south, the larger ones in Al-Abdally in the north;
- institutional schemes, which are operated by the government through the Public Authority for Agricultural Affairs and Fish Resources (PAAF);
- company-owned schemes such as the United Company for Agricultural Production, located in Al-Sulaibiya in the centre of the country.

The cost of irrigation development for small schemes (< 10 ha), equipped with micro-irrigation including one well and a pump, amounts to $US 19 000/ha. The cost falls as the irrigation scheme size increases and for large schemes (> 30 ha) it is about $US 15 000/ha. Annual operation and maintenance costs per ha are estimated at 2% of the investment costs.

There are no water charges for groundwater use. Farmers are charged for desalinated water use and the charge varies from $US 0.9/m^3 for small schemes to $US 1.5/m^3 for large schemes. The treated sewage water charge is $US 0.07/m^3.

The major irrigated crops are vegetables and fodder (Figure 5). Alfalfa is the largest open field crop and its production is about 100 tons/ha per year. The production of tomatoes in open fields is 40 tons/ha per year, while the production in greenhouses can reach 200 tons/ha per year.

Impervious layers exist at various depths in the Al-Wafra creating waterlogging in some areas. In 1994 this was estimated at 2 840 ha, due to poor natural drainage. On-farm drainage systems have not yet been developed, but some studies related to this subject are being conducted by PAAF and MEW. Small-scale subsurface drainage systems were installed in some public gardens (2 ha). The area salinized by irrigation has been estimated at 4 080 ha in 1994.

FIGURE 5
Irrigated crops f/p (total: 4 770 ha in 1994)

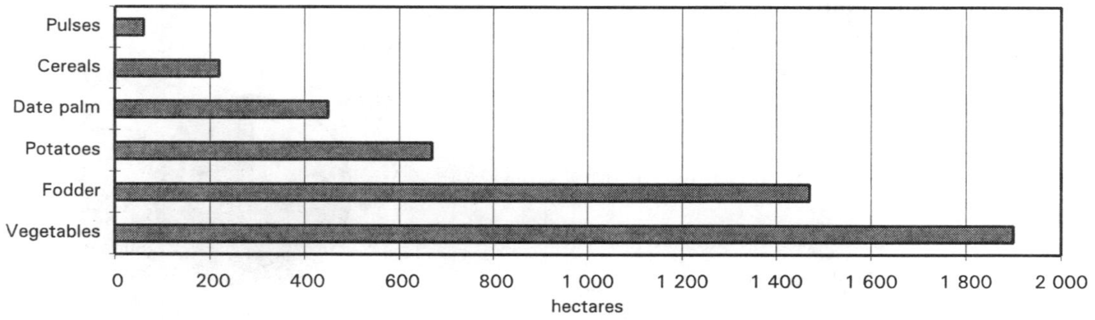

INSTITUTIONAL ENVIRONMENT

The main institutions involved in water resources management are:

- the Public Authority for Agricultural Affairs and Fish Resources (PAAF), established in 1983, with:
  - ◊ the Soil and Water Division: responsible for the design and evaluation of farm irrigation systems, testing irrigation equipment, crop water requirement research, monitoring of groundwater quality and quantity, and water resources planning;
  - ◊ the Landscape and Greenery Department: responsible for irrigation designs for highways and forestry areas;
- the Ministry of Electricity and Water (MEW), established in 1962: responsible for studies, development, exploration, monitoring and giving licences for drilling and using groundwater;
- the Ministry of Public Works (MPW), established in 1962: responsible for sewage water networks and collection reservoirs, wastewater treatment and utilization. Also responsible for the delivery of treated sewage effluent to farms and public gardens;
- the Kuwait Institute for Scientific Research (KISR): in charge of research related to water resources;
- the Environmental Protection Council: in charge of monitoring water quality.

TRENDS IN WATER RESOURCES MANAGEMENT

Water is the main limiting factor for the expansion of agriculture in Kuwait. Since water is subsidized, the government is aiming at maximizing water use efficiency through the promotion of modern irrigation systems and the increase of productivity by introducing crop varieties that are better adapted to Kuwait's environment.

Studies have been carried out to find other sources of irrigation water that are economically feasible and environmentally safe, such as industrial wastewater reuse, injection of treated wastewater into the aquifers and the use of desalinated seawater for irrigation.

Waterlogging and salinization problems are increasing at an alarming rate, which stresses the urgent need to study drainage requirements, both for agricultural and landscaping areas, and to convince the farmers/users of the need for adequate drainage facilities.

MAIN SOURCES OF INFORMATION

KISR [Kuwait Institute for Scientific Research]. 1994. Geohydrological studies of Al-Wafra and Al-Abdally farm areas, Volume 1. Prepared by the Hydrology Department, Water Resources Division. Kuwait.

PAAF [Public Authority for Agricultural Affairs and Fish Resources]. 1994. Soil and water (brief description). Prepared by the Technical Committee of the Soil and Water Division and the Landscape and Greenery Department.

Senay, Y. 1981. Geohydrology. In: Geology and groundwater hydrology of the State of Kuwait. Ministry of Electricity and Water.

Kyrgyzstan

GEOGRAPHY AND POPULATION

Kyrgyzstan is a mountainous, landlocked country in Central Asia, bordered on the north by Kazakhstan, in the east and south-east by China, in the south-west by Tajikistan and in the west by Uzbekistan. It became independent in 1991 and has a total area of 198 500 km^2. About 94% of the country is located at more than 1 000 metres above sea level and 40% above 3 000 metres, covered with glaciers and snow. The highest mountain is Victory Peak (7 429 m), which stands in the immense Mustag massif, one of the world's largest glaciers, covering 1 579 km^2. To the south are the Pamirs, another towering range of mountains.

In 1990, the cultivated area was estimated at 1.32 million ha, which is almost 7% of the total area of the country. About 1.28 million ha consisted of annual crops and 40 600 ha consisted of permanent crops. Agriculture is highly mechanized. In 1990, there were 176 collective farms and 290 state farms with about 29 000 tractors.

TABLE 1
Basic statistics and population

Physical areas:			
Area of the country	1995	19 850 000	ha
Cultivable area		-	ha
Cultivated area	1990	1 323 400	ha
- annual crops	1990	1 282 800	ha
- permanent crops	1990	40 600	ha
Population:			
Total population	1995	4 745 000	inhabitants
Population density	1995	24	inhab./km^2
Rural population	1995	61	%
Water supply coverage:			
Urban population	1990	86	%
Rural population	1990	75	%

The total population is about 4.7 million (1995), of which 61% is rural. The natural annual demographic growth rate is estimated at 2.2%. However, taking into consideration migration it is estimated at 1.8%. In 1991, 35% of the labour force was engaged in agriculture. In 1994, agriculture accounted for 34% of the country's GDP.

CLIMATE AND WATER RESOURCES

Climate

The climate in Kyrgyzstan is of the continental type with hot summers and cold winters, during which frost occurs in all parts of the country. Average annual precipitation is estimated at 316 mm, varying from 150 mm in the plains to over 1 000 mm in the mountains.

Water resources

As a result of tens of thousands of years of glaciation, there are some 3 000 lakes (many of which are located at an elevation of over 3 000 metres), continuously filled by rivers flowing down from the mountains. Average natural surface water flow in the country is estimated at 47.23 km^3/year, all internally produced, of which 28.97 km^3 or 61% flows through the Amu Darya and Syr Darya that drain into the Aral Sea. Of these 47.23 km^3/year, the former Soviet Unionallocated 11.60 km^3 to Kyrgyzstan, while the rest was allocated to the surrounding

TABLE 2
Water: sources and use

Renewable water resources:			
Average precipitation		316	mm/yr
		62.73	km^3/yr
Internal renewable water resources		47.23	km^3/yr
Total (actual) renewable water resources	1995	11.60	km^3/yr
Dependency ratio	1995	0	%
Total (actual) renewable water resources per inhabitant	1995	2 445	m^3/yr
Total dam capacity	1990	21 500	$10^6\ m^3$
Water withdrawal:			
- agricultural	1990	10 465	$10^6\ m^3/yr$
- domestic	1990	259	$10^6\ m^3/yr$
- industrial	1990	312	$10^6\ m^3/yr$
Total water withdrawal		11 036	$10^6\ m^3/yr$
per inhabitant	1990	2 527	m^3/yr
as % of total (actual) renewable water resources		95.1	%
Other water withdrawal	1990	137	$10^6\ m^3/yr$
Average groundwater depletion		-	$10^6\ m^3/yr$
Wastewater - Non-conventional water sources:			
Wastewater:			
- produced wastewater	1990	115	$10^6\ m^3/yr$
- treated wastewater	1990	11	$10^6\ m^3/yr$
- reused treated wastewater	1990	9	$10^6\ m^3/yr$
Desalinated water		-	$10^6\ m^3/yr$

republics of Kazakhstan, Uzbekistan, Turkmenistan and Tajikistan. This allocation is apparently still endorsed by the five states of Central Asia. Out of the Kyrgyzstan allocation, 5.30 km^3 originate from the Amu Darya and Syr Darya. Fresh groundwater resources, usually of very high quality, have been estimated at 13 km^3/year and constitute the baseflow of surface runoff.

Dams

In 1990, total dam capacity was 21.5 km^3. The largest dam is the Toktogul dam on the Syr Darya river with a reservoir capacity of 19.5 km^3. However, due to its location downstream near the border with Uzbekistan, it does not play an important role in the irrigation of areas within Kyrgyzstan. The same applies to the Kirov dam which has a capacity of 0.55 km^3.

FIGURE 1
Water withdrawal (total: 11.036 km^3 in 1990)

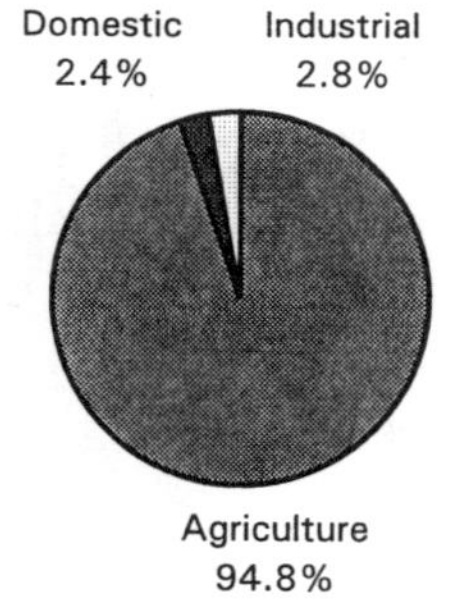

Water withdrawal

In 1990, total water withdrawal was about 11 km^3, which is almost the total allocation of Kyrgyzstan (11.6 km^3/year) (Figure 1). Average annual surface water availability for irrigation during the period 1985-1992 was about 10.77 km^3, although the water requirement had been evaluated at 10.83 km^3, leading to an irrigation water deficit for the whole country of 0.06 km^3. However, in some provinces there is a fairly severe water shortage, while in other provinces there is a surplus. Surface water is mainly used for irrigation purposes, while groundwater is mainly used for domestic and industrial purposes.

TABLE 3
Irrigation and drainage

Irrigation potential		1990	2 247 300	ha
Irrigation:				
1. Full or partial control irrigation: equipped area		1990	1 072 600	ha
- surface irrigation		1990	931 588	ha
- sprinkler irrigation		1990	141 000	ha
- micro-irrigation		1990	12	ha
% of area irrigated from groundwater		1990	0.2	%
% of area irrigated from surface water		1990	99.8	%
% of area irrigated from non-conventional sources		1990	0.0	%
% of equipped area actually irrigated			-	%
2. Spate irrigation area			-	ha
3. Equipped wetland and inland valley bottoms (i.v.b.)			-	ha
Total irrigation (1+2+3)		**1990**	**1 072 600**	**ha**
- as % of cultivated area			81	%
4. Flood recession cropping area			-	ha
Total water managed area (1+2+3+4)		**1990**	**1 072 600**	**ha**
- as % of cultivated area			81	%
- increase over last 10 years			-	%
- power irrigated area as % of water managed area			-	%
Full or partial control irrigation schemes:	Criteria			
Large-scale schemes	> 5 000 ha	1990	643 200	ha
Medium-scale schemes		1990	229 400	ha
Small-scale schemes	< 1 000 ha	1990	200 000	ha
Total number of households in irrigation		1990	705 825	
Irrigated crops:				
Total irrigated grain production			-	tons
as % of total grain production			-	%
Harvested crops under irrigation (full or partial control)			**-**	**ha**
- permanent crops: total			-	ha
- annual crops: total			-	ha
. fodder crops (perennial and annual)		1992	396 900	ha
. cereals		1992	288 300	ha
. vegetables (including potatoes)		1992	27 800	ha
. cotton, tobacco		1992	40 900	ha
. other annual crops			-	ha
Drainage - Environment:				
Drained area		1990	149 200	ha
as % of cultivated area			11	%
- drained areas in full or partial control irrigated areas			-	ha
- drained areas in equipped wetland and i.v.b			-	ha
- other drained areas			-	ha
- total drained area with subsurface drains			-	ha
- total drained area with surface drains			-	ha
Flood-protected area			-	ha
Area salinized by irrigation		1993	123 400	ha
Population affected by water-borne diseases			-	inhabitants

IRRIGATION AND DRAINAGE DEVELOPMENT

Irrigation is a key input into the agricultural sector of Kyrgyzstan, reaching over 80% of the cultivated area. It has also been critical in maintaining and stabilizing agricultural production. Irrigation potential has been estimated at about 2.25 million ha. In 1990, the total water managed area, all equipped for fully or partially controlled irrigation, was estimated at 1.07 million ha, which is almost 48% of the irrigation potential.

Sprinkler irrigation is practised on 141 000 ha and only 12 ha of micro-irrigation are reported. On the remaining part, representing 87% of the total area, surface irrigation methods are used (Figure 2). Out of the total irrigated area, only 20% has dependable water

FIGURE 2
Irrigation techniques f/p
(total: 1 072 600 ha in 1990)

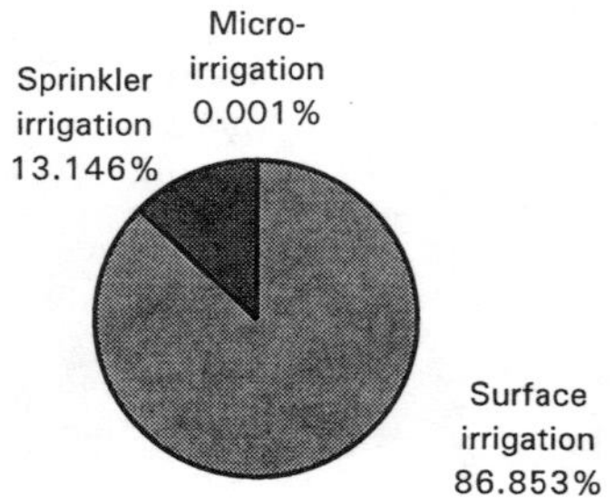

FIGURE 3
Origin of irrigation water f/p
(total: 1 072 600 ha in 1990)

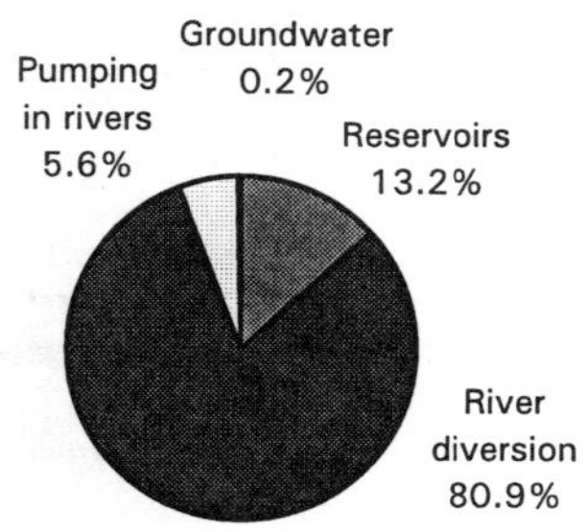

either from storage dams or groundwater. The remaining 80% is supplied through the diversion of mountain streams (Figure 3).

The main systems, particularly those downstream of large storage dams, are well maintained. However, the distribution systems within the state and collective farms are generally poorly designed, built and maintained. Seepage and leakage losses in the distribution system are considerable, resulting in a conveyance/distribution efficiency estimated at 55%.

In 1990, there were in total 1 346 irrigation schemes, the characteristics of which are presented in the table below.

Characteristics of the irrigation schemes

Size	Number of schemes	Total area of schemes (ha)	Average area of scheme (ha)	Number of beneficiaries	Number of benefic./ha
Small (<1000 ha)	1 174	200 000	170	486 400	2.4
Medium (1000-5000 ha)	101	229 400	2 270	648 400	2.8
Large (>5000 ha)	71	643 200	9 060	1 688 500	2.6
Total	1 346	1 072 600	800	2 823 300	2.6

Large schemes represent 60% of the irrigated area (Figure 4). They are state or collective farms. With the government programme to privatize agricultural production it is expected, however, that the number of small-scale farms (even less than 50 ha) will increase considerably.

The average cost of irrigation development varies from $US 1 080/ha for small schemes to $US 2 340/ha for medium schemes and $US 3 530/ha for large schemes. Average annual operation and maintenance costs are estimated at $US 26, 55 and 80/ha respectively.

FIGURE 4
Typology of f/p control irrigation schemes
(total: 1 072 600 ha in 1990)

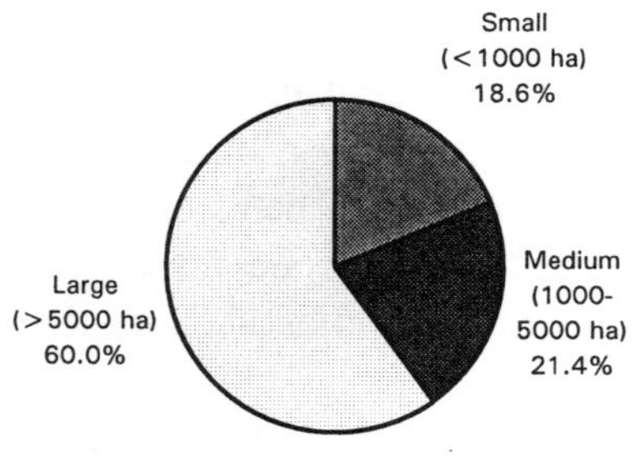

FIGURE 5
Major irrigated crops in 1992

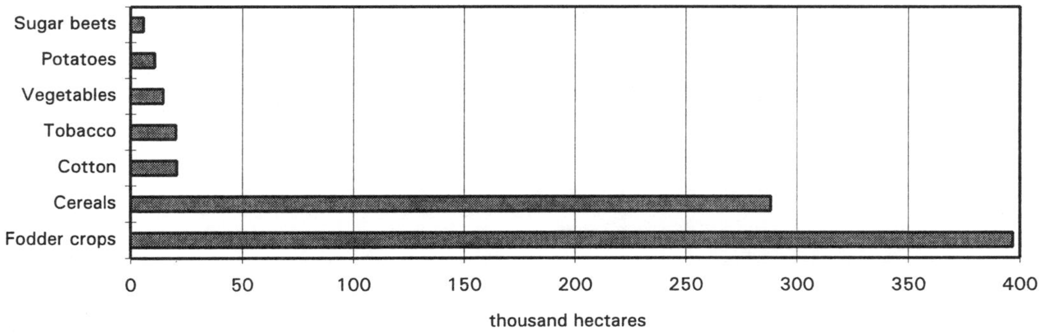

In the past years, farmers were not charged for water, although the land tax is two or three times higher on irrigated land than it is on non-irrigated land of similar quality. However, the financial situation has changed dramatically and the Ministry of Irrigation (MOI) is no longer able to cover irrigation costs by general tax revenues. In 1992-93, a water fee was imposed on state and collective farms, a service charge for delivering water. Only 20-30% of the farms paid the fee. According to the draft budget for 1995, water users would have to pay 78% of the maintenance costs of the irrigation system through this water fee and the state budget would pay 22%. Over half the equipped area is said to need rehabilitation.

The major irrigated crops are fodder crops and cereals (Figure 5). Although yields are generally low for irrigated land by world standards, they are about five times higher than yields in non-irrigated areas.

Waterlogging, drainage and salinity problems

In 1993, 123 400 ha had salinity problems, of which 63 400 ha were slightly saline, 34 200 ha moderately saline and 25 800 ha highly saline. The waterlogged area was estimated at 89 200 ha in 1990, while the drained lands amounted to 149 200 ha.

Salinity and drainage problems are likely to increase in the near future. With the very restricted budget facing MOI, it is unlikely that it will be able to maintain and operate the existing drainage system effectively, nor improve or extend it. Most drainage and salinity problems are in the northern part of the country, in the Chui province.

INSTITUTIONAL ENVIRONMENT

The Ministry of Water Resources/Economy and the Ministry of Agriculture are the main institutions involved in water resources, irrigation and drainage planning and development.

In the past, irrigation systems were designed and operated to deliver water to the large state or collective farms and it was a relatively easy task for MOI to deliver water to each farm. However, the increasing number of small farms following the privatization programmes, will need new legal institutions to interface with the farmers and provide adequate support.

Article 18 of the new water law includes specific provisions for establishing water users associations that receive water from MOI and allocate it among members of the association, many of whom will be farmers. They would have legal standing, be independent of the government, and be able to collect taxes from their members, borrow funds, and take appropriate action to maintain and upgrade 'their' part of the irrigation system, which are the on-farm systems formerly operated by the state and collective farms.

TRENDS IN WATER RESOURCES MANAGEMENT

According to the agricultural sector review of the World Bank (1995), the following three key issues need immediate attention to secure the sustainability of irrigation in Kyrgyzstan.

- New institutional arrangements must be made at farm level to manage and maintain the distribution of water within the former state and collective farms as farming units are privatized. For this, the need to create water users' associations has been felt.

- The financing of operation and maintenance of the existing systems must be secured and obtained largely from water users. The new water law has a section devoted to water fees and taxes. There are charges for water use, for the service of providing water (collection, transport, distribution and purification) and for the discharge of polluting substances into water. There are also provisions for increased fees if water consumption rates exceed these forecast levels and there are fee exemptions for the use of water-saving technologies and other water conservation measures.

- Environmental degradation of the irrigation systems and the irrigated lands must be guarded against through increased efforts to improve drainage and to reduce salinity and soil erosion. For this, a programme to improve irrigation efficiency and reduce water applications, especially in the higher lands, is needed. The steep slopes of the irrigated lands in the mountain areas with shallow soils should enable the conversion to pressurized irrigation methods, especially where gravity pressure is feasible.

MAIN SOURCES OF INFORMATION

World Bank. 1993. Kyrgyzstan: The transition to a market economy. A World Bank Country Study. Washington D.C.

World Bank. 1995. The Kyrgyz Republic: Agricultural Sector Review. Report No 12989-KG. Washington D.C.

Lebanon

GEOGRAPHY AND POPULATION

Lebanon, with a total area of 10 400 km^2, is situated east of the Mediterranean Sea and stretches about 210 km along the coast and 50 km inland. It is bordered by Syria in the north and east and by Israel in the south. Administratively it is divided into six Mohafazats or provinces.

Topographically, Lebanon can be divided from west to east into four parallel parts:

- a flat, narrow coastal strip parallel to the sea;
- the Lebanon Mountains chain, the highest crest of which is just over 3 000 metres;
- the Bekaa Valley at a height of around 900 metres;
- the Anti-Lebanon Mountains chain, which rises to 2 800 metres, in the east.

TABLE 1
Basic statistics and population

Physical areas:			
Area of the country	1995	1 040 000	ha
Cultivable area	1994	360 000	ha
Cultivated area	1994	189 206	ha
- annual crops	1994	104 120	ha
- permanent crops	1994	85 086	ha
Population:			
Total population	1995	3 009 000	inhabitants
Population density	1995	289	inhab./km^2
Rural population	1995	13	%
Water supply coverage:			
Urban population	1991	95	%
Rural population	1991	85	%

The cultivable area is estimated at 360 000 ha, or 35% of the total area. During the period 1992-94, the total cultivated area was estimated at 189 206 ha, of which 104 120 ha consisted of annual crops and 85 086 ha consisted of permanent crops, mainly fruit trees and olives. The Ministry of Agriculture is planning to start up a national agricultural census in 1996, the last one being carried out in 1970. According to the 1970 census there were 140 000 farm holdings with 63% having less than 2 ha of land, which means that agriculture is characterized by land fragmentation. However, there are indications of a decrease in the number of very small farms and in 1985 it was reported that about 46% of the farm holdings had less than 2 ha of land.

The total population is about 3 million (1995), of which only 13% is rural. The annual demographic growth rate is estimated at 2%. The agricultural labour force declined from 25% in 1967 to less than 9% in 1990. However, agriculture remains an important source of income in rural areas and although it is difficult to estimate the number of full-time farmers, most families have agriculture as a part-time activity. Seasonal labour represents between 30 and 40% of the present agricultural labour force. In 1992, agriculture accounted for an estimated 8.8% of GDP and for 30% of total export earnings.

TABLE 2
Water: sources and use

Renewable water resources:			
Average precipitation		827	mm/yr
		8.6	km^3/yr
Internal renewable water resources		4.800	km^3/yr
Total (actual) renewable water resources	1995	4.407	km^3/yr
Dependency ratio	1995	0.8	%
Total (actual) renewable water resources per inhabitant	1995	1 465	m^3/yr
Total dam capacity	1995	250	$10^6\ m^3$
Water withdrawal:			
- agricultural	1994	875	$10^6\ m^3/yr$
- domestic	1994	368	$10^6\ m^3/yr$
- industrial	1994	50	$10^6\ m^3/yr$
Total water withdrawal		1 293	$10^6\ m^3/yr$
per inhabitant	1994	444	m^3/yr
as % of total (actual) renewable water resources		29.3	%
Other water withdrawal	1994	700	$10^6\ m^3/yr$
Average groundwater depletion		-	$10^6\ m^3/yr$
Wastewater - Non-conventional water sources:			
Wastewater:			
- produced wastewater	1991	165	$10^6\ m^3/yr$
- treated wastewater	1991	4	$10^6\ m^3/yr$
- reused treated wastewater	1991	2	$10^6\ m^3/yr$
Desalinated water		-	$10^6\ m^3/yr$

CLIMATE AND WATER RESOURCES

Climate

The climate of Lebanon is typically Mediterranean, with heavy rains in the winter season (January to May) and dry and arid conditions in the remaining 7 months of the year. However, the influence of the Mediterranean Sea, the topographic features, as well as the Syrian Desert in the north create a variety of micro-climates within the country with contrasting temperatures and rainfall distribution. The average annual temperature is 20°C on the coast (ranging from 13°C in winter to 27°C in summer), 16°C in the Beeka valley (ranging from 5°C in winter to 26°C in summer) and less than 10°C at higher elevations in the mountain zones (ranging from 0°C in winter to 18°C in summer). Average annual rainfall is estimated at 823 mm, varying from 600 to 900 mm along the coastal zones to 1 400 mm on the high mountains and decreasing to 400 mm in the eastern parts and less than 200 mm in the north-east. Above 2 000 metres, precipitation is essentially niveous and helps to sustain a base yield for about 2 000 springs during the dry period. Rainfall occurs on eighty to ninety days a year, mainly between October and April. About 75% of the annual stream flow occurs in the five-month period from January to May, 16% from June to July and only 9% in the remaining five months from August to December.

Water resources

In total, there are about 40 major streams in Lebanon and, based on the hydrographic system, the country can be divided into five regions:

- the El Assi (Orontes) river basin in the north. The El Assi flows into Syria in the north-east of the country;
- the Litani river basin in the east and south. The Litani reaches the sea in the south-west of the country;

TABLE 3
Irrigation and drainage

Irrigation potential		1992	177 500	ha
Irrigation:				
1. Full or partial control irrigation: equipped area		1993	87 500	ha
- surface irrigation		1993	53 500	ha
- sprinkler irrigation		1993	21 000	ha
- micro-irrigation		1993	13 000	ha
% of area irrigated from groundwater		1993	45.7	%
% of area irrigated from surface water		1993	54.3	%
% of area irrigated from non-conventional sources		1993	0.0	%
% of equipped area actually irrigated			-	%
2. Spate irrigation area			-	ha
3. Equipped wetland and inland valley bottoms (i.v.b.)			-	ha
Total irrigation (1+2+3)		**1993**	**87 500**	**ha**
- as % of cultivated area			46	%
4. Flood recession cropping area			-	ha
Total water managed area (1+2+3+4)		**1993**	**87 500**	**ha**
- as % of cultivated area			46	%
- increase over last 10 years		1983-93	30	%
- power irrigated area as % of water managed area		1993	56.6	%
Full or partial control irrigation schemes:	Criteria			
Large-scale schemes	> 1 000 ha	1993	33 000	ha
Medium-scale schemes		1993	28 000	ha
Small-scale schemes	< 100 ha	1993	26 500	ha
Total number of households in irrigation		1993	58 500	
Irrigated crops:				
Total irrigated grain production		1994	32 600	tons
as % of total grain production		1994	39	%
Harvested crops under irrigation (full or partial control)		**1994**	**87 500**	**ha**
- permanent crops: total		1992	29 844	ha
- annual crops: total		1994	57 656	ha
. vegetables		1994	35 475	ha
. potatoes		1994	10 365	ha
. cereals (wheat, barley, maize)		1994	5 711	ha
. sugar beet		1994	3 800	ha
. other annual crops		1994	2 305	ha
Drainage - Environment:				
Drained area		1993	10 800	ha
as % of cultivated area			6	%
- drained areas in full or partial control irrigated areas			-	ha
- drained areas in equipped wetland and i.v.b			-	ha
- other drained areas			-	ha
- total drained area with subsurface drains			0	ha
- total drained area with surface drains		1993	10 800	ha
Flood-protected area		1995	1 500	ha
Area salinized by irrigation			-	ha
Population affected by water-borne diseases			-	inhabitants

- the Hasbani river basin in the south-east. The Hasbani, which flows into Israel in the south-east of the country, is a tributary of the Jordan river;
- all the remaining major coastal river basins. The northern El Kebir river basin is shared with Syria, the river itself forming part of the border between the two countries before flowing into the sea;
- all the remaining small in-between scattered and isolated subcatchments with no noticeable surface streamflow, like the endorheic catchments and isolated coastal 'pockets'.

Lebanon has a relatively favourable position as far as its rainfall and water resources are concerned, but constraints for development consist of the limited water availability during the seven dry summer months. Annual internal renewable water resources are estimated at about 4.8 km^3. Annual surface runoff is estimated at 4.1 km^3 and groundwater recharge at 3.2 km^3, of which 2.5 km^3 constitutes the baseflow of the rivers. About 1 km^3 of this flow comes from over 2 000 springs with about 10-15 l/s of average unit yield, sustaining a perennial flow for 17 of the total of 40 major streams in the country.

Water flows

Lebanon being at a higher elevation than its neighbours has practically no incoming surface water flow. A contribution of 74 million m^3/year to the El Kebir river, to the north, is estimated to be generated by the 707 km^2 bordering Syrian catchment areas. There might also be some groundwater inflow from these areas, but no figures on quantities are available. Surface water flow to Syria is estimated at 510 million m^3/year through the El-Assi (Orontes)) river and the bordering El Kebir river. A recent (informal) agreement between Lebanon and Syria on the Orontes river has led to a share of 80 million m^3/year for Lebanon and the remainder for Syria. Surface water flow to Israel is estimated at 160 million m^3/year, of which about 138 million m^3 through the Hasbani river including a contribution of 30 million m^3 from its tributary, the Wazzani spring. Annual groundwater outflow is estimated at 1 030 million m^3, of which 130 million m^3 to Syria, 180 million m^3 to Israel and 720 million m^3 to the sea.

The relative importance of groundwater flow to the sea and the difficulties related to its control, added to the difficult geological conditions of most of the investigated sites for storage dams, make the manageable resources of Lebanon certainly much lower than the global figure of 4.8 km^3/year. The most realistic figure recognized does not exceed 2.2-2.5 km^3/year.

Dams

The Karaoun dam on the Upper Litani river is the largest, with a storage capacity of 220 million m^3 and an effective storage of 160 million m^3 from year to year. It regulates the downstream flow of the Litani river for power generation and irrigation. The Bisri dam on the Awali river is currently at the final design stage for a storage capacity of 128 million m^3 and is intended mainly for supplying water to Greater Beirut. The Kardalé dam on the middle reach of the Litani river, also to give a storage capacity of 128 million m^3, has been postponed at the preliminary design stage, in view of the prevailing adverse security situation in the southern border region.

The Green Plan, which is a public authority established in 1963 responsible for the development of water reservoirs, and the private sector, have already developed hundreds of small earth and concrete storage pounds, with a maximum per unit capacity of 0.2 million m^3. During the period 1964-1992 the Green Plan led to a total of 3.5 million m^3 of earth pounds and 0.35 million m^3 of concrete pounds.

The Litani River Authority implemented three hillside stock ponds in the early 1970s for a total storage capacity of about 1.8 million m^3.

Wastewater

About 165 million m^3 of wastewater was produced in 1991, of which 130 million m^3 of domestic origin and 35 million m^3 of industrial origin. There is infiltration from water from cesspools to groundwater and a direct outflow of sewerage water to the natural watercourses, especially in the inland villages and communities where it is commonly directed to watercourses. In 1991, the quantity of treated wastewater was roughly estimated at 4 million m^3, the quantity reused at 2 million m^3, for some informal irrigation. Some illicit irrigation from untreated wastewater is practised.

Seepage from the water supply networks, of between 35 and 50%, is almost all infiltrated to the aquifers and extracted again via tubewells, especially in the Greater Beirut Metropolitan area.

FIGURE 1
Water withdrawal (total: 1 293 million m^3 in 1993)

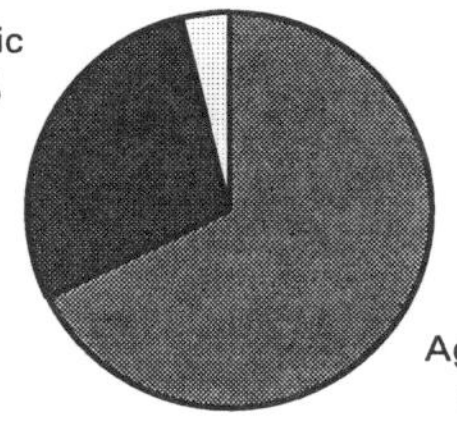

Water withdrawal

In 1994, water withdrawal for agricultural, domestic and industrial purposes was estimated at 1 293 million m^3, of which almost 68% for agricultural purposes (Figure 1). The assessment of agricultural water use is based on a water use of 11 200 m^3/ha per year from surface water and 8 575 m^3/ha per year from groundwater. In 1991, about 700 million m^3 was estimated to be used for hydropower, with direct restitution to the natural river courses.

IRRIGATION AND DRAINAGE DEVELOPMENT

Irrigation potential, based on soil and water resources, is estimated at 177 500 ha. In 1993, the total area equipped for irrigation was estimated at 87 500 ha, of which 67 500 ha for perennial irrigation and 20 000 ha for seasonal irrigation (spring).

Surface irrigation, mainly basin and furrow irrigation, is practised on 53 500 ha. It usually comprises diversion structures or simple intakes on streams or springs, open concrete main canals and earth or concrete secondary canals. Sprinkler irrigation is practised on 21 000 ha, especially for potatoes and sugar beet in the Bekaa Central Plain. Micro-irrigation is practised on 13 000 ha, especially in North Bekaa (Qaa region) and in the coastal region (Figure 2).

FIGURE 2
Irrigation techniques f/p
(total: 87 500 ha in 1993)

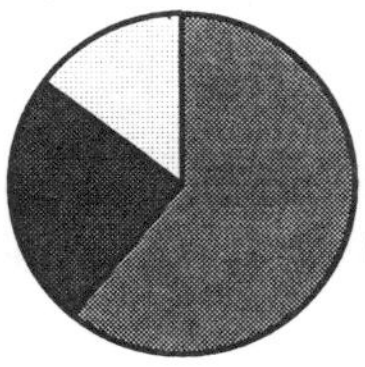

The main source of irrigation water is the Litani river and the Litani-Awali complex of water resources. In 1993, it was estimated that 54.3% was irrigated from surface water and 45.7% from groundwater (artesian wells, recharge wells and springs) (Figure 3). The use of groundwater for irrigation has increased in the past few years in view of the delay in the implementation of governmental schemes. Individual farmers in the schemes who face water shortages are increasingly relying on supplementary supply from groundwater by means of private wells and in 1992-95 about

2 000 wells were added to an overall total of more than 10 000 wells, especially in the southern coastal hills and in North and Middle Bekaa Central Plain.

FIGURE 3
Origin of irrigation water f/p
(total: 87 500 ha in 1993)

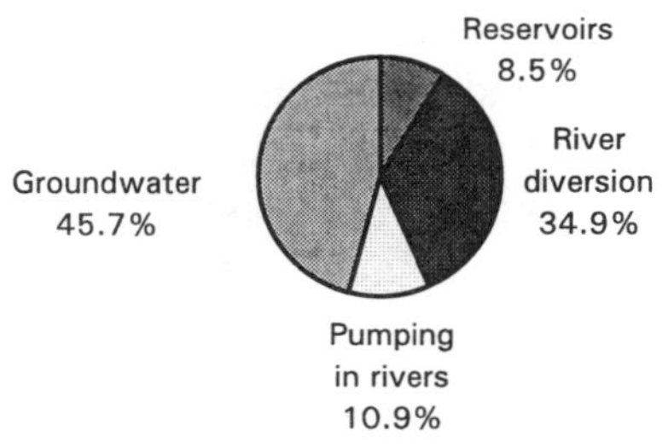

The public irrigation subsector, essentially unchanged since 1970, consists of about 5 large-scale schemes (> 1 000 ha) and 50 medium and small-scale schemes (Figure 4). The total number of households in the irrigation schemes is estimated at 58 500. Most of the schemes are old, poorly maintained and in an advanced state of deterioration. It is estimated that most of the area irrigated by surface water needs rehabilitation. The private irrigation subsector continues to develop to a much larger and dynamic extent, based essentially on tubewells.

FIGURE 4
Types of irrigation schemes f/p
(total: 87 500 ha in 1994)

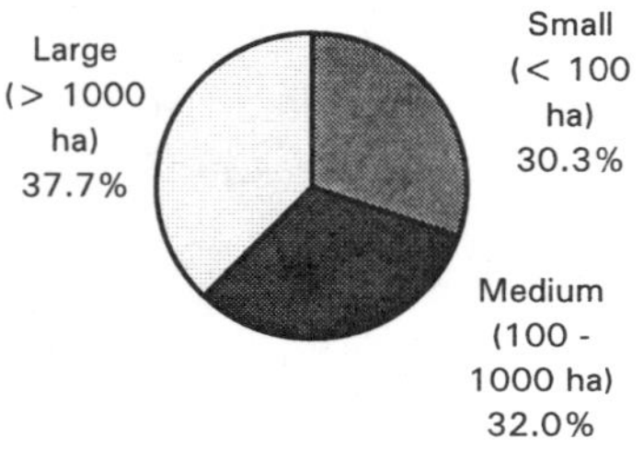

The average cost for irrigation development ranges from $US 2 500/ha for small schemes, $US 3 750/ha for medium schemes and between $US 4 000 and 7 000/ha for large schemes. Estimates of operation and maintenance costs are $US 40/ha per year for small schemes with gravity surface irrigation. As regards medium schemes, these cost from $US 100/ha per year for grabity surface irrigation to $US 600/ha per year for private wells and, for large schemes, from $US 400/ha per year for private pumping in rivers to $US 600/ha per year for tubewells.

In governmental irrigation systems, irrigation water is charged at a flat rate per planted area, except in the modern pressurized irrigation schemes of the Litani River Authority in South Bekaa and Saida-Jezzine where volumetric metering is provided. At present, for the Qasmieh-Ras El Ain coastal scheme, water charges are fixed at $US 260/ha for farmers who irrigate by gravity directly from the canal. In the Danneyeh scheme and the Akkar scheme, in the north, water charges range from $US 30 to 125/ha.

In the period 1992-94 the major irrigated crops were vegetables, fruit trees, potatoes and, to a lesser extent, cereals (maize, wheat, barley) and sugar beet (Figure 5). The average yield for irrigated wheat and barley was estimated at 5.0 tons/ha, as against 2.2 tons/ha for rainfed wheat and barley.

The amount of agricultural land suffering from drainage problems is fairly limited and is mainly in South Bekaa (about 5 000 ha) and in the Boquaia Plain in Akkar (about 4 000 ha). The area benefiting from improved drainage through open ditches and river calibration was estimated at about 10 800 ha in 1993. River calibration is also done to protect against flood damage, especially over the Litani river, upstream of Karaoun Lake, where the drainage and

FIGURE 5
Irrigated crops f/p (total: 87 500 ha during 1992-94)

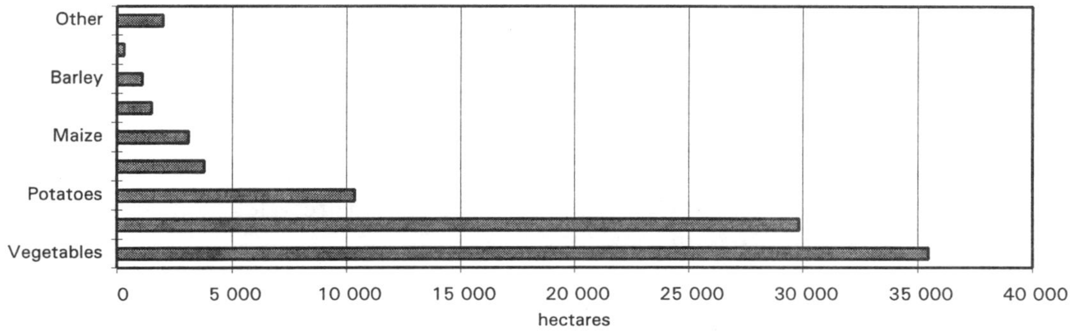

calibration works realized in the 1970s helped to alleviate the flood damage on about 1 500 ha.

INSTITUTIONAL ENVIRONMENT

The institutions with direct decision-making authority for irrigation and water resources development and management are the Ministry of Hydraulic and Electrical Resources (MHER) and, under its responsibility, the Litani River Authority (LRA) which is in charge of implementing the two most important ongoing projects, in South Bekaa (21 500 ha) and Southern Lebanon (30 000 ha), and of exploiting the potential Qasmieh - Ras El Ain coastal irrigation scheme (4 000 ha). Also under the responsibility of MHER, there are 25 local irrigation authorities, each exploiting a small or medium irrigation project on areas of between 10 and 300 ha, and 126 local irrigation committees, which were instituted by Ministerial Decrees during the period 1984-1990. An institutional plan is under way to merge these authorities and committees into five main authorities, at the Mohafazat, or provincial, level (excluding Beirut).

The Council for Development and Reconstruction (CDR) is, at present, the most powerful institution in irrigation rehabilitation and development. It manages a $US 58 million loan from IBRD for the rehabilitation and modernization of the 5 large irrigation schemes and about 28 medium schemes, for a total area of about 43 000 ha. It also manages a $US 13 million loan from IFAD for the on-farm support activities connected to the above projects.

Responsibilities for agricultural development lie within the following administrations: the Ministry of Agriculture (MOA) and, under its responsibility, the Agricultural Research Institute (IRAL) and the Green Plan (GP). MOA has recently issued institutional regulations to allow autonomy to the GP, with responsibility for land preservation, land reclamation, rural roads construction, small hydraulic development of hillside stock ponds, and schemes and farm-level infrastructure works. In addition, MOA has recently established a new autonomous institution, called 'The Institution for Alternate Crops', centring on the Baalbeck-Hermel traditional region of illicit crops.

Water laws and regulations on water are mostly outdated and, at present, water rights constitute a constant source of disputes. Groundwater is tapped by thousands of informal private wells with no licences, no water metering and no charges or taxes for the tapped volumes.

Many sectorial and regional water resources planning studies are under way by CDR, MHER and LRA. The establishment of the National Water Master Plan is also foreseen. The most important decree for regional water resources planning is the Decree No 14522 of 10 May 1970 which organizes the allocation of the available water resources south of the Beirut river up to the southern international borders and up to the 800 metres elevation on the western skirts. This decree, however, requires updating.

TRENDS IN WATER RESOURCES MANAGEMENT

The development of potential future irrigation aims at increasing the actual area of 87 500 ha up to the potential soil and water ceiling of 177 500 ha, concentrated mainly in South Lebanon, the Bekaa Plain and the northern coastal areas (Akkar).

The irrigation potential in Lebanon is linked to the physical mobilization of water and to the rehabilitation and modernization of irrigation infrastructures. An increase in the irrigated area can be achieved from surface water resources through the construction of storage dams and inter-regional transfers, for example the Khardalé dam over the middle Litani river (now postponed) and the 'Canal 800' conveyor for the irrigation of 15 000 ha in South Lebanon. More than 83 sites for possible dam construction, with a total capacity of 873 million m^3, have already been prospected and are recommended for further investigation.

Future drainage development involves completing and achieving the calibration of the Litani river and its seven tributaries in the South Bekaa Plain, in order to reclaim about 1 500 ha of the waterlogged area, and to facilitate the drainage works in another risky area of 3 500 ha which is also exposed to frequent floods from rivers. Environmental issues, such as the preservation of marshy lowlands for migratory birds, should also be given consideration.

MAIN SOURCES OF INFORMATION

CDR/MHER/BTD-CADRES-Consulting Engineers. 1994. Rapid initial assessment of small and medium irrigation schemes in Lebanon. CDR, Beirut.

FAO. 1994. Irrigation rehabilitation and modernization project: Preparation report. FAO Investment Centre/World Bank Cooperative Programme Report No. 5/94 CP-LEB 8. Rome.

FAO. 1995. Agricultural infrastructure development project: Preparation report. FAO Investment Centre/World Bank Cooperative Programme Report No. 106/95 CP-LEB 9. Rome.

Geadah, Adib. 1993. Projet d'irrigation de la Bekaa-Sud: Etude de factibilité, document de travail en vue de la reprise des activités. Litani River Authority, Beirut.

International Bechtel Inc. and Dar Al-Handasah (Shair and Partners). September 91 - June 1993. Recovery planning for reconstruction and development of Lebanon (Phases 1 and 2). Report No. L9125/1-2. Beirut.

Jaber, Bassam. 1994. The water problem in Lebanon. Conference on the problems of water in the Middle East. Centre for strategic studies, research and documentation, Beirut.

Jaber, Bassam. 1995. The water resources in Lebanon. Conference on the environmental management for sustainable development in Lebanon. UNEP/LNCSR, Beirut.

Litani River Authority, Directorate of Studies. 1993. The Master Plan and the 15-year plan for the equipment and the exploitation of the Litani River Basin. Litani River Authority, Beirut.

Saadé Riad. 1995. 42ème rapport annuel sur la production agricole au Liban. Centre de Recherches et d'Etudes pour l'Agriculture au Liban (CREAL), Beirut.

World Bank. 1994. Irrigation rehabilitation and modernization project: Staff appraisal report. Report No. 13012-LE. Washington DC.

Libya

GEOGRAPHY AND POPULATION

The Libyan Arab Jamahiriya has a total area of about 1.76 million km^2. It is bordered in the east by Egypt and Sudan, on the south by Chad and Niger, in the west by Algeria and Tunisia, and in the north by the Mediterranean Sea. Four physiographic regions can be distinguished:

- the Coastal Plains, that run along the Libyan coast and vary in width;
- the Northern Mountains, that run close to the coastal plains and include the Jabal Nafusah in the west and the Jabal al Akhdar in the east;
- the Internal Depressions, that cover the centre of Libya and include several oases.
- the Southern and Western Mountains.

About 95% of the country is desert. The cultivable area was estimated at 3.80 million ha, which is only slightly over 2% of the total area. In 1987, the total cultivated area was estimated at 2.28 million ha, or 60% of the cultivable area, of which 1.93 million ha consisted of annual crops and 0.35 million ha consisted of permanent crops.

TABLE 1
Basic statistics and population

Physical areas:			
Area of the country	1995	175 954 000	ha
Cultivable area	1992	3 800 000	ha
Cultivated area	1987	2 278 648	ha
- annual crops	1987	1 933 648	ha
- permanent crops	1987	345 000	ha
Population:			
Total population	1995	5 407 000	inhabitants
Population density	1995	3	inhab./km^2
Rural population	1995	14	%
Water supply coverage:			
Urban population	1990	100	%
Rural population	1990	80	%

The total population is about 5.4 million (1995), of which 14% is rural. The annual demographic growth rate was 4.1% between 1980-91. The average population density is 3 inhabitants/km^2, but varies between 50 inhabitants/km^3 in the northern regions of Tripolitania and Cyrenaica to less than 1 inhabitant/km^3 elsewhere. Agriculture contributes less than 5% to GDP, although it provides employment for approximately 13% of the active population.

CLIMATE AND WATER RESOURCES

Climate

The climatic conditions are influenced by the Mediterranean sea to the north and the Sahara desert to the south, resulting in an abrupt transition from one kind of weather to another. The following broad climatic divisions can be made:

- the Mediterranean coastal strip with dry summers and relatively wet winters;
- the Jabal Nafusah and Jabal Akhdar highlands experiencing a plateau climate with higher rainfall and humidity and low winter temperatures, including snow on the hills;

TABLE 2
Water: sources and use

Renewable water resources:			
Average precipitation		26	mm/yr
		46.1	km^3/yr
Internal renewable water resources		0.6	km^3/yr
Total (actual) renewable water resources	1995	0.6	km^3/yr
Dependency ratio	1995	0	%
Total (actual) renewable water resources per inhabitant	1995	111	m^3/yr
Total dam capacity	1992	387	10^6 m^3
Water withdrawal:			
- agricultural	1994	4 000	10^6 m^3/yr
- domestic	1994	500	10^6 m^3/yr
- industrial	1994	100	10^6 m^3/yr
Total water withdrawal		4 600	10^6 m^3/yr
per inhabitant	1994	880	m^3/yr
as % of total (actual) renewable water resources		767	%
Other water withdrawal		-	10^6 m^3/yr
Average groundwater depletion		-	10^6 m^3/yr
Wastewater - Non-conventional water sources:			
Wastewater:			
- produced wastewater		-	10^6 m^3/yr
- treated wastewater		-	10^6 m^3/yr
- reused treated wastewater	1990	100	10^6 m^3/yr
Desalinated water	1988	70	10^6 m^3/yr

- moving southwards to the interior, pre-desert and desert climatic conditions prevail, with torrid temperatures and large daily thermal amplitudes. Rain is rare and irregular and diminishes progressively towards zero.

The average annual rainfall is 26 mm, with more than 100 mm/year over only 7% of the land surface. The highest rainfall occurs in the northern Tripoli region (Jabal Nafusah and Jifarah Plain) and in the northern Benghazi region (Jabal al Akhdar) - these two areas being the only ones where the average annual rainfall exceeds the minimum value (250-300 mm) considered necessary to sustain rainfed agriculture. Rainfall occurs during the winter months, but great variability is observed over space and time (year to year).

Surface water

The total mean annual runoff calculated or measured at the entrance of the wadis in the plains is estimated at 200 million m^3/year, but part of it either evaporates or contributes to the recharge of the aquifers. Therefore the surface water resources are roughly estimated at 100 million m^3/year. Sixteen dams, with a total storage capacity of 387 million m^3 and with an expected average annual volume of water controlled in the order of 60 million m^3, had been constructed by 1991. Additional dams are planned, to achieve a total storage capacity of 686 million m^3. This difference between the average annual runoff and the storage capacity of the dams is so that the runoff water of exceptionally wet years can be stored.

Groundwater

Currently, aquifers are only recharged only in the northern regions, namely in the north-western zone, Jabal Nafusah and Jifarah Plain, and in the north-eastern zone, Jabal al Akhdar. Renewable groundwater resources are estimated at 800 to 1 000 million m^3/year, but part (perhaps 50%) now flows out either to the sea or to evaporative areas (*sabkhas*). Not all the

TABLE 3
Irrigation and drainage

Irrigation potential	1992	750 000	ha
Irrigation:			
1. Full or partial control irrigation: equipped area	1990	470 000	ha
- surface irrigation		-	ha
- sprinkler irrigation		-	ha
- micro-irrigation		-	ha
% of area irrigated from groundwater		-	%
% of area irrigated from surface water		-	%
% of area irrigated from non-conventional sources		-	%
% of equipped area actually irrigated	1990	51	%
2. Spate irrigation area		-	ha
3. Equipped wetland and inland valley bottoms (i.v.b.)		-	ha
Total irrigation (1+2+3)	**1990**	**470 000**	**ha**
- as % of cultivated area		21	%
4. Flood recession cropping area		-	ha
Total water managed area (1+2+3+4)	**1990**	**470 000**	**ha**
- as % of cultivated area		21	%
- increase over last 10 years		-	%
- power irrigated area as % of water managed area		-	%
Full or partial control irrigation schemes:	Criteria		
Large-scale schemes	> - ha	-	ha
Medium-scale schemes		-	ha
Small-scale schemes	< - ha	-	ha
Total number of households in irrigation		-	
Irrigated crops:			
Total irrigated grain production		-	tons
as % of total grain production		-	%
Harvested crops under irrigation (full or partial control)		-	**ha**
- permanent crops: total		-	ha
- annual crops: total		-	ha
.		-	ha
.		-	ha
.		-	ha
.		-	ha
. other annual crops		-	ha
Drainage - Environment:			
Drained area		-	ha
as % of cultivated area		-	%
- drained areas in full or partial control irrigated areas		-	ha
- drained areas in equipped wetland and i.v.b		-	ha
- other drained areas		-	ha
- total drained area with subsurface drains		-	ha
- total drained area with surface drains		-	ha
Flood-protected area		-	ha
Area salinized by irrigation		-	ha
Population affected by water-borne diseases		-	inhabitants

renewable groundwater can be abstracted without affecting the environment, because of the deterioration of water quality by saline water encroachment. For this reason, the safe yield has been estimated at 500 million m^3/year. South of the 29th parallel, an important development of Palaeozoic and Mesozoic continental sandstone enabled water to be stored safely during the long period of the late Quaternary, before the climate turned extremely arid. Most water used in Libya comes from these huge fossil reserves.

Through the Great Manmade River Project about 2 km^3/year of fossil water is expected to be transported from the desert to the coastal areas, mainly for irrigation but part will be used for the water supply of the major cities.

Desalinated water and treated wastewater

There have been several attempts during the last 20 years to introduce and expand sea water desalination plants and wastewater treatment facilities. However, so far restrictions have been imposed by the high cost of energy and spare parts. A number of desalination plants of different sizes have been built near large municipal centres and industrial complexes. The total capacity of installed plants is approximately 140 million m^3/year, but sections of them are either not in use or only partly operational. It is estimated that only 70 million m^3 of water is desalinated each year. The present level of wastewater treatment is estimated at about 100 million m^3/year.

Water withdrawal

Total water withdrawal for agricultural, domestic and industrial purposes was estimated at 4.6 km^3 in 1994, which is almost eight times the annual renewable water resources. About 87% of total water withdrawal is used for agricultural purposes (Figure 1). All desalinated water is currently used for domestic and industrial purposes and all treated wastewater for agricultural purposes.

FIGURE 1
Water withdrawal (total: 4.6 km^3 in 1994

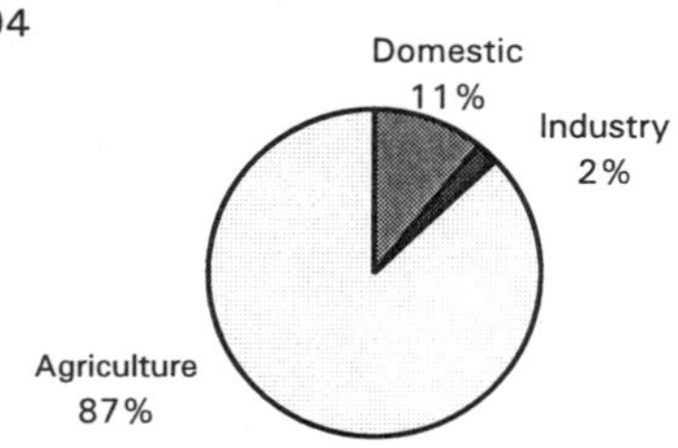

IRRIGATION AND DRAINAGE DEVELOPMENT

Irrigation potential has been estimated at 750 000 ha. However, the development of this potential would have to rely mainly on the use of fossil water. Considering renewable water resources, it is estimated that a maximum of 40 000 ha could be irrigated in the coastal areas.

The total water managed area is approximately 470 000 ha, all equipped for full or partial control irrigation. Sprinkler irrigation is practised on almost the entire area, because of the sandy soils prevailing in most areas of Libya. Of the total area of 470 000 ha, only 240 000 ha was actually irrigated according to figures from several years ago, distributed as listed in the table below.

Regional distribution of irrigation

Zone	Geographic region	Area actually irrigated (ha)	Estimated water use ($10^6 m^3$/yr)*	Year (approx.)
Murzuq Basin	South-west	50 000	551	1985
Jifarah Plain	North-west	100 000	500	1985
Jabal Nafusah; Suf al Jin; Western Sirte; Hammada el Hamra	North-west and North	25 000	241	1985
Al Jabal al Akhdar	North-east	40 000	150	1985
Al Kufrah-As Sarir	South-east	25 000	423	1990
Total		240 000	1 865	

* *Figures probably under-estimated.*

There are three different categories of farming in the irrigation subsector:

- smallholders, generally on 1 to 5 ha plots, receive substantial State support for water equipment, energy, and agricultural inputs. This type of farming represents approximately

30 to 40% of the total irrigated area but is mostly concentrated in the traditional development areas, i.e. the Jifarah Plain, the Jabal al Akhdar, and the Murzuq Basin;

- large-scale state farming, mainly located in the southern areas, where new irrigation schemes have been set up based on highly productive deep wells supplying water to blocks divided into small plots and cultivated by small-scale farmers; and

- large-scale state farming, mainly located in the desert areas (usually pivot systems), operated by state technicians and workers.

At present, no water fees are imposed on water users.

INSTITUTIONAL ENVIRONMENT

Responsibility for all water resources assessment and monitoring rests with the General Water Authority, while the Secretariat of Agriculture and Animal Wealth is responsible for the development of irrigated agriculture and the implementation of major projects. A special Authority - The Great Manmade River Water Utilization Authority - is responsible for the use for agricultural purposes of the water transported from the desert to the coast. The Secretariat of Municipalities takes care of the water supply to urban settlements.

TRENDS IN WATER RESOURCES MANAGEMENT

Over-extraction of groundwater in the coastal regions has led to a continuing decline in the groundwater level and sea water intrusion is estimated to be advancing at a rate of 100 to 250 m/year. If this over-extraction is not stopped or reversed, it is expected that these intrusions will lead to the contamination and pollution of all productive aquifers by the year 2000.

Irrigation development in Libya is linked to the implementation of the project to transport fossil water from the aquifer below the desert. Through the Great Manmade River Project, consisting of five phases, the total volume of water to be transferred and redistributed within the country will amount to about 2 km^3/year for a period of at least 50 years (Figure 2):

FIGURE 2
Planned water transfer through the Great Manmade River Project

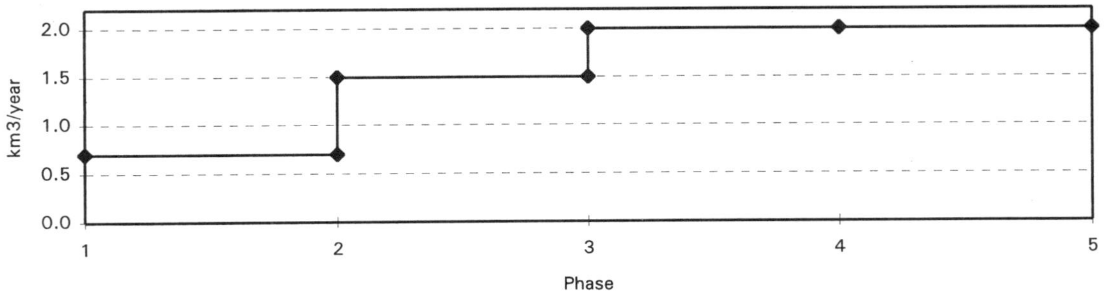

- *Phase 1*, which has been completed recently, is expected to supply the north-central and north-eastern zones extending from Benghazi to Sirt with a total of 700 million m^3/year at a continuous flow rate of 2.0 million m^3/day. Water will be produced by two well fields, Sarir and Tazirbu;
- *Phase 2*, which is under construction, will deliver 800 million m^3/year at a rate of 2.5 million m^3/day to the north-western part of the country (Jifarah Plain) from more than 500 wells, distributed in several well fields located in the northern and north-eastern part of the Murzuq basin;
- *Phase 3* will add about 500 million m^3/year to Phase 1 at a rate of 1.6 million m^3/day from an additional well field within the Kufra basin;
- *Phases 4 and 5* will not involve any additional water production. Instead, the conveyance lines of Phase 1 will be extended farther to the east to reach Tobruk, and farther to the west to link with the Phase 2 pipelines.

Part of the water transported will be used for the water supply of the major cities on the coast: Tripoli, Benghazi, Sirt and Misratah.

In addition to the development of both groundwater and surface water, a further increase in water supplies could be achieved through expanding the reuse of treated wastewater and/or adopting desalination technology.

MAIN SOURCES OF INFORMATION

Pallas, P. 1980. Water resources of the Socialist People's Libyan Arab Jamahiriya. In: The Geology of Libya. Proceedings of the Second Symposium on the Geology of Libya. Academic Press, London.

Saad A. Alghariani. 1993. Satisfying future water demands of northern Libya.

Salem, O.M. 1992. The Great Manmade River project. In: Water Resources Development, 8(4).

Malta

GEOGRAPHY AND POPULATION

The Maltese Archipelago is located in the central part of the Mediterranean Sea about 90 km south of Sicily (Italy), 300 km east of Tunisia and 350 km north of Libya. It consists of 3 main islands, Malta, Gozo and Comino and some islets (Cominotto, Filfla) without inhabitants.

The total area of the archipelago is 316 km^2. The available agricultural land has steadily decreased during recent decades, due to urban and tourist development. Between 1983 and 1991 it decreased by almost 7% from 11 491 ha to 10 721 ha. All available agricultural land is cultivated. Pressures are also heightened on already overcharged infrastructures such as water and electricity by the presence of more than 1 million tourists a year.

TABLE 1
Basic statistics and population

Physical areas:			
Area of the country	1995	31 600	ha
Cultivable area	1991	10 721	ha
Cultivated area	1991	10 721	ha
- annual crops	1991	10 091	ha
- permanent crops	1991	630	ha
Population:			
Total population	1995	366 000	inhabitants
Population density	1995	1 158	inhab./km^2
Rural population	1995	11	%
Water supply coverage:			
Urban population	1995	100	%
Rural population	1995	100	%

The total population is 366 000 (1995), of which 11% is rural. With a population density of 1 158 inhabitants/km^2 it is one of the most densely populated countries in the world. About 2% of the labour force are full-time farmers and another 15% are part-time farmers. The rural population is decreasing because of a reduction in the number of full-time farmers and a preference for urban settlement. On the other hand, old and abandoned farmhouses are being reinstated or converted to residential farmhouses by non-farming families who prefer to live in the countryside. In 1993, agriculture, including fishing, accounted for only 3% of GDP.

CLIMATE AND WATER RESOURCES

Climate

Due to its geographical location the archipelago enjoys a typical Mediterranean climate characterized by hot, dry summers and mild, humid winters. The rainy season is between September and March with the last rain in April. Average annual rainfall is around 524 mm and the temperature varies between 7°C and 15°C in January to between 25°C and 35°C in August. A record temperature of 42°C was recorded in August 1995.

Water resources

Despite the relatively low rainfall and the arid appearance of the Maltese Islands, local catchment characteristics are very favourable for the storage of rainwater and the hydrological cycle provides a generous supply of freshwater which undoubtedly contributed to the early settlement of the inhabitants.

TABLE 2
Water: sources and use

Renewable water resources:			
Average precipitation		524	mm/yr
		0.1656	km^3/yr
Internal renewable water resources		0.0155	km^3/yr
Total (actual) renewable water resources	1995	0.0155	km^3/yr
Dependency ratio	1995	0	%
Total (actual) renewable water resources per inhabitant	1995	42	m^3/yr
Total dam capacity	1993	0.154	$10^6\ m^3$
Water withdrawal:			
- agricultural	1995	6.6	$10^6\ m^3/yr$
- domestic	1995	48.6	$10^6\ m^3/yr$
- industrial	1995	0.5	$10^6\ m^3/yr$
Total water withdrawal		55.7	$10^6\ m^3/yr$
per inhabitant	1995	152	m^3/yr
as % of total (actual) renewable water resources		359	%
Other water withdrawal		-	$10^6\ m^3/yr$
Average groundwater depletion	1995	17.92	$10^6\ m^3/yr$
Wastewater - Non-conventional water sources:			
Wastewater:			
- produced wastewater	1991	23.7	$10^6\ m^3/yr$
- treated wastewater	1993	1.8	$10^6\ m^3/yr$
- reused treated wastewater	1993	1.6	$10^6\ m^3/yr$
Desalinated water	1995	31.4	$10^6\ m^3/yr$

Surface water resources

Total surface water resources are estimated at 0.5 million m^3/year. Structurally, Malta tilts gently to the east giving rise to a topography that is high along the western shores and gently slopes down to sea level along the eastern shores. This implies that the surface drainage lines cross the entire width of the island from their source close to the western shore before reaching the sea on the east. This favourable topography, combined with the good water storage capacity of the soil, excellent infiltration characteristics and effective runoff interception by numerous dams and cisterns, gives the surface water maximum time to seep into the ground and thus minimizes runoff losses to the sea. A greater amount of surface runoff, however, is lost from urban areas via sewers or directly to the sea especially in coastal towns and villages. Attempts are in hand to tackle the storm water wastage problem for a more effective use of surface runoff.

Morphologically, the Maltese Islands are divided into two main units by the Victoria Lines fault, that crosses the northern part of the island of Malta from west to east. North of this fault, Malta is broken up into a number of horsts and grabens by less pronounced faults. Drainage is parallel to the general strike of the horst-graben system and the few intermittent streams flow into the bays to the north-east. The second morphologic unit lies south of the Victoria Lines fault, where two main drainage systems are found. The main one converges into the Valletta basin by a system of east- or north-east-trending streams, while the second one converges into Marsaxlokk bay to the south-east.

Groundwater resources

The renewable groundwater potential on the Maltese Islands is estimated as being approximately 40 million m^3/year. In order not to deplete the storage capacity of the main aquifer without causing salt water intrusion, only 15 million m^3/year of groundwater would be potentially extractable. Based on the 1995 figures of the Water Services Corporation,

TABLE 3
Irrigation and drainage

Irrigation potential	1995	2 000	ha
Irrigation:			
1. Full or partial control irrigation: equipped area	1990	763	ha
- surface irrigation	1990	113	ha
- sprinkler irrigation	1990	150	ha
- micro-irrigation	1990	500	ha
% of area irrigated from groundwater	1990	63.3	%
% of area irrigated from surface water	1990	0.0	%
% of area irrigated from non-conventional sources	1990	36.7	%
% of equipped area actually irrigated	1990	94.8	%
2. Spate irrigation area		-	ha
3. Equipped wetland and inland valley bottoms (i.v.b.)		-	ha
Total irrigation (1+2+3)	**1990**	**763**	**ha**
- as % of cultivated area		7	%
4. Flood recession cropping area		-	ha
Total water managed area (1+2+3+4)	**1990**	**763**	**ha**
- as % of cultivated area		7	%
- increase over last 10 years		-	%
- power irrigated area as % of water managed area		-	%
Full or partial control irrigation schemes: Criteria			
Large-scale schemes = 280 ha	1990	280	ha
Medium-scale schemes	1990	0	ha
Small-scale schemes < 3 ha	1990	483	ha
Total number of households in irrigation	1990	3 000	
Irrigated crops:			
Total irrigated grain production	1990	0	tons
as % of total grain production	1990	0	%
Harvested crops under irrigation (full or partial control)	**1990**	**1 807**	**ha**
- permanent crops: total	1990	0	ha
- annual crops: total	1990	1 807	ha
. melons	1990	550	ha
. tomatoes	1990	540	ha
. potatoes	1990	377	ha
. pumpkins/marrow	1990	170	ha
. cauliflower	1990	170	ha
Drainage - Environment:			
Drained area		-	ha
as % of cultivated area		-	%
- drained areas in full or partial control irrigated areas		-	ha
- drained areas in equipped wetland and i.v.b		-	ha
- other drained areas		-	ha
- total drained area with subsurface drains		-	ha
- total drained area with surface drains		-	ha
Flood-protected area		-	ha
Area salinized by irrigation		-	ha
Population affected by water-borne diseases		-	inhabitants

however, 19.75 million m^3/year were extracted from 13 pumping stations, approximately 160 boreholes in Malta and Gozo, and about 2800 registered private wells (the latter extracting an estimated total of 2.44 million m^3/year). This means that groundwater depletion does in fact take place. Moreover, there is significant extraction from illegal and unregistered wells (probably up to 2.97 million m^3/year), leading to a total groundwater extraction of 22.72 million m^3/year.

Dams

Most runoff occurs after heavy torrential rain. This is the only time when surface water flows, for a few days at most, along the beds of the major valleys. To retain this storm discharge, a large number of small dams have been constructed across the drainage lines. They also serve the purpose of reducing the rate of soil erosion. Open reservoirs have been constructed along recently made roads to minimize runoff. Total dam capacity is estimated at 154 000 m^3.

Desalinated water and treated wastewater

At present 31.4 million m^3/year of desalinated water are being produced from four sea water Reverse Osmosis Plants and one brackish water Reverse Osmosis Plant, but this is a rather expensive procedure. In 1993, of the total produced wastewater estimated at 23.7 million m^3, about 1.82 million m^3 was treated and 1.56 million m^3 of this was reused.

Water withdrawal

Total water withdrawal, was estimated at 55.68 million m^3 in 1995, of which 87% for domestic purposes (Figure 1). Of the total quantity of groundwater used (22.72 million m^3), 17.20 million m^3 were used for domestic purposes, 5.41 million m^3 for agricultural and the remaining 0.11 million m^3 for industrial purposes. The desalinated water (31.40 million m^3) was all used for the provision of potable water in the public supply, which is equal to 65% of the total potable water supply. Of the total reused treated wastewater (1.56 million m^3); 1.22 million m^3 was reused in agriculture and 0.34 million m^3 in industry (Figure 2).

FIGURE 1

Water withdrawal (total: 55.68 million m^3 in 1995)

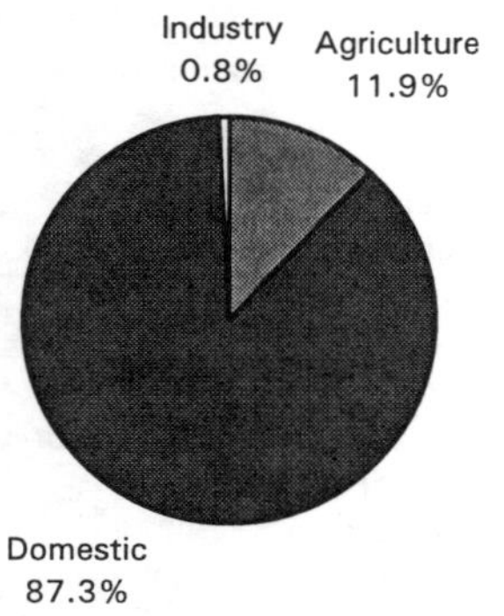

FIGURE 2

Origin of water used by sector (total: 55.68 million m^3 in 1995)

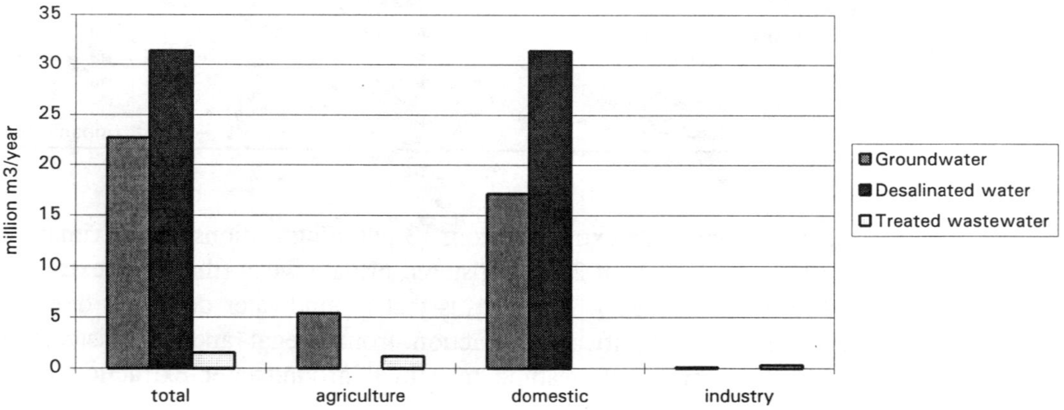

IRRIGATION AND DRAINAGE DEVELOPMENT

In 1990, the total water managed area, all equipped for full or partial control irrigation, was estimated at 763 ha, which is about 7% of the agricultural land. Of this area, 280 ha are equipped for irrigation by treated sewage water from a sewage treatment plant (Figure 3), which was completed in 1983 to provide 7 000 m^3 water per day, but at present 240 ha are actually irrigated per year using 1.22 million m^3 of treated wastewater. Tenders have been issued to upgrade this sewage treatment plant to produce 17 000 m^3 per day for irrigation and industrial purposes. Work is expected to be completed by early 1997.

FIGURE 3
Origin of irrigation water f/p (total: 763 ha in 1995)

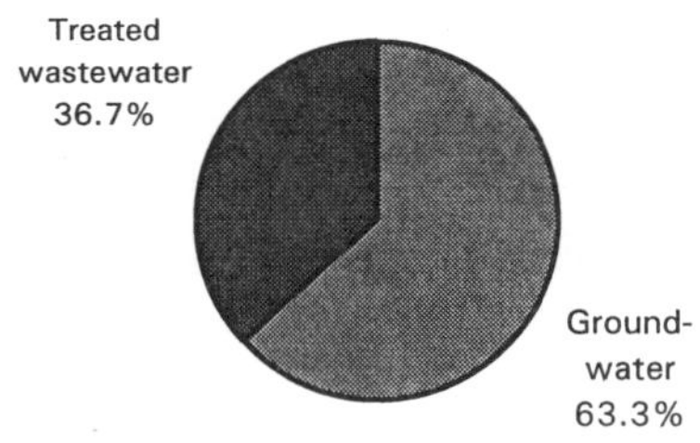

Plans are also under way to construct three other sewage treatment plants, two in Malta and one in Gozo, with a total capacity of 73 000 m^3 per day to treat all the sewage produced in the country by the year 2000. Hence the total available treated wastewater would be about 90 000 m^3 per day. Since the maximum daily irrigation water requirements are estimated at 60 m^3/ha per day (July/August), the potential area for irrigation from treated wastewater would be 1 500 ha if all the wastewater was reused for irrigation. The potential area to be irrigated from treated wastewater could even be increased up to 2 500 ha if wastewater could be stored in winter (when irrigation water requirements are lower) for use during the summer. However, this solution is financially prohibitive. Thus, the total irrigation potential is estimated at 2 000 ha, of which 500 ha irrigated by groundwater and 1 500 ha irrigated by treated wastewater. Nevertheless, there are several limitations to reaching this potential including fund availability, problems of access to the fields, the size and fragmentation of farm holdings, farm labour demand, marketing, water charges and the Groundwater Protection Zones. From preliminary cost estimates of water distribution networks to supply treated wastewater to farmers from the new sewage treatment plants, the capital investment required would amount to $US 11 000/ha.

According to the Agricultural Census, 483 ha is the area actually irrigated by groundwater, while the Water Services Corporation (WSC) recorded an area of only about 269 ha irrigated by groundwater. The difference can be explained by looking at the history of groundwater extraction in Malta. Groundwater abstraction is administered and regulated by WSC and the law states that nobody is allowed to sink shafts to exploit the groundwater since this is used as potable water. However, before this legislation, which was enacted in 1943, several farmers had sunk shafts to abstract water. The thousands of sunk shafts are the property of the farmers themselves and though, under the law, the abstraction of water is controlled by WSC, the shafts are administered individually by farmers. Each shaft may irrigate only a hectare or two, depending on the area cultivated by the farmer when the shaft was sunk some 60 or 70 years ago. But individual records of present water abstraction are not kept and these shafts are scattered over all rural areas, but predominantly in the north of the country. The number of private wells officially registered by WSC is about 2800, to which an important number of unregistered wells should be added. Total groundwater withdrawal from the wells is estimated at 5.41 million m^3/year, of which 2.44 million m^3 extracted by registered wells and an estimated 2.97 million m^3/year by unregistered wells.

FIGURE 4
Irrigation techniques f/p (total: 763 ha in 1995)

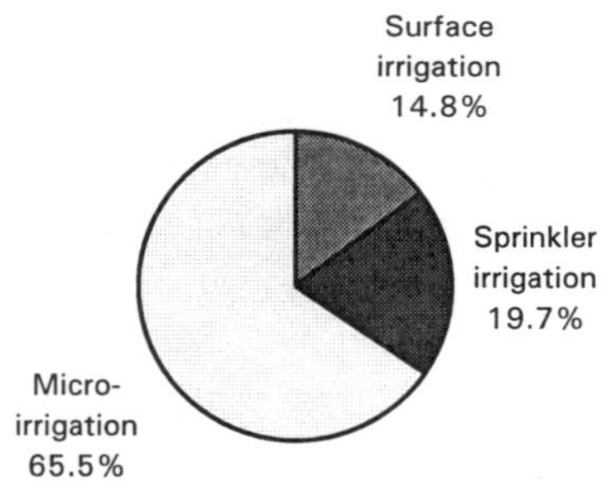

Apart from the treated wastewater and groundwater used for irrigation, water harvesting practices are also widely spread all over the islands. Throughout the centuries, several farmers have built or dug small reservoirs in the rock to collect rainwater to be used mainly as supplementary irrigation. Recently, several reservoirs with capacities of 100 - 2 000 m^3 were constructed for irrigation in spring or early summer, some with financial assistance from the government. Though this water collection may seem insignificant, it is in fact very important to provide a supplementary source of water for Maltese agriculture and it covers an area estimated at 1 953 ha. For example, probably over 1 000 ha of the 1 400 hectares spring potato crop which is planted in the rainy season in December and January, and harvested in May and June, is irrigated with this source of water in the absence of rainfall. Similarly, unquantified large areas of vegetable production may also receive additional irrigation from this surface runoff water from September to early June.

On the areas irrigated by treated wastewater, the distribution network for surface irrigation was originally constructed of concrete channels which created problems and brought complaints from farmers wanting their share of irrigation water. Though a report was drawn up by a consultancy firm to improve this system, funds were never made available to execute the works. Fortunately, the farmers have themselves invested heavily in installing micro-irrigation equipment to irrigate cash crops. Part of the area is also equipped with sprinklers to irrigate potatoes in spring since the potato crop is more responsive to sprinkler irrigation. Out of a total managed area of 763 ha, it is estimated that 500 ha are equipped with micro-irrigation systems, 150 ha with sprinkler irrigation systems, while on the remaining 113 ha surface irrigation is practised (Figure 4). The cost of irrigation development is in the range of $US 1 600/ha for micro-irrigation, while the operation and maintenance costs are about $US 800/ha per year.

There are three categories of farming in the irrigation subsector and an estimated 3 000 farmers, both full-time and part-time, are involved:

- Most of the irrigated farms are normally leased to farmers and operated by individuals. The source of water is often shared with others, since both the land and water rights are inherited with the lease resulting in land fragmentation. Very often the irrigated farm is scattered over various localities giving rise to problems of access to the fields, conveyance of water and laying of irrigation networks and schemes. Some farmers frequently have shares of water rights from various groundwater sources which may further complicate the irrigation scheduling. There are no water charges for abstraction of groundwater from these private boreholes.

- The irrigation scheme which is supplied with treated wastewater from the sewage treatment plant is run by a government agency. Treated water is supplied to five government reservoirs and by means of a channel system is distributed to farmers for a nominal fee of $US 100/ha. This was considered as a social project to increase revenue in the farming community. The running of the individual farms is however entirely the responsibility of the farmer.

- Water from government-owned boreholes producing 0.09 million m^3/year second class water (high nitrate or high salinity) is offered to farmers in the vicinity at $US 0.11/$m^3$, in preference to using it for domestic supply.

The major crops under full or partial control irrigation are melons, tomatoes, potatoes, pumpkins/ marrow and cauliflower (Figure 5), with a total cropped area of 1 807 ha (average cropping intensity is 2.5). In addition, the areas benefiting from water harvesting techniques (1 953 ha) are generally cropped once a year. With the introduction of micro-irrigation and improved farming practices, the yields of some products significantly increased during the period 1990-95. Yields of melons increased from 6 tons/ha in 1990 to an estimated 20 tons/ha in 1995.

FIGURE 5
Irrigated crops f/p (total: 1 807 ha in 1990, crop intensity 2.5)

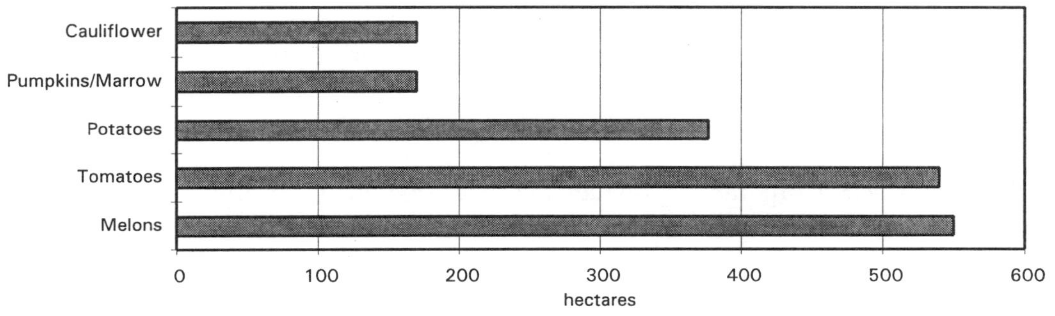

Revenue from one hectare of fully irrigated land is calculated to be $US 24 750, which is at least ten times the revenue from dry farming. The main reason is that vegetable production during the totally dry period, from May to the end of August, may only be carried out under constant irrigation. Moreover, owing to favourable temperatures, with full irrigation a cropping intensity of 3 may be achieved. In the case of treated wastewater irrigation scheme, cropping intensity averages about 2.

Drainage is not practiced in Malta since infiltration rates are high, averaging 74 mm/hr. The few small wetland areas adjacent to the sea have been declared conservation areas, two of which are bird sanctuaries.

There are no salinization problems since any salt accumulation in the top layer of the soil is washed down through the shallow soils with regular frequent irrigation and torrential rains in the winter. Figure 6 shows clearly that any salt accumulation as a result of irrigation with 4 000 μs/cm wastewater during the dry summer is removed during winter by torrential rains. Moreover both groundwater aquifers are deep and do not rise to the top soil.

FIGURE 6
Average soil salinity

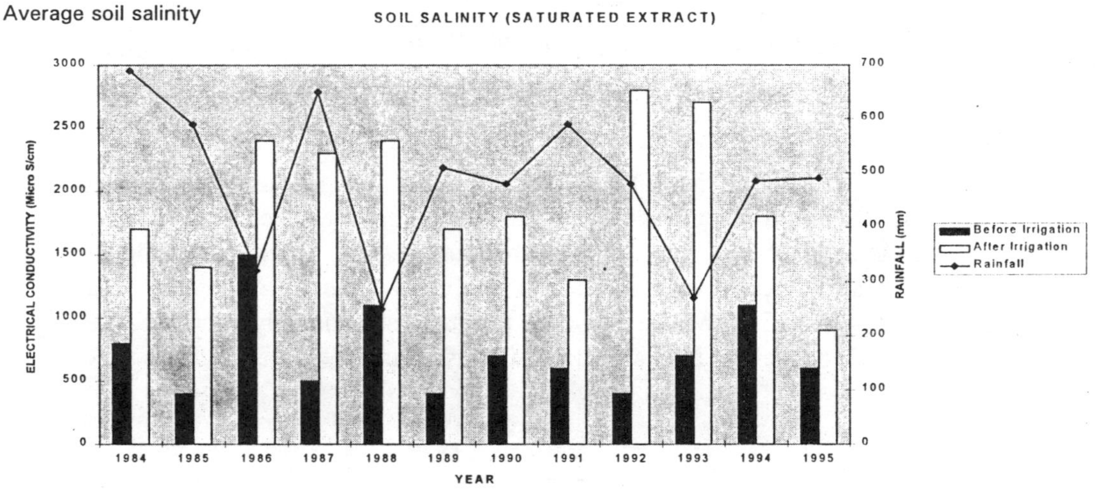

INSTITUTIONAL ENVIRONMENT

The two main institutions involved in water resources management are:

The Ministry of the Environment:

- The Water Services Corporation provides the legal support to an effective national water resources management which manages the supply and demand of potable and non-potable water. The Water Services Corporation Act, enacted in 1991, serves as the legislative backbone and incorporates all the legal tools and institutional framework to deal with every situation in the water sector in Malta and to coordinate all the activities concerned in full awareness of the environmental constraints and national socio-economic requirements. The Corporation will also be responsible, once the relative parts of the Act come into force, for wastewater collection and treatment.

- The Drainage Department is responsible for the design, construction and maintenance of the sewage system and sewage treatment plants.

- The Secretariat for the Environment is responsible for quality control of the treated wastewater.

The Ministry of Food, Agriculture and Fisheries:

- The Land and Water Division is in charge of promoting irrigated agriculture and responsible for the distribution of treated wastewater.

TRENDS IN WATER RESOURCES MANAGEMENT

As a consequence of the groundwater depletion, a steady salinization of the groundwater sources has taken place over the last 20 years. For this reason, other alternative water supplies, like desalinated sea water and treated wastewater, will have to continue to be adopted so as to stop and reverse this depletion..

The commitment of Malta to treat all sewage water produced by the year 2000 would make treated wastewater available to irrigate about 1 500 ha, in addition to 500 ha irrigated by groundwater. However, the main constraints to achieving this potential are the funding of the irrigation network, realistic water charges and marketing prices.

Domestic use of groundwater has top priority and hence further development of this source for irrigation is very limited. Through a more efficient use of water by means of micro-irrigation, there would be great potential for an expansion in irrigated areas. The government is financially assisting farmers to buy irrigation equipment by offering grants and subsidizing interest rates under the Financial Assistance Policy.

MAIN SOURCES OF INFORMATION

BRGM. 1992. Study of the fresh water resources in Malta. Report for the Water Services Corporation.

Central Office of Statistics. 1995. Maltese population estimates 1971-94 and Census of agriculture 1990-91.

Cowiconsult. 1992. Sewerage Master Plan for Malta and Gozo. Ministry of the Environment.

FAO. 1992. Malta agricultural policy and the E.C. membership challenges and opportunities. Report prepared by Van As, I. and Quick, D.

Gauci, Vince. 1994. Sewage effluent production 1993. Annual report, Waste Recycling Department.

Mitschoff, Josef. 1990. Upgrading and modernization of Sant'Antnin sewage effluent irrigation system. Ministry for Development of Infrastructure.

Water Services Corporation. 1996. Annual Report 1995.

Mauritania

GEOGRAPHY AND POPULATION

Mauritania, located in the north-west of the African continent, has a total area of about 1 million km^2. It is bordered in the north-west by Western Sahara, in the north-east by Algeria, in the east and south-east by Mali, in the south-west by Senegal and in the west by the Atlantic Ocean.

The cultivable area has been estimated at about 20 million ha, or almost 20% of the total area. In 1989, the total cultivated area was 204 000 ha, of which 199 000 ha consisted of annual crops and about 5 000 ha consisted of permanent crops.

TABLE 1
Basic statistics and population

Physical areas:			
Area of the country	1995	102 552 000	ha
Cultivable area	1985	20 129 689	ha
Cultivated area	1989	204 000	ha
- annual crops	1989	199 000	ha
- permanent crops	1994	5 000	ha
Population:			
Total population	1995	2 274 000	inhabitants
Population density	1995	2	inhab./km^2
Rural population	1995	46	%
Water supply coverage:			
Urban population	1990	67	%
Rural population	1990	65	%

The total population is about 2.3 million (1995), of which 46% is rural. The population density is about 2 inhabitants/km^2, but this figure varies from 0.1 in Tiris Zemmour to 18.5 in the Gorgol region. The annual demographic growth rate is about 2.7%. Agriculture, including livestock, accounts for 20% of GDP.

CLIMATE AND WATER RESOURCES

Climate

Average annual rainfall is 99 mm, but this varies from less than 20 mm in the north to more than 500 mm in the south-east (Sélibaby). The rainy season lasts only 3 months. Average minimum and maximum temperature vary from 16°C in January (Bir Mougrein) to 36°C in June (Kiffa).

Water resources

The total (actual) renewable surface water resources are estimated at 11.1 km^3/year and the internal renewable surface water resources at 0.1 km^3/year. The hydrographic system is dominated by the Senegal river which runs along the southern border and its tributaries on the right bank downstream of Bakel. There are some wadis with sporadic flows during the rainy season. Total groundwater resources have been estimated at 3.7 km^3, which refers to the continuous aquifers only. The yields of the alluvial groundwater in the wadis and the intermittent aquifer systems have not yet been quantified. Internal renewable groundwater resources are estimated at 0.3 km^3/year.

TABLE 2
Water: sources and use

Renewable water resources:			
Average precipitation		99	mm/yr
		101.9	km^3/yr
Internal renewable water resources		0.4	km^3/yr
Total (actual) renewable water resources	1995	11.4	km^3/yr
Dependency ratio	1995	96.5	%
Total (actual) renewable water resources per inhabitant	1995	5 013	m^3/yr
Total dam capacity	1994	900	$10^6 m^3$
Water withdrawal:			
- agricultural	1985	1 500	$10^6 m^3/yr$
- domestic	1985	101	$10^6 m^3/yr$
- industrial	1978	29	$10^6 m^3/yr$
Total water withdrawal		1 630	$10^6 m^3/yr$
per inhabitant	1985	923	m^3/yr
as % of total (actual) renewable water resources		14.3	%
Other water withdrawal		-	$10^6 m^3/yr$
Average groundwater depletion		-	$10^6 m^3/yr$
Wastewater - Non-conventional water sources:			
Wastewater:			
- produced wastewater		-	$10^6 m^3/yr$
- treated wastewater		-	$10^6 m^3/yr$
- reused treated wastewater		-	$10^6 m^3/yr$
Desalinated water	1990	1.7	$10^6 m^3/yr$

FIGURE 1
Water withdrawal (total: 1.63 km^3 in 1985)

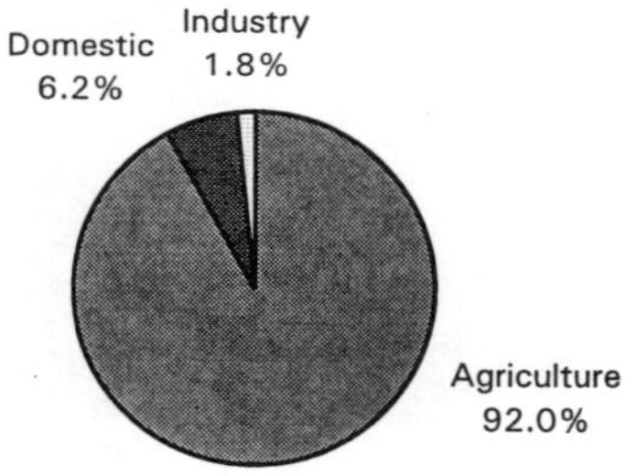

FIGURE 2
Distribution of the water managed areas (total: 113 200 ha in 1994)

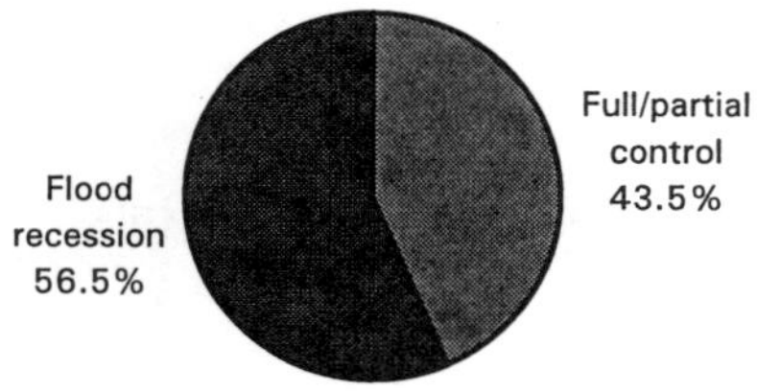

In 1985, total water withdrawal was 1.63 km^3, of which 92% for agricultural purposes (Figure 1).

It is estimated, that in the year 2000 about 3 750 water points will be needed to supply drinking water to the rural population, while at present 2 118 water points have been constructed.

IRRIGATION AND DRAINAGE DEVELOPMENT

Irrigation potential, considering soil resources, has been estimated at 257 000 ha, of which 110 000 ha for flood recession cropping. Considering the available water resources as well, the irrigation potential in the Senegal basin is estimated at 165 000 ha, of which 125 000 ha along

TABLE 3
Irrigation and drainage

Irrigation potential		1994	170 000	ha
Irrigation:				
1. Full or partial control irrigation: equipped area		1994	49 200	ha
- surface irrigation			-	ha
- sprinkler irrigation			-	ha
- micro-irrigation			-	ha
% of area irrigated from groundwater		1994	9.6	%
% of area irrigated from surface water		1994	90.4	%
% of area irrigated from non-conventional sources		1994	0.0	%
% of equipped area actually irrigated		1994	54.4	%
2. Spate irrigation area			-	ha
3. Equipped wetland and inland valley bottoms (i.v.b.)			-	ha
Total irrigation (1+2+3)		**1994**	**49 200**	**ha**
- as % of cultivated area			24	%
4. Flood recession cropping area		1988	64 000	ha
Total water managed area (1+2+3+4)		**1994**	**113 200**	**ha**
- as % of cultivated area			55	%
- increase over last 10 years			-	%
- power irrigated area as % of water managed area		1994	41.6	%
Full or partial control irrigation schemes:	Criteria			
Large-scale schemes	> 200 ha	1994	6 217	ha
Medium-scale schemes		1994	3 037	ha
Small-scale schemes	< 50 ha	1994	39 967	ha
Total number of households in irrigation			-	
Irrigated crops:				
Total irrigated grain production		1993	136 498	tons
as % of total grain production		1993	66	%
Harvested crops under irrigation (full or partial control)		**1993**	**134 899**	**ha**
- permanent crops: total		1993	4 751	ha
- annual crops: total		1993	130 148	ha
. sorghum		1993	62 488	ha
. vegetables/polyculture		1993	40 820	ha
. rice		1993	22 408	ha
. other annual crops (maize)		1993	4 432	ha
Drainage - Environment:				
Drained area		1994	12 784	ha
as % of cultivated area			6	%
- drained areas in full or partial control irrigated areas			-	ha
- drained areas in equipped wetland and i.v.b			-	ha
- other drained areas			-	ha
- total drained area with subsurface drains			-	ha
- total drained area with surface drains			-	ha
Flood-protected area			-	ha
Area salinized by irrigation			-	ha
Population affected by water-borne diseases			-	inhabitants

the main Senegal river and 40 000 ha in the Gorgol and Karakoro tributary areas. In addition, there are around 5 000 ha of oases.

In 1994, the water managed area was evaluated at 113 200 ha, of which 49 200 ha consisted of full or partial control irrigation schemes. Flood recession cropping covered the remaining 64 000 ha (Figure 2).

The area equipped for full or partial control irrigation consists mainly of small schemes, with an area of less than 50 ha each (Figure 3). The total area can be divided into 4 571 ha of oases and 44 629 ha of schemes in the provinces of the valley of the Senegal river and its tributaries (Figure 4):

FIGURE 3
Typology of f/p control irrigation schemes (total: 49 200 ha in 1994)

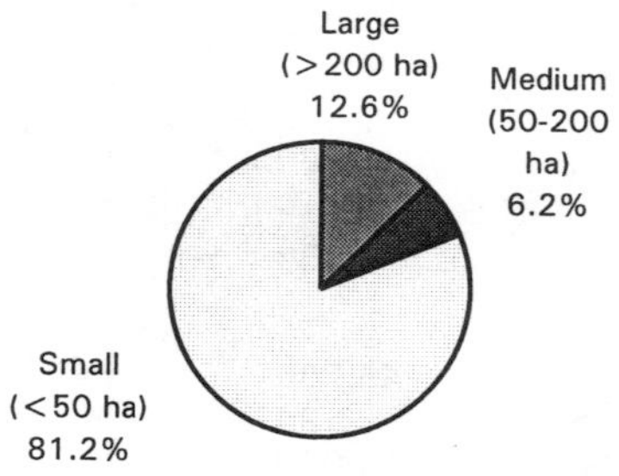

FIGURE 4
Distribution of the f/p control irrigation schemes (total: 49 200 ha in 1994)

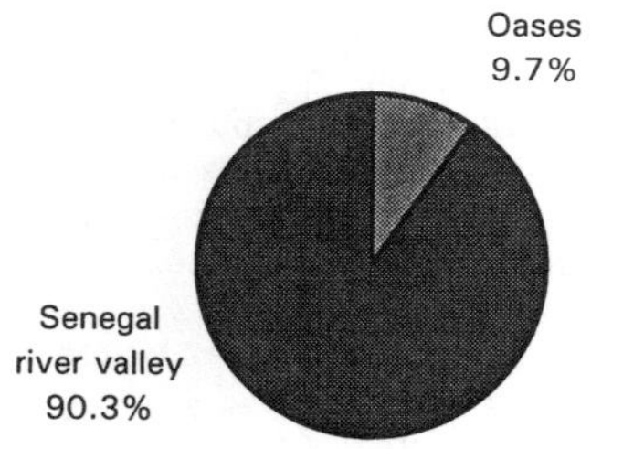

- In total there are 218 oases, irrigated by groundwater. There are 31 400 wells in the oases, in the majority of which the water is extracted manually (Figure 5). About 400 000 people live in the oases, but only 14 700 families possess the 4 751 ha of palm trees (more than one million productive palm trees) and cultivate around 244 ha of annual crops under the palm trees. The oases are very small, more than 46% being less than 10 ha (Figure 6). The whole area equipped for irrigation is irrigated all year round.

FIGURE 5
Methods of groundwater extraction in the oases (total: 4 751 ha in 1993)

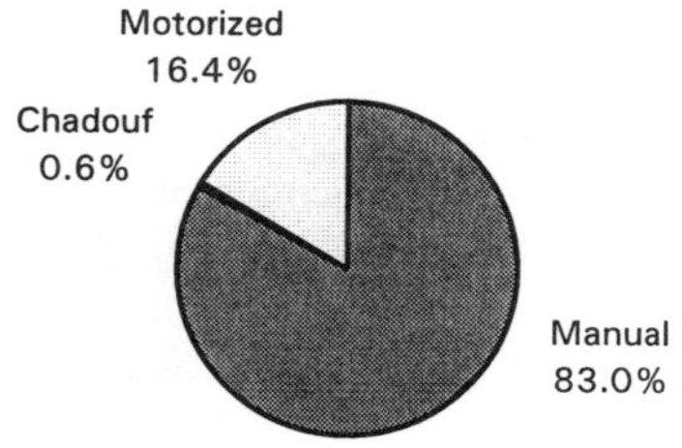

FIGURE 6
Typology of oases (total: 4 751 ha in 1993)

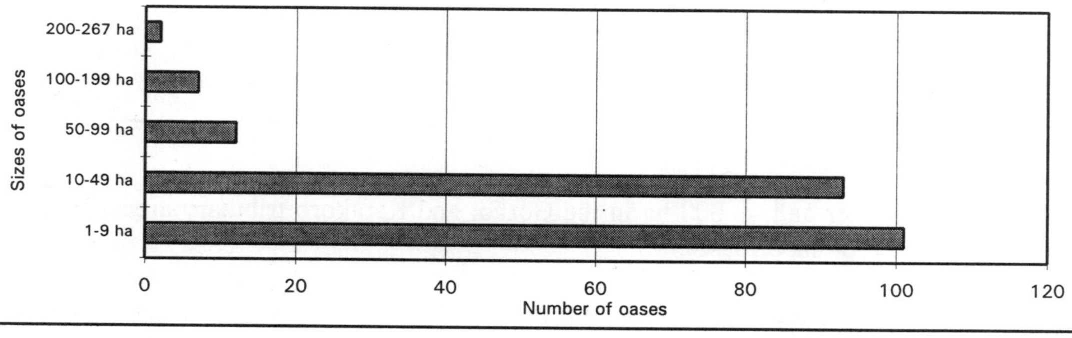

- The schemes in the valley of the Senegal river and its tributaries are only partly irrigated each year (22 000 ha only in 1994). Large areas need rehabilitation, as almost 38 000 ha are considered to be in a quite bad or bad state, according to a recent survey. All these schemes are irrigated by surface water, mainly through pumping from the river (Figure 7).

FIGURE 7
Origin of irrigation water f/p
(total: 49 200 ha in 1994)

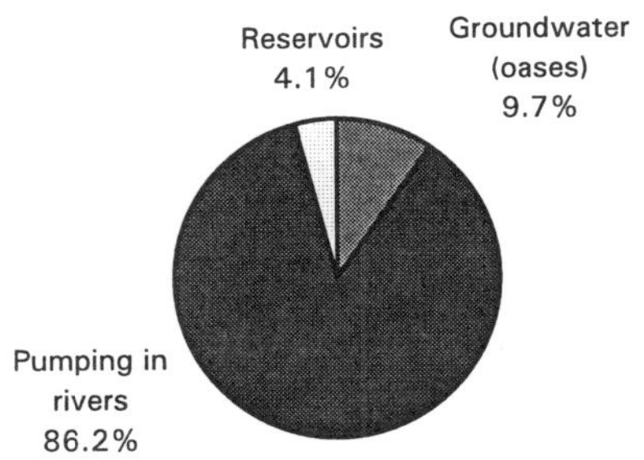

FIGURE 8
Typology of flood recession cropping
(total: 64 000 ha in 1988)

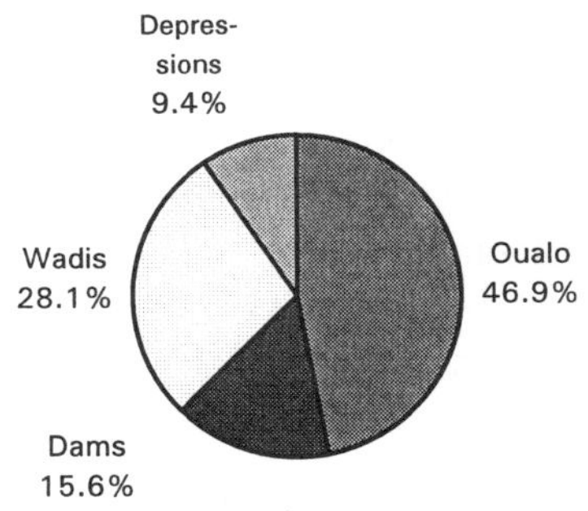

The main irrigation technique used in the full or partial control irrigation schemes is surface irrigation. The average cost of irrigation development varies between $US 2 200/ha for small schemes to about $US 12 500/ha for large schemes.

Different types of flood recession cropping can be distinguished (Figure 8):

- the oualo in the major bed of the Senegal and Gorgol river valleys. The areas vary between 20 000 ha and 40 000 ha depending on the year, with an average of 30 000 ha;
- flood recession cropping around the large dams, with a total area of 10 000 ha, of which 4 000 ha in Lake Rkiz, 3 500 ha in Achram and 2 500 ha in Foum Gleita;
- flood recession cropping in the beds of the wadis on areas of less than 1 000 ha, with a total area varying between 16 000 and 20 000 ha;
- topographic depressions inundated by surface runoff water (6 000 ha). Estimates on the total area vary between 2 000 and 10 000 ha.

The major irrigated crops are sorghum, vegetables and rice. Palm trees grow in the oases. Mixed farming, with vegetable growing is important in the flood recession zones (Figure 9).

FIGURE 9
Irrigated crops (total: 134 899 ha in 1993)

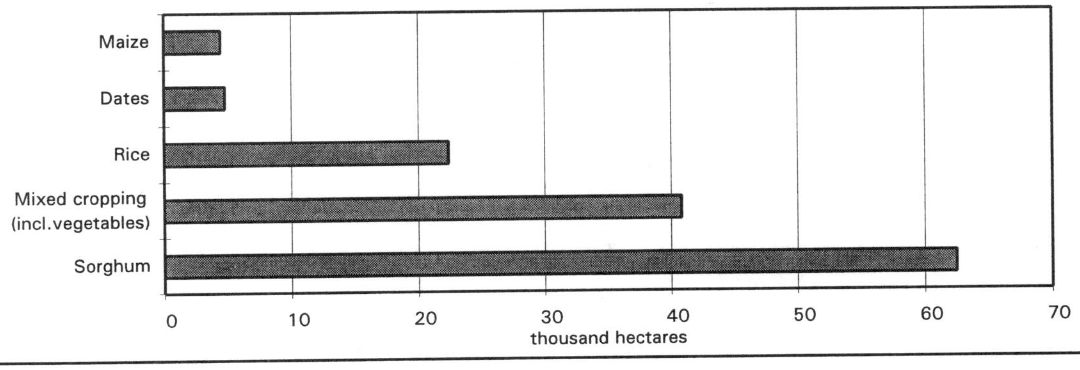

INSTITUTIONAL ENVIRONMENT

The public institutions intervening in hydraulic and agricultural development are:

The Ministry of Rural Development and Environment:

- Société nationale de développement rural (SONADER) (rural development);
- Direction de l'environnement et de l'aménagement rural (DEAR) (environment and rural management);
- Direction du développement des ressources agro-pastorales (DRAP) (agri-pastoral resources development;
- Cellule de planification (Planning Unit).

The Ministry of Hydraulics:

- Direction de l'hydraulique (Hydraulics Department);
- Cellule nationale de l'Office de mise en valeur du Sénégal (OMVS) (National Desk of the Senegal River Development Office).

Decree No 90020 of 19 January 1990 establishes the clauses of the reorganization of the land properties, putting the accent on the individualization of land tenure, in three phases: authorization to cultivate, provisional concession and permanent concession.

TRENDS IN WATER RESOURCES MANAGEMENT

The major constraints to the exploitation of the water resources and the development of irrigation in Mauritania can be summarized as follows:

- high costs of irrigation development and especially of earth structures;
- accessibility (river and oases);
- weak support given to the private sector with insufficient investment to sustain the productive capital;
- marketing problems;
- inadequate organization of the operation of the schemes, especially of the communal schemes;
- high costs of extraction of the groundwater resources;
- salinization of the soils (delta and lower river);
- high evapotranspiration (2 000 to 3 000 mm/year from south to north);
- insufficient and irregular rainfall and surface flow in the wadis.

MAIN SOURCES OF INFORMATION

Cellule de planification, Ministère du développement rural et de l'environnement. 1994. Le développement rural en chiffres.

FAO/DDC. 1988. Programme national de construction des barrages en terre.

FAO/DDC. 1990. Projet de développement intégré de la vallée du fleuve Sénégal. Document de travail (volet irrigation).

GERSAR. 1994. Etudes d'application des schémas directeurs du fleuve Sénégal et du Delta.

Ministère du développement rural et de l'environnement. 1993. Annuaire statistique des oasis.

PNUD. 1990. Schéma directeur pour la mise en valeur des ressources en eau. Rapport du projet PNUD/DTCD/MAU/87/008.

SONADER. 1993. Enquête périmètres irrigués.

SONADER. 1994. Rapport d'activités 1993.

Morocco

GEOGRAPHY AND POPULATION

Morocco, located in the north-west of the African continent, has a total area[1] of 446 500 km². It is bordered in the west by the Atlantic Ocean, in the north by the Mediterranean Sea, in the north-east by Algeria and in the south-east and south by Mauritania.

The country can be divided into four physiographic units:

- The Coastal Plains, that extend along the entire Moroccan coastline. They are narrow on the Mediterranean Coast and wide on the Atlantic Coast. These plains are crossed by the majority of the rivers and valleys of the country;
- The Northern Hills, that run parallel to the Mediterranean Sea and are called the El-Reef Mountains with their peak reaching 2 456 metres above sea level;
- The Central Hills, that run along the middle of the country and extend from north-east to south-west. They consist of the mountain ranges of the Central, Upper and Lower Atlas, which run almost parallel to one another. The peak is in the Upper Atlas at 4 165 metres above sea level;
- The Desert Hills, that are extensions of the southern slopes of the Upper and Lower Atlas Mountains.

The cultivable area has been estimated at 8 million hectares, which is 18% of the total area. In 1993, the total cultivated area was 7.23 million ha, of which 6.57 million ha consisted of annual crops and 0.66 million ha consisted of permanent crops.

TABLE 1
Basic statistics and population

Physical areas:			
Area of the country	1995	44 650 000	ha
Cultivable area	1992	8 000 000	ha
Cultivated area	1993	7 232 700	ha
- annual crops	1993	6 568 300	ha
- permanent crops	1993	664 400	ha
Population:			
Total population	1995	27 028 000	inhabitants
Population density	1995	61	inhab./km²
Rural population	1995	52	%
Water supply coverage:			
Urban population	1990	100	%
Rural population	1990	18	%

The total population is 27 million (1995), of which 52% is rural. The average population density is 61 inhabitants/km², but on the coastal areas the population density is highest. It is low in the mountain and desert zones, where it is concentrated around the oases. In 1993, agriculture, including forestry and fisheries, accounted for 14.3% of GDP, employing about 47% of the labour force. The irrigated areas represent 17% of the cultivated areas, but contribute about 45% to agricultural earnings.

1 Following the practice of the United Nations, this figure does not include the area of the Western Sahara, which is about 266 000 km². This does not imply any position taken by FAO concerning the legal or constitutional status of the Western Sahara.

TABLE 2
Water: sources and use

Renewable water resources:			
Average precipitation		336	mm/yr
		150.0	km^3/yr
Internal renewable water resources		30.0	km^3/yr
Total (actual) renewable water resources	1995	30.0	km^3/yr
Dependency ratio	1995	0	%
Total (actual) renewable water resources per inhabitant	1995	1 110	m^3/yr
Total dam capacity	1992	11 000	10^6 m^3
Water withdrawal:			
- agricultural	1991	10 180	10^6 m^3/yr
- domestic	1992	543	10^6 m^3/yr
- industrial	1992	322	10^6 m^3/yr
Total water withdrawal		11 045	10^6 m^3/yr
per inhabitant	1991	436	m^3/yr
as % of total (actual) renewable water resources		36.8	%
Other water withdrawal	1990	380	10^6 m^3/yr
Average groundwater depletion		-	10^6 m^3/yr
Wastewater - Non-conventional water sources:			
Wastewater:			
- produced wastewater	1994	370	10^6 m^3/yr
- treated wastewater		-	10^6 m^3/yr
- reused treated wastewater		-	10^6 m^3/yr
Desalinated water	1992	3.4	10^6 m^3/yr

CLIMATE AND WATER RESOURCES

Climate

Average annual rainfall is 340 mm, but varies from more than 450 mm in the north, where rainfed agriculture is possible, to less than 150 mm towards the south-east, where irrigation is absolutely necessary. Over 50% of the precipitation is concentrated on only 15% of the country's area.

Water resources

The water resources have been evaluated at 30 km^3/year, out of which 16 km^3 of surface water and 5 km^3 of groundwater are considered to represent water development potential (Figure 1). The most important rivers are equipped with dams, allowing surface water to be stored for use during the dry seasons. In 1990, 34 dams were operational, with a total dam capacity of 11 km^3. Of these dams, 13 were used in the schemes operated by the regional agricultural development offices (ORMVA or Office Régional de Mise en Valeur Agricole).

Water withdrawal

In 1992, water withdrawal was estimated at about 11 km^3, of which over 92% for agricultural purposes (Figure 2). Of this total of 11 km^3, 7.5 km^3 was surface water and 3.5 km^3 groundwater. In 1990, 236 000 water points were counted in the rural areas (91% wells, 8% springs, 1% surface water points), which was equal to about 1 water point to 50 inhabitants, of which 16% were operated by mechanical or motorized water extraction techniques. A health survey showed that 84% of water points delivered non potable water. About 10% of the rural population receives water in a more or less satisfactory way through collective systems and 8% benefits from a water distribution network, which leads to a water supply coverage of 18% in rural areas. A major effort is being undertaken, starting in 1996,

TABLE 3
Irrigation and drainage

Irrigation potential		1993	1 653 000	ha
Irrigation:				
1. Full or partial control irrigation: equipped area		1989	1 093 200	ha
- surface irrigation		1989	986 000	ha
- sprinkler irrigation		1989	103 200	ha
- micro-irrigation		1989	4 000	ha
% of area irrigated from groundwater		1989	31.1	%
% of area irrigated from surface water		1989	68.3	%
% of area irrigated from non-conventional sources		1989	0.6	%
% of equipped area actually irrigated			-	%
2. Spate irrigation area		1989	165 000	ha
3. Equipped wetland and inland valley bottoms (i.v.b.)			-	ha
Total irrigation (1+2+3)		**1989**	**1 258 200**	**ha**
- as % of cultivated area			17	%
4. Flood recession cropping area			-	ha
Total water managed area (1+2+3+4)		**1989**	**1 258 200**	**ha**
- as % of cultivated area			17	%
- increase over last 10 years			-	%
- power irrigated area as % of water managed area			-	%
Full or partial control irrigation schemes:	Criteria			
Large-scale schemes	> - ha	1989	431 650	ha
Medium-scale schemes				ha
Small-scale schemes	< - ha	1989	661 550	ha
Total number of households in irrigation			-	
Irrigated crops:				
Total irrigated grain production		1990	1 025 000	tons
as % of total grain production		1990	15	%
Harvested crops under irrigation (full or partial control)		**1990**	**1 073 000**	**ha**
- permanent crops: total		1990	186 000	ha
- annual crops: total		1990	887 000	ha
. cereals		1990	410 000	ha
. vegetables		1990	140 000	ha
. fodder crops		1990	100 000	ha
. pulses		1990	89 000	ha
. other annual crops (sugar beet/cane, cotton, ...)		1990	148 000	ha
Drainage - Environment:				
Drained area			-	ha
as % of cultivated area			-	%
- drained areas in full or partial control irrigated areas			-	ha
- drained areas in equipped wetland and i.v.b			-	ha
- other drained areas			-	ha
- total drained area with subsurface drains			-	ha
- total drained area with surface drains			-	ha
Flood-protected area			-	ha
Area salinized by irrigation			-	ha
Population affected by water-borne diseases			-	inhabitants

FIGURE 1
Distribution of the renewable water resources over the major hydrological zones
(total: 30 km^3/year)

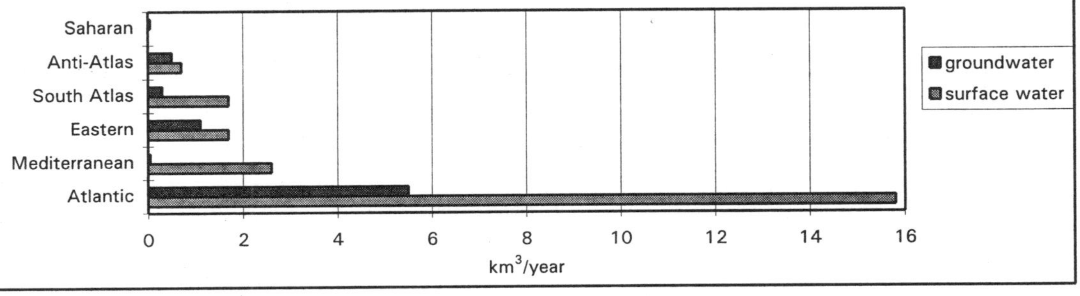

FIGURE 2
Water withdrawal (total: 11.05 km^3 in 1992)

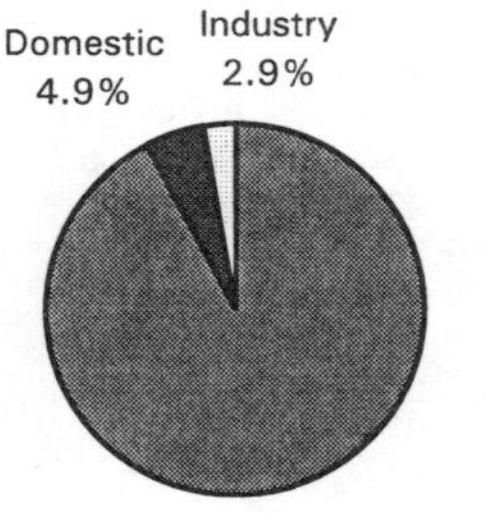

FIGURE 3
Distribution of the water managed areas (total: 1 258 200 ha in 1989)

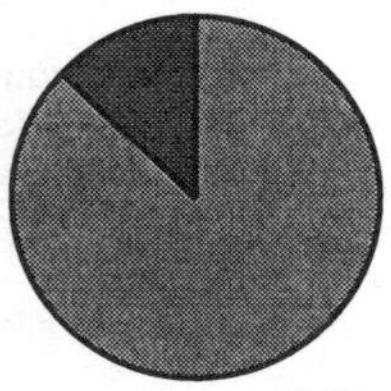

FIGURE 4
Origin of irrigation water f/p (total 1 093 200 ha in 1989)

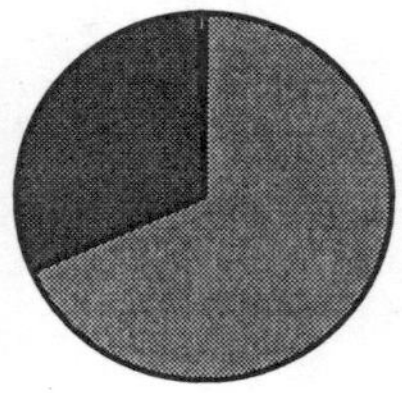

FIGURE 5
Irrigation techniques f/p (total: 1 093 200 ha in 1989)

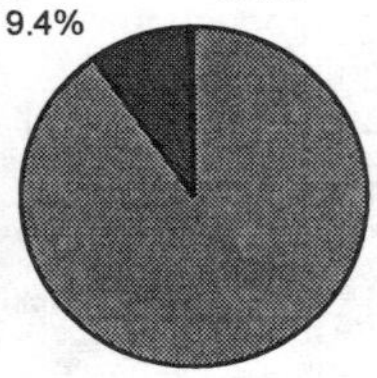

to increase this percentage within the next eight years. Hydro-electricity produced 1 500 GWh in 1991, or 30% of the total energy production of the country. In 1990, 230 million m^3 of water were used for hydropower and 150 million m^3 for environmental protection (wadis).

IRRIGATION AND DRAINAGE DEVELOPMENT

Irrigation is a strategic sector in Morocco. The water managed areas, in total about 1.26 million ha, represent only 17% of the cultivated area, but 76% of the irrigation potential area estimated at 1.65 million ha. Spate irrigation covers 165 000 ha, the remaining 1.09 million ha being full or partial control irrigation schemes (Figure 3), of which about two-thirds are irrigated by surface water (Figure 4). In 1995, about 7 000 ha were estimated to be irrigated by untreated wastewater, using about 60 million m^3/year. Surface irrigation is the principal irrigation technique, sprinkler irrigation being practised only on a few large schemes (Figure 5). Cereals are the major irrigated crops (Figure 6).

Large schemes

Since the 1960s emphasis has been put on the construction of dams and on the development of large schemes (referred to as 'grande hydraulique') (Figure 7). The schemes (with areas

FIGURE 6
Irrigated crops f/p (total: 1 073 000 ha in 1990)

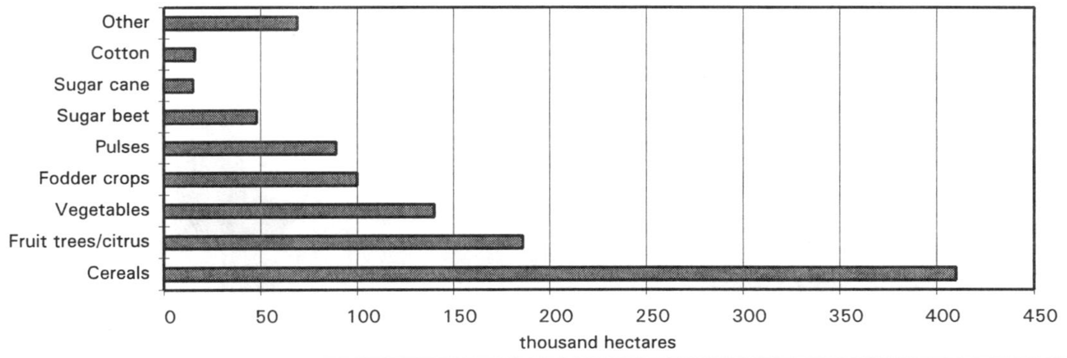

FIGURE 7
Evolution of the areas of the large schemes

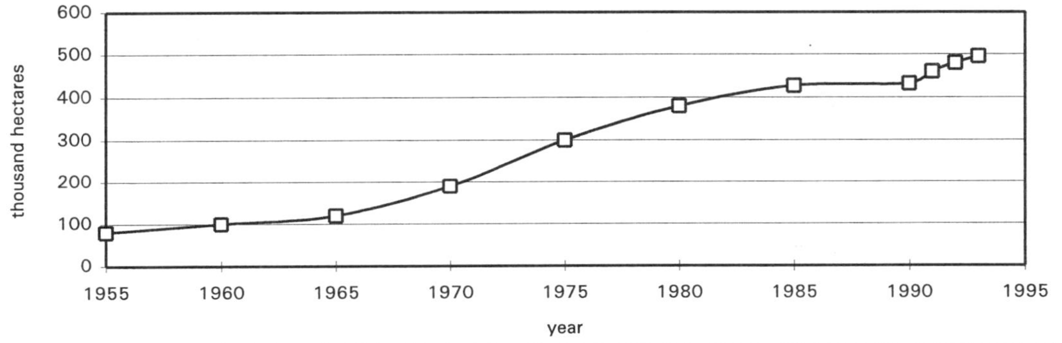

> 30 000 ha) are managed in a decentralized manner by the ORMVA. In total there are nine schemes over a total area of 431 650 ha in 1989 and 496 000 ha in 1993. Sprinkler irrigation is practised on 23.5% of the ORMVA schemes.

The agricultural investments code of 1969 (currently under revision) defines the contribution of the farmers to the costs of the equipment and to the operation and maintenance charges of the irrigation network. The annual operation costs of the large schemes vary from $US 6 to 12/ha, and the maintenance cost of the external equipment from $US 6 to 17/ha. Energy costs represent a large part of the total costs. Sprinkler irrigation has an extra cost of $US 144/ha due to the energy needed.

Medium and small schemes

In addition to the above areas covered by the large schemes, there are about 826 550 ha of medium and small schemes (referred to as 'petite et moyenne hydraulique'), including the spate irrigation area. Out of this total area, 416 550 ha are located within the zones managed by the ORMVA (Figure 8). About 397 000 ha are irrigated by perennial water, 265 000 ha by seasonal water and 165 000 ha by spate water (Figure 9). The schemes in the mountains mostly use springwater, distributed by gravity.

In all schemes, the water distribution is based on rotating turns. The measurement unit most frequently used in surface irrigation is the 'farm stream', which corresponds to a discharge of 30 litres/sec. This unit also forms the basis for the dimensions of the canals in the large schemes.

FIGURE 8
Distribution of the water managed areas
(total: 1 258 200 ha in 1989)

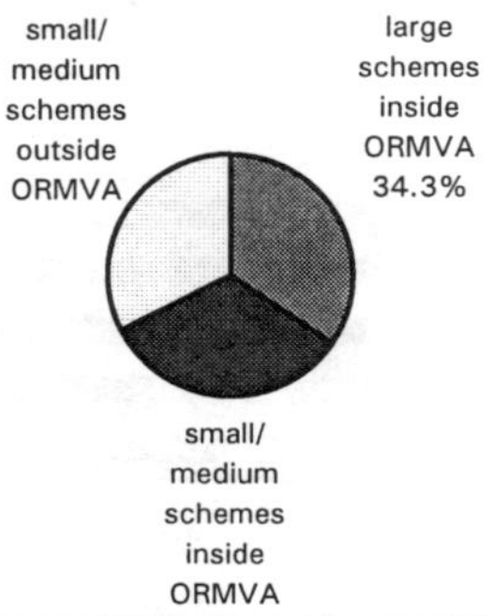

FIGURE 9
Distribution of the small and medium schemes according to water availability
(total: 826 550 ha in 1989)

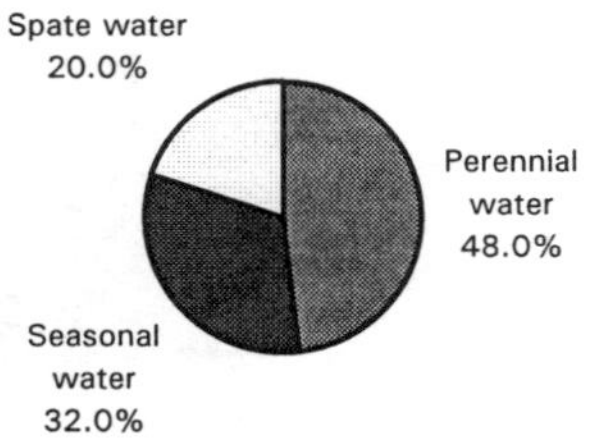

INSTITUTIONAL ENVIRONMENT

The Superior Council of Water and Climate (Conseil supérieur de l'eau et du climat) is the principal institution involved in the water resources management subsector. It has the mandate to coordinate the development of the water resources by examining the development policies of the sector, approving the regional master plans related to the development of the water resources (prepared by the Directorate of Rural Equipment), resolving conflicts over the allocation of the water resources and establishing policies for water quality conservation. The General Directorate for Hydraulics (DGH or Direction générale de l'hydraulique) is in charge of the secretariat of the Council and brings together the main services concerned in this sector, elected representatives, socio-professional organizations, local authorities and representatives of the different types of water users.

The main organizations involved in the drinking water supply subsector are:

- The DGH, which is part of the Ministry of Public Works. It is in charge of water supply at basin level and is responsible for research and the exploitation of the water resources;

- The National Office for Drinking Water (ONEP or Office national de l'eau potable), which is placed under the Ministry of Public Works. It is in charge of water distribution control in urban areas and in some rural municipalities. It plans, builds and operates the installations for treatment and transport from the primary sources, i.e. reservoirs and primary canals;

- 16 autonomous, inter-communal state-owned water companies, which are placed under the Ministry of Interior and supervised by the Directorate for state-owned companies and services conceded by this Ministry. They are in charge of water distibution in the municipalities;

- The Ministry of Public Health (MSP or Minstère de la santé publique) which, together with ONEP, is in charge of quality control for water resources for drinking water supply networks in the towns and villages to which it provides water.

The main organizations involved in the irrigated agriculture sector are:

- The Ministry of Agriculture and Agricultural Development (MAMVA or Ministère de l'agriculture et de la mise en valeur agricole), which is in charge of the supervision of new investments, in particular the extension, rehabilitation and maintenance of all the large and medium schemes;

- The DGH, which is in charge of providing irrigation water for the large schemes. It constructs and maintains the large hydraulic structures like dams, river diversion structures and projects for the exploitation of groundwater;

- The Agricultural Engineering Service (AGR or Administration du génie rural), which is responsible for the management of the irrigation schemes:

    ◊ The ORMVA, which are public but financially autonomous entities placed under the MAMVA, and which are responsible for the planning and management of the water resources for agriculture and the design, construction and management of the large schemes. They are also responsible for the small and medium schemes within their geographical jurisdiction;
    ◊ Outside the areas controlled by the ORMVA, the provincial Directorates for Agriculture are in charge of the promotion and management of the small and medium irrigation schemes, in reality mainly limited to extension activities.

On the 20 September 1995, a new Water Law became effective.

TRENDS IN WATER RESOURCES MANAGEMENT

The surface water resources are limited and must be saved in order to be able to satisfy the water needs for drinking, industrial and agricultural purposes in the 21st century. While globally the water demand is satisfied, certain regions already suffer from water scarcity, especially during dry years.

The multiplication of the number of dams is one way chosen to increase water availability. Siltation of dams is a problem, at present estimated at 50 million m^3 per year. The capacity already lost in 1990 was estimated at 800 million m^3, which is 7% of total capacity. A programme for the protection of dams against siltation has been set up. Another way chosen to increase water availability is increasing groundwater extraction. However the cost of groundwater extraction is very high and a number of aquifers are already over-exploited. Government policies are moving towards demand management.

A National Irrigation Programme for the year 2000 (PNI 2000 or Programme national d'irrigation en 2000) was adopted in 1992, with the following objectives:

- to equip by the year 2000 the whole area controlled by existing dams and dams under construction (about 250 000 ha);

- to improve the performances of the old irrigation schemes through modernization and/or rehabilitation of equipment (about 200 000 ha);

Land ownership, which is characterized by very small properties (< 5 ha), land fragmentation, the absence of land ownership deeds and security, is being reviewed together with its implications for schemes and resources management.

MAIN SOURCES OF INFORMATION

AIGR [Amicale des ingénieurs du génie rural]. 1991. Annuaire 1990 du Génie rural.

BRGM/SOGREAH/SCET/CINAM. 1991. Etude du schéma directeur national d'approvisionnement en eau potable des populations rurales. Mission d'évaluation globale de la situation actuelle d'AEP du milieu rural, PNUD.

Conseil supérieur de l'eau. 1991 Aménagement des bassins versants et protection des barrages contre l'envasement.

Conseil supérieur de l'eau. 1993. Aménagement hydro-agricole: situation actuelle et perspectives.

Direction des statistiques. 1994. Annuaire statistique du Maroc 1993.

FAO-CP. 1996. Maroc: Programme d'approvisionnement groupé en eau potable. Rapport de préparation, 2 volumes. Rome. Report N° 96/062-CP-MOR.

Jellali, Mohammed. 1996. Développement des ressources en eau au Maroc: situation actuelle et perspectives. Direction générale de l'hydraulique, Ministère des travaux publics.

Lahlou, O. 1988. L'aménagement hydro-agricole, cas du périmètre irrigué marocain.

Nadifi, K. and Wahabi, R. 1993. Water resources in Morroco: the state of art and future prospects. Communication presented at the 'Symposium régional sur l'utilisation et la conservation des ressources en eau'.

SCET-AGRI&GERSAR. 1987. The Moroccan irrigation potential.

SCET-AGRI&GERSAR/World Bank/UNDP/French Government. 1984. Study on options and investment priorities in irrigation development. Moroccan country case study, 2 volumes.

UNDP. 1985. Options and investment priorities in irrigation development. Report No. MOR/SR2.

World Bank. 1993. Morocco irrigated areas agricultural services project. Report No. 11947-MOR.

World Bank. 1995. Le secteur de l'eau au Maroc. Report No. 12649-MOR.

Oman

GEOGRAPHY AND POPULATION

The Sultanate of Oman occupies the south-eastern corner of the Arabian Peninsula and has a total area of 312 500 km^2. It is bordered in the north-west by the United Arab Emirates, in the west by Saudi Arabia and in the south-west by Yemen. A detached area of Oman, separated from the rest of the country by the United Arab Emirates, lies at the tip of the Musandam Peninsula, on the southern shore of the Strait of Hormuz. The country has a coastalline of almost 1 700 km, from the Strait of Hormuz in the north to the borders of the Republic of Yemen in the south-west, overlooking three seas: the Persian Gulf, the Gulf of Oman and the Arabian Sea.

Oman can be divided into the following physiographic regions:

- The whole coastal plain. The most important parts are the Batinah Plain in the north, which is the principal agricultural area, and the Salalah Plain in the south. The elevation ranges between zero near the sea to 500 metres further inland.
- The mountain ranges, which occupy 15% of the total area of the country. The mountain range that runs in the north close to the Batinah Plain is the Jebel Al Akhdar with a peak at 3 000 metres. Other mountains are located in the Dhofar province, in the extreme southern part of the country, with peaks from 1 000 to 2 000 metres.
- The internal regions. Between the coastal plain and the mountains in the north and south lie the internal regions, consisting of several plains with elevations not exceeding 500 metres.

The cultivable area has been estimated at 2.2 million ha, which is 7% of the total area of the country. The cultivated area was 61 550 ha in 1993, of which 18 550 ha consisted of annual crops and 43 000 ha consisted of permanent crops. Over half the agricultural area is located in the Batinah Plain in the north which has a total area representing about 3% of the area of the country.

TABLE 1

Basic statistics and population

Physical areas:			
Area of the country	1995	31 250 000	ha
Cultivable area	1990	2 200 000	ha
Cultivated area	1993	61 550	ha
- annual crops	1993	18 550	ha
- permanent crops	1993	43 000	ha
Population:			
Total population	1995	2 163 000	inhabitants
Population density	1995	7	inhab./km^2
Rural population	1995	87	%
Water supply coverage:			
Urban population	1990	87	%
Rural population	1990	42	%

The total population is about 2.16 million (1995), of which 87% is rural according to United Nations estimates. According to the national population census of 1993, 28% of the total population was rural. The difference between the two figures is explained by the fact that the UN standards for Oman consider as rural all the inhabitants of the country, except those of the two cities: Muscat and Matrah. The annual demographic growth rate is estimated at 3.7%. While agriculture and fisheries employed about 37% of the total labour force in 1993, they accounted for only 3.3% of GDP.

TABLE 2
Water: sources and use

Renewable water resources:			
Average precipitation		54	mm/yr
		17.0	km3/yr
Internal renewable water resources		0.985	km3/yr
Total (actual) renewable water resources	1995	0.985	km3/yr
Dependency ratio	1995	0	%
Total (actual) renewable water resources per inhabitant	1995	455	m3/yr
Total dam capacity	1996	58	106 m3
Water withdrawal:			
- agricultural	1991	1 148	106 m3/yr
- domestic	1991	56	106 m3/yr
- industrial	1991	19	106 m3/yr
Total water withdrawal		1 223	106 m3/yr
per inhabitant	1991	728	m3/yr
as % of total (actual) renewable water resources		124	%
Other water withdrawal		-	106 m3/yr
Average groundwater depletion		-	106 m3/yr
Wastewater - Non-conventional water sources:			
Wastewater:			
- produced wastewater	1991	58	106 m3/yr
- treated wastewater	1991	28	106 m3/yr
- reused treated wastewater	1991	26	106 m3/yr
Desalinated water	1995	34	106 m3/yr

CLIMATE AND WATER RESOURCES

Climate

The climate differs from one region to another. It is hot and humid during summer in the coastal areas and hot and dry in the interior regions with the exception of some higher lands and the southern Dhofar region, where the climate remains moderate throughout the year. In the north and centre of Oman rainfall occurs during the winter (November-April), while in the south and some internal parts of the country it is a result of seasonal summer storms (June-September). Average annual rainfall has been estimated at 55 mm, varying from less than 20 mm in the internal desert regions to over 300 mm in the mountain areas.

Water resources

A great deal of uncertainty lies in the assessment of Oman's water resources. Internal renewable water resources have been evaluated at 985 million m^3/year. Surface water resources are scarce. In nearly all wadis, surface runoff occurs only for some hours or up to a few days after a storm, in the form of rapidly rising and falling flood flows. Since 1985, 15 major recharge dams have been constructed together with many smaller structures, in order to retain a portion of the peak flows, thus allowing more opportunity for groundwater recharge. In addition, several flood control dams produce significant recharge benefits. In 1996, the total dam capacity is 58 million m^3. Groundwater recharge is estimated at 955 million m^3/year.

Non-conventional water sources

In 1995, the total produced wastewater was estimated at 58 million m^3. Only 28 million m^3 was treated, of which 26 million m^3 was reused. Also in 1995, the quantity of desalinated water was 34 million m^3.

TABLE 3
Irrigation and drainage

Irrigation potential		-	ha
Irrigation:			
1. Full or partial control irrigation: equipped area	1993	61 550	ha
- surface irrigation	1993	57 820	ha
- sprinkler irrigation	1993	1 640	ha
- micro-irrigation	1993	2 090	ha
% of area irrigated from groundwater	1993	100	%
% of area irrigated from surface water	1993	0	%
% of area irrigated from non-conventional sources	1993	0	%
% of equipped area actually irrigated	1993	100	%
2. Spate irrigation area		-	ha
3. Equipped wetland and inland valley bottoms (i.v.b.)		-	ha
Total irrigation (1+2+3)	1993	61 550	ha
- as % of cultivated area		100	%
4. Flood recession cropping area		-	ha
Total water managed area (1+2+3+4)	1993	61 550	ha
- as % of cultivated area		100	%
- increase over last 10 years		-	%
- power irrigated area as % of water managed area	1993	100	%
Full or partial control irrigation schemes: Criteria			
Large-scale schemes > - ha		-	ha
Medium-scale schemes		-	ha
Small-scale schemes < - ha		-	ha
Total number of households in irrigation	1993	62 411	
Irrigated crops:			
Total irrigated grain production	1994	18 900	tons
as % of total grain production	1994	100	%
Harvested crops under irrigation (full or partial control)	1993	70 930	ha
- permanent crops: total	1993	43 000	ha
- annual crops: total	1993	27 930	ha
. fodder crops (mainly alfalfa)	1993	17 330	ha
. vegetables	1993	5 700	ha
. cereals	1993	4 900	ha
Drainage - Environment:			
Drained area		-	ha
as % of cultivated area		-	%
- drained areas in full or partial control irrigated areas		-	ha
- drained areas in equipped wetland and i.v.b		-	ha
- other drained areas		-	ha
- total drained area with subsurface drains		-	ha
- total drained area with surface drains		-	ha
Flood-protected area		-	ha
Area salinized by irrigation		-	ha
Population affected by water-borne diseases		-	inhabitants

Water withdrawal

In 1995, total water withdrawal was 1 223 million m^3, of which almost 94% for agricultural purposes (Figure 1). The treated wastewater was reused mainly for the irrigation of trees along the roads, while the desalinated water was used for domestic purposes. At present, groundwater depletion is thus estimated at around 240 million m^3/year.

FIGURE 1
Water withdrawal (total: 1.2 km^3 in 1991)

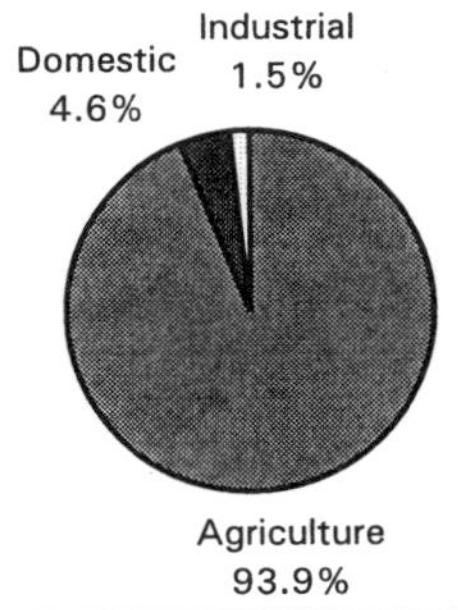

IRRIGATION AND DRAINAGE DEVELOPMENT

All agriculture in Oman is irrigated and since the 1970s the equipped area increased from about 28 000 ha to 61 550 ha in 1993 (Figure 2), of which 34 930 ha, or almost 57%, is located in the Al Batinah province in the north. Although 2.2 million ha are considered to be suitable for agriculture, there are no figure on the irrigation potential, as no reliable data are available on groundwater availability in the deep aquifers. At present groundwater depletion already takes place, especially in coastal areas, leading to sea water intrusion and a deterioration in the water quality.

FIGURE 2
Evolution of the irrigated area

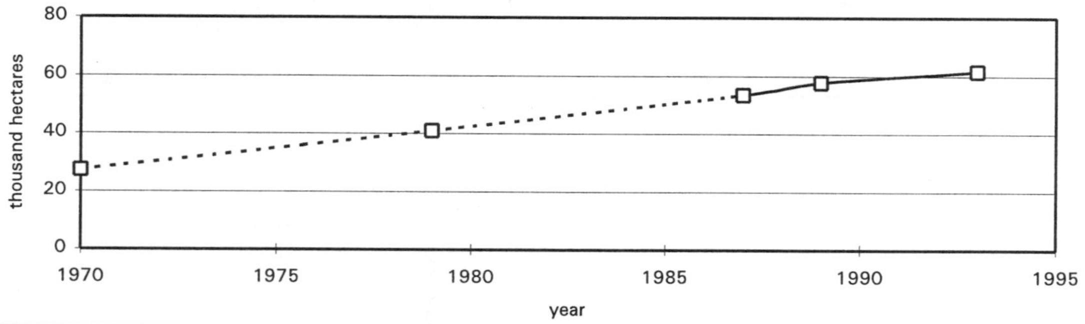

The falaj system ('aflaj' in the plural) is the traditional method developed centuries ago for supplying water for irrigation and domestic purposes. Many of the systems currently in use are estimated to be over a thousand years old. The falaj comprises the entire system: the source, which might be a qanat, a spring or the upper reaches of flowing wadis from which water is diverted; the conveyance system, which is usually an open earth or cement-lined ditch; and the delivery system. The falaj has assumed social significance, and well-established rules of usage, maintenance and administration have evolved.

FIGURE 3
Sources of water of the households involved in irrigation (total: 62 411 households in 1993)

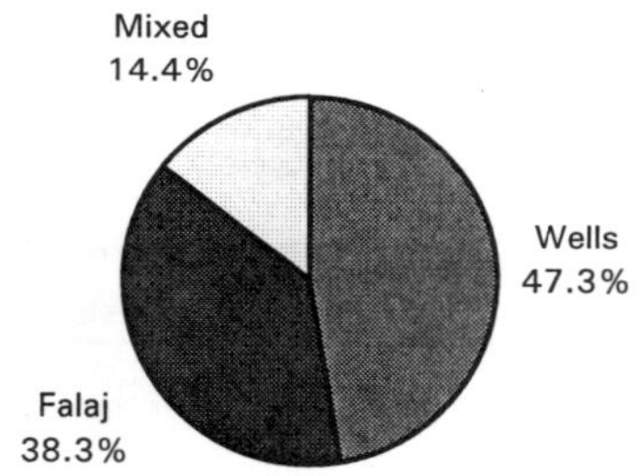

Originally, the falaj developed where higher elevation water sources such as springs, qanats or surface water could be intercepted by diversion or small catchment dams and then conveyed by gravity to the point of use. More recently, however, dug wells have been used to supplement the falaj water. This is especially the case in the coastal areas where many hand-dug wells and tubewells have been constructed. For 47% of the total number of 62 411 households involved in irrigation, wells are now the main source of water, 39% rely on falaj water, while the remaining 14% have access to both sources (Figure 3).

FIGURE 4
Irrigation techniques f/p (total: 61 550 ha in 1993)

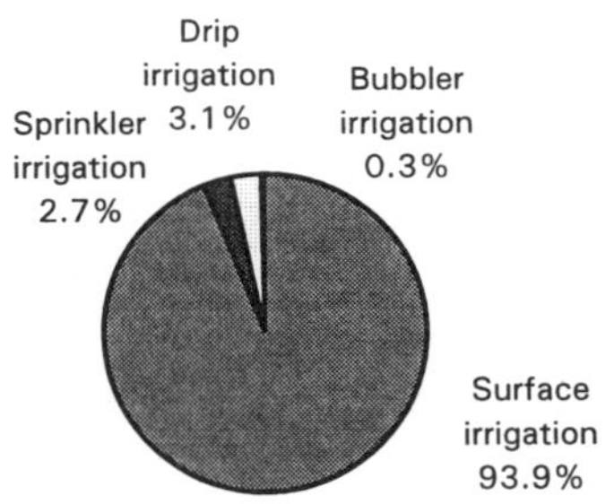

Of the total area of 61 550 ha equipped for irrigation, all of which is power irrigated using groundwater (wells, falaj), only 1 640 ha, or 2.7%, benefit from sprinkler irrigation and 2 090 ha, or 3.4%, from micro-irrigation techniques (Figure 4). Although the Ministry of Agriculture and Fisheries (MAF) is making efforts to introduce modern irrigation techniques, the traditional flood system remains the most common irrigation technique. In order to encourage farmers to take up the new techniques, MAF has approved a financial subsidy varying between 75% for small-scale schemes (less than 10 feddans or 4.2 ha), 50% for medium-scale schemes and 25% for large-scale schemes (more than 50 feddans or 21 ha). Most of the area consists of small schemes.

In 1996, the cost of irrigation development was estimated at $US 3 250/ha for medium and large schemes and $US 4 415/ha for small schemes. These costs represent the average cost of installing sprinkler irrigation and micro-irrigation systems. The average annual operation and maintenance costs are $US 845, 1 170 and 1 820/ha for large, medium and small schemes respectively.

Date palm is the main crop grown in Oman, occupying about half the total cropped area. Other crops are fodder crops (mainly alfalfa), other fruit trees (citrus, bananas, mangoes, coconuts) vegetables and cereals (mainly barley, wheat and sorghum) (Figure 5).

FIGURE 5
Irrigated crops f/p (total: 70 930 ha in 1993)

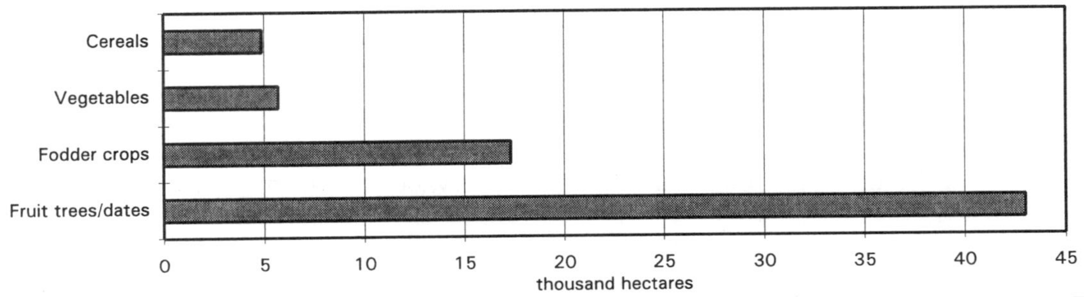

No reliable information on the area salinized by irrigation is available. A study done in 1994 on the salinity of soils in general in Oman, states that an area of 11.7 million ha, which is 35% of the total area of Oman, is affected by salinity. No drainage is practised.

INSTITUTIONAL ENVIRONMENT

The Ministry of Water Resources (MWR) is in charge of water resources assessment, whereas the Ministry of Agriculture and Fisheries (MAF) is in charge of irrigation.

In 1988, Royal Decree No. 83/88 declared the water resources of Oman a national resource. This is the most far-reaching and important piece of legislation on water resources. Oman has several laws on water resources and the main measures taken for water management and conservation are:

- no wells may be constructed within 3.5 km of the mother well of the falaj;
- permits are required for the construction of new wells, for deepening existing wells, for changes in use and for installing a pump;
- all drilling and well digging contractors are required to register with MWR on a yearly basis;
- MWR has the cooperation of other government agencies such as the Ministry of Interior and the Royal Oman Police in dealing with offenders.

TRENDS IN WATER RESOURCES ASSESSMENT

Three broadly-based programmes have been set up by the government for:

- the improvement of data collection;
- a detailed assessment of the water resources;
- a study of water demand and its spatial distribution.

In addition to the above measures taken for water management and conservation, the government has recently initiated programmes to relocate some of the large-scale farms in the Batinah and Salalah Plains, where the water resources are over-utilized, to areas with under-utilized water resources. Several water conservation initiatives have been developed, like leakage control in municipal water supply schemes and the improvement of irrigation methods through subsidy programmes. Public awareness of water resources issues has created a general and focused understanding of the overall situation and of the specific contribution each citizen can make.

The main issues and strategies that the government will address in the coming years are:

- creating and cultivating conservation awareness;
- matching water use to water availability;
- establishing an integrated programme for the conservation and management of the resources at basin level;
- controlling saline intrusion by reducing abstraction below the long-term recharge;
- adopting improved irrigation techniques and selecting appropriate crops to reduce agricultural water use;
- controlling urban water losses;
- increasing the use of treated wastewater and desalinated water;
- protecting the groundwater resources in qualitative as well as quantitative terms;
- constructing new groundwater recharge dams.

MAIN SOURCES OF INFORMATION

Department of Agricultural Statistics. 1995. Agricultural Census 1992-93. Ministry of Agriculture and Fisheries.

Ministry of Agriculture and Fisheries. [?]. Development and optimization of the use of water resources in the Sultanate of Oman.

Ministry of Water Resources. 1991. National Water resources Master Plan. Oman.

World Bank. 1988. Sultanate of Oman: Recent economic developments and prospects. Report No 6899-OM. Washington DC.

Pakistan

GEOGRAPHY AND POPULATION

Pakistan, with a total area of 796 100 km^2, is located in Southern Asia. It is bordered by India in the east, China in the north-east, Afghanistan in the north and north-west, Iran in the south-west and the Arabian Sea to the south. Pakistan is divided into four provinces, namely the Punjab, Sindh, North West Frontier Province (NWFP) and Balochistan.

The country can be divided into five physiographic regions:

- The Himalayan mountain ranges in the north-western part on the border with India and China. The highest peak, the Godwin-Austin (7 610 m) is part of the Trans-Himalayan Range;
- The Hindu Kush and the Western Mountains in the north on the border with Afghanistan. The Tirichmir (7 690 m) is the highest peak in the Hindu Kush range;
- The Potwar Plateau, just south of Islamabad. The elevation varies from 300 to 600 m. South of the Potwar Plateau is the Salt Range;
- The Indus Plain, stretching from the Salt Range to the Arabian Sea. This flat plain is largely made up of alluvium, over 300 m deep, deposited by the Indus river and its tributaries;
- The Balochistan Plateau in the south-west of the country, with an average altitude of about 600 m. Dry hills run across the plateau from north-east to south-west. A large part of the north-west is desert.

The total cultivable area was estimated at 29.9 million hectares in 1990, which is 37% of the total area, mainly concentrated in the Indus plain. In 1990, the total cultivated area was estimated at almost 16.6 million hectares, or 55% of the cultivable area, of which 16.1 million ha consisted of annual crops and almost 0.5 million ha consisted of permanent crops.

TABLE 1
Basic statistics and population

Physical areas:			
Area of the country	1995	79 610 000	ha
Cultivable area	1990	29 900 000	ha
Cultivated area	1990	16 556 000	ha
- annual crops	1990	16 100 000	ha
- permanent crops	1990	456 000	ha
Population:			
Total population	1995	140 947 000	inhabitants
Population density	1995	177	inhab./km^2
Rural population	1995	65	%
Water supply coverage:			
Urban population	1990	82	%
Rural population	1990	42	%

The total population is 141 million (1995), of which 65% is rural. Average population density is 177 inhabitants per km^2, but the population is mainly concentrated in the Indus plain. Average annual population growth is estimated at about 3.2%. In 1992, agriculture accounted for 26% of GDP and provided employment to 48% of the labour force. It supplies most of the country's food, but is also the source of raw materials for major domestic industries, particularly for cotton products which account for 80% of export earnings.

TABLE 2
Water: sources and use

Renewable water resources:			
Average precipitation		494	mm/yr
		393.3	km^3/yr
Internal renewable water resources		248.0	km^3/yr
Total (actual) renewable water resources	1995	418.27	km^3/yr
Dependency ratio	1995	40.7	%
Total (actual) renewable water resources per inhabitant	1995	2 968	m^3/yr
Total dam capacity		-	$10^6\ m^3$
Water withdrawal:			
- agricultural	1991	150 600	$10^6\ m^3/yr$
- domestic	1991	2 500	$10^6\ m^3/yr$
- industrial	1991	2 500	$10^6\ m^3/yr$
Total water withdrawal		155 600	$10^6\ m^3/yr$
per inhabitant	1991	1 277	m^3/yr
as % of total (actual) renewable water resources		37.2	%
Other water withdrawal		-	$10^6\ m^3/yr$
Average groundwater depletion		-	$10^6\ m^3/yr$
Wastewater - Non-conventional water sources:			
Wastewater:			
- produced wastewater		-	$10^6\ m^3/yr$
- treated wastewater		-	$10^6\ m^3/yr$
- reused treated wastewater		-	$10^6\ m^3/yr$
Desalinated water		-	$10^6\ m^3/yr$

CLIMATE AND WATER RESOURCES

Climate

Pakistan lies in the subtropical arid zone and most of the country is subjected to a semi-arid climate. June is the hottest month in the plains and July in the mountainous areas, with temperatures over 38°C, while the mean monthly minimum is only 4°C in December/January. Average annual precipitation is estimated at 494 mm, but is uneven over much of the Indus basin. It reaches up to 1 500 mm in the north. Most of the rainfall in Pakistan originates from summer monsoons.

Water resources

Pakistan can be divided into three hydrological units:

- the Indus basin, covering more than 566 000 km^2 (or 71% of the territory), comprising the whole of the provinces of the Punjab, Sindh and NWFP and the eastern part of Balochistan. The Indus river has 2 main tributaries, the Kabul on the right bank and the Panjnad on the left bank. The Panjnad is the resulting flow of five main rivers (literally Punjab means 'five waters'): the Jhelum and Chenab, known as the western rivers, and the Ravi, Beas, and Sutlej known as the eastern rivers.
- the Karan desert in the west of Balochistan (in the west of the country), which is an endorheic basin covering 15% of the territory. The Mashkel and Marjen rivers are the principal source of water in the basin. The water is discharged in the Hamun-i-Mashkel lake, in the south-west at the border with Iran.
- the arid Makran coast along the Arabian Sea covering 14% of the territory in its south-western part (Balochistan province). The Hob, Porali, Hingol and Dasht are the principal rivers of this coastal zone.

TABLE 3
Irrigation and drainage

Irrigation potential		-	ha
Irrigation:			
1. Full or partial control irrigation: equipped area	1990	14 327 000	ha
- surface irrigation		-	ha
- sprinkler irrigation		-	ha
- micro-irrigation		-	ha
% of area irrigated from groundwater	1989	34.0	%
% of area irrigated from surface water	1989	66.0	%
% of area irrigated from non-conventional sources	1989	0.0	%
% of equipped area actually irrigated		-	%
2. Spate irrigation area	1990	1 402 448	ha
3. Equipped wetland and inland valley bottoms (i.v.b.)		-	ha
Total irrigation (1+2+3)	**1990**	**15 729 448**	**ha**
- as % of cultivated area		95	%
4. Flood recession cropping area	1990	1 230 552	ha
Total water managed area (1+2+3+4)	**1990**	**16 960 000**	**ha**
- as % of cultivated area		102	%
- increase over last 10 years	1980-90	14	%
- power irrigated area as % of water managed area		-	%
Full or partial control irrigation schemes:	Criteria		
Large-scale schemes	> - ha	-	ha
Medium-scale schemes		-	ha
Small-scale schemes	< - ha	-	ha
Total number of households in irrigation		-	
Irrigated crops:			
Total irrigated grain production		-	tons
as % of total grain production		-	%
Harvested crops under irrigation (full or partial control)		-	**ha**
- permanent crops: total		-	ha
- annual crops: total		-	ha
. wheat	1992	6 497 400	ha
. cotton		-	ha
.		-	ha
.		-	ha
. other annual crops		-	ha
Drainage - Environment:			
Drained area	1992	5 100 165	ha
as % of cultivated area		31	%
- drained areas in full or partial control irrigated areas	1992	5 100 165	ha
- drained areas in equipped wetland and i.v.b		-	ha
- other drained areas		-	ha
- total drained area with subsurface drains		-	ha
- total drained area with surface drains		-	ha
Flood-protected area		-	ha
Area salinized by irrigation		-	ha
Population affected by water-borne diseases		-	inhabitants

The internal renewable water resources are estimated at 248 km^3/year. Surface runoff is estimated at 243 km^3, while groundwater resources are about 55 km^3, most being the baseflow of the river system.

The Indus basin has a total drainage area of 1.06 million km^2, of which 56% lies in Pakistan, and the other 44% in China, Afghanistan and India. Because of the importance of irrigation in the Indus plain, the water balance of the Indus basin has been carefully studied, which is not the case for the other basins. Therefore most of the results found refer only to the Indus basin. The mean annual inflow into the country through the western rivers (the Indus, including the Kabul tributary, the Jhelum and the Chenab) amounted to 170.27 km^3 in 1995. The mean annual natural inflow into the country through the eastern rivers (the Ravi, the Beas and the Sutlej) is estimated at 11.1 km^3, but this is reserved for India, according to the 1960 Indus Water Treaty.

Given the seasonal nature of the Himalayan runoff, roughly 85% of annual flows are in the *Kharif* season (summer), and only 15% in the *Rabi* season (winter).

Dams and hydropower

In 1986 there were 40 dams with a height of over 15 metres in Pakistan. In 1992, 12 other dams were under construction. The two major dams are Tarbela (13.7 km^3 of maximum reservoir capacity and 12.0 km^3 of active reservoir capacity) and Mangla (7.2 km^3 and 5.9 km^3 respectively), both being also equipped for electricity generation. In 1994, hydropower represented 49.9% of total installed capacity, and 47.5% of total energy generation, Tarbela dam alone represents 36.7% of the total. The gross theoretical hydropower potential was estimated in 1991 at 150 000 GWh/year.

FIGURE 1
Water withdrawal (total: 155.6 km^3 in 1991)

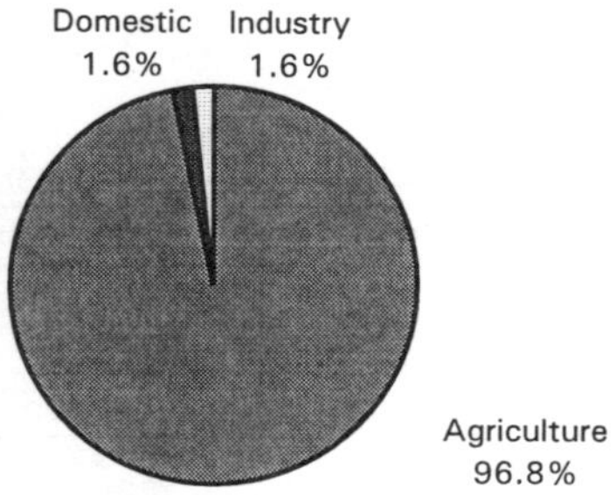

Water withdrawal

Total water withdrawal in 1991 was estimated at 156 km^3, of which 97% for agricultural purposes. (Figure 1). Groundwater abstraction for agriculture has been roughly estimated at 55 km^3/year, which is approximately the volume of groundwater renewable resources. However, in some areas, development appears to have reached the point where groundwater is being mined. Most urban and rural water is supplied from groundwater. Over 50% of the village water supply is obtained through hand pumps installed by private households. In saline groundwater areas, irrigation canals are the main source of domestic water.

FIGURE 2
Distribution of the water managed area (total: 16.96 million ha in 1990)

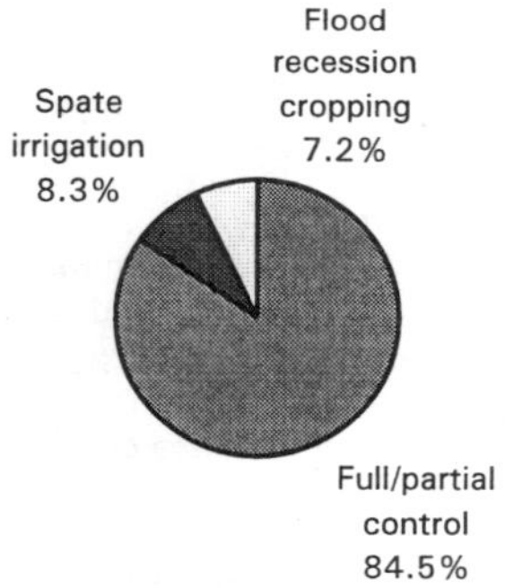

IRRIGATION AND DRAINAGE DEVELOPMENT

The water managed areas in Pakistan are estimated at 16 960 000 ha in 1990, and can be divided according to the following classification (Figure 2):

Full or partial control irrigation schemes, covering a total area of 14 327 000 ha comprising:

- the Indus Basin Irrigation System (IBIS), which is by far the largest irrigation system in Pakistan, covering areas in all provinces. In 1993, irrigated areas in the IBIS were estimated at 13 972 500 ha (Figure 3);
- in NWFP, about 7 800 ha are irrigated through pump lifts. They are maintained by the Provincial Irrigation Department (PID);

- in northern parts of NWFP, irrigation is practised by means of contour channels off-taking from the locally available water sources, often steep side streams or springs. Most of these schemes which cover about 26 700 ha are owned and operated directly by the beneficiaries through traditional social organizations;
- in Balochistan, about 50 000 ha are irrigated from about 800 karezes (tunnels or underground channel tapping an aquifer) and perennial springs. They are generally small group-operated schemes, with a size ranging between 50 and 400 ha;
- in Balochistan, about 130 000 ha of small group-operated schemes are irrigated from infiltration galleries or small weirs in rivers;
- in Balochistan, about 130 000 ha are irrigated by individuals pumping from tubewells and 10 000 ha from open wells.

FIGURE 3
Irrigated areas in the Indus Basin Irrigation System, by province (total: 13.97 million ha in 1993)

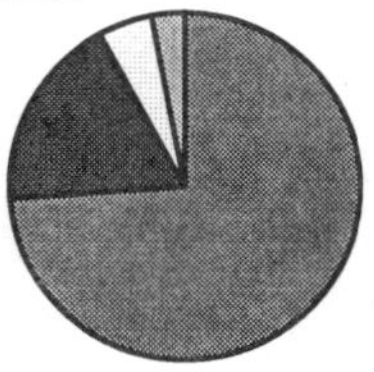

FIGURE 4
Flood recession cropping areas, by province (total: 1.23 million ha in 1990)

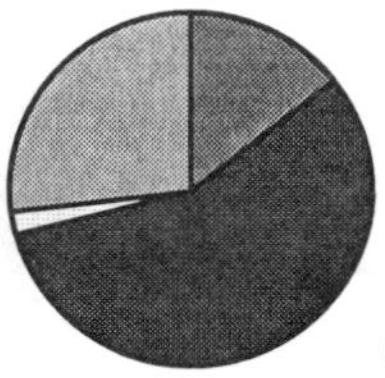

Spate irrigation, covering a total area of 1 402 448 ha. In Pakistan, these areas are known as *Rod Kohi* in NWFP and Punjab, or *Bandat* in Balochistan, and are often called flood irrigation. This kind of irrigation relies on the floods of the hill torrents. Wherever possible, the runoff is harnessed for irrigation by weirs or temporary diversion structures. Farmers divert the spate flow onto their fields by constructing breachable earth bunds (called *gandas*) across the rivers, or by constructing stone/gravel spurs leading towards the centre of the river. Captured water flows from field to field and when the soil profile is saturated, the lower bund is breached to release water into another field. Annual average cropping intensity is 20%.

Flood recession cropping areas, covering a total area of 1 230 552 ha (Figure 4). In Pakistan these areas are known as *Sailaba*, and are often called falling flood irrigation areas. Sailaba cultivation is carried out on extensive tracts of land along the rivers and hill streams subject to annual inundation. It utilizes the moisture retained in the root zone after the flood subsides together with sub-irrigation due to the capillary rise of groundwater and any rain.

Apart form these water managed areas, some attempts have been made to develop water harvesting, which is known in Pakistan as *Khushkaba*.

According to the usual Pakistani classification (Figure 5) irrigation consists of:

- government canals: 11 310 000 ha in 1990, of which 74% in the Punjab and 20% in the Sindh province;
- private canals: 430 000 ha, of which 86% are in NWFP;
- tubewells: 4 260 000 ha, of which 92% in Punjab province;

FIGURE 5
Origin of irrigation water in the water managed areas
(total: 16.96 million ha in 1990)

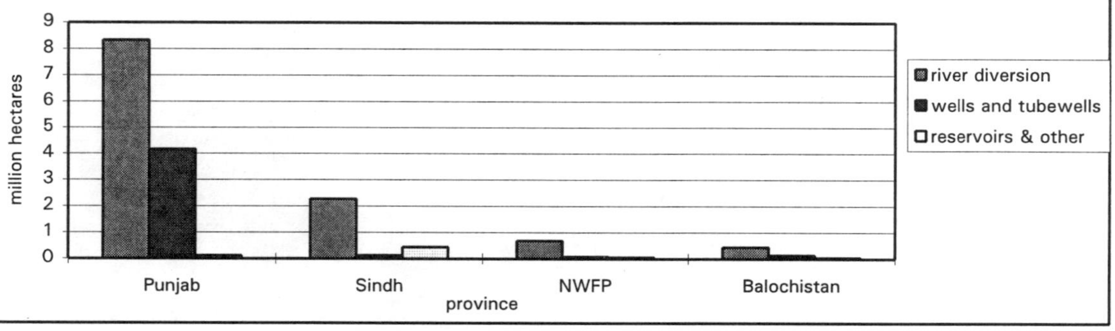

FIGURE 6
Tubewells for irrigation (total: 305 231 in 1989)

- open wells: 280 000 ha, of which 82% in Punjab province;
- tanks: 60 000 ha, all of them in Punjab;
- other means: 620 000 ha, of which 73% in Sindh province.

In 1989, 305 231 tubewells were used for irrigation in Pakistan, most of them being privately owned and using diesel (Figure 6).

The total water managed area of 16.96 million ha is higher than the total cultivated area (16.56 million ha). This is due to the fact that not all the water managed areas are actually cultivated. This is especially the case for spate irrigation and flood recession cropping areas. The main irrigated crops are wheat, rice and cotton. Due to inadequate water availability in winter (storage capacity is too small) and at the beginning and end of summer, cropping intensity is exceptionally low.

The Indus Basin Irrigation System

The 4000 year old Indus civilization has its roots in irrigated agriculture. The abandoned creeks of the meandering river, inundated during the flood season, are supposed to have served as the irrigation channels of the ancient systems. The first controlled all-year irrigation began in 1859 with the completion of the Upper Bari Doab Canal emanating from the Madhopur headworks on the Ravi river. The Sukkur barrage, completed in 1932, is considered as the first modern hydraulic structure on the downstream Indus river.

Although irrigation takes place in other areas of Pakistan, information on the history and development of irrigation generally refers to the Indus Basin Irrigation System (IBIS). Development of irrigation in the Indus basin has progressed in the form of discrete barrage-controlled systems. With water rather than land being the main constraint, the irrigation systems were generally designed to use the available river supplies for bringing the largest possible areas under crops, with minimum water provided to bring the crops to maturity.

The IBIS is characterized by its supply-based structure. The was designed to distribute water with minimum human interference. There are few structures to regulate canal flow. No escapes are provided at the tail end of the system and the surplus flows have to be absorbed within the systems. Drain construction has not kept pace with requirements, while infrastructure development has often obstructed natural drainage flows.

With almost 14 million ha, the IBIS is the largest contiguous irrigation system in the world. It consists of an extensive network of barrages, canals and watercourses. The total length of the canals is about 61 000 km with communal watercourses, farm channels, and field ditches covering another 1.6 million km. In the Indus system, river water is diverted by barrages and weirs into main canals and subsequently branch canals, distributaries and minors. The flow to the farm is delivered by the watercourses (there are over 107 000 watercourses) which are supplied through outlets (*moghas*) from the distributaries and minors. The mogha is designed to allow a discharge that self-adjusts to variations in the parent canal. Within the watercourse command (an area ranging from 80 to 280 ha), farmers receive water proportional to their land holding. The entire discharge of the watercourse is given to one farm for a specified period on a seven day rotation. The rotation schedule, called *warabandi*, is established by the Provincial Irrigation Department, unless the farmers can reach a mutual agreement.

Operation and maintenance

The public sector operates the irrigation systems above the moghas. Each season, the Water and Power Development Authority (WAPDA) of the Federal Government estimates water availability for the following season. Provincial Irrigation Departments (PID) inform WAPDA of provincial water demands at specific locations. WAPDA releases water from the reservoirs to meet demands as closely as possible. The limited reservoir capacity of the systems does not allow the full regulation of rivers for irrigation.

The operation and maintenance expenditure is collected by levying water charges and/or drainage taxes. In Punjab and NWFP, water charges are assessed by PID. In Sindh and Balochistan, they are assessed by the Provincial Revenue Department (PRD). Currently, water and drainage charges are not linked to operation and maintenance needs. They are, moreover, collected in all regions by PRD, and are deemed to be part of provincial revenues. The gap between operation and maintenance expenditures and recoveries through water charges is high (44%) and increasing. The difficulties faced in cost recovery have resulted in very poor operation and maintenance which, together with deliveries at less than the designed levels and illegal diversion, has led to major inequalities in the distribution of surface water. In reality, water often does not reach users toward the tail end of system, which can partly explain the increasing groundwater extraction.

In March 1991, an agreement was reached between the provinces on the apportionment of the Indus waters to replace a much older agreement. The new agreement has released the provincial canal systems from the need to be in operation all the time so as to protect or establish future rights. Now that the supplies have been apportioned, including the formula for sharing any surplus river flows, the provincial systems are free to move toward more efficient water use.

The first Water Users Associations (WUA) were created in 1981 under the World Bank-supported On-Farm Water Management Programme. They were formed at the watercourse level, with a primary objective of rehabilitating the watercourses. By 1991, there were some 17 000 WUA, representing about 16% of all watercourses, with a membership of 85 000 farmers.

Irrigation water withdrawal and water losses

Over the past 20 years, groundwater use has been a major factor in increasing agricultural production. Groundwater tubewells not only supply additional water but have provided flexibility to match surface water supplies with crop water requirements. However, because of uncontrolled and rapid private sector development of groundwater (6% annual growth), there is a danger of excessive lowering of water tables and intrusion of saline water into freshwater aquifers. Within the IBIS, total water availability at the farm gate has significantly increased in the last 15 years (Figure 7), and changed slightly in its composition, with a higher use of groundwater extracted by tubewells. In 1975, surface water represented 70% of the total water available, groundwater provided through private tubewells 22.5% and groundwater provided through public tubewells 7.5%. In 1990, the figures were 63%, 27% and 10% respectively.

FIGURE 7
Overall water availability at the farm gate

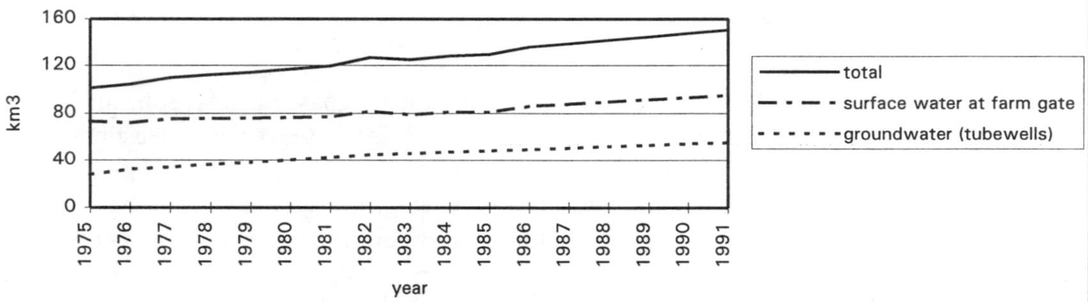

The Water Resources Section of the Planning and Development Division has estimated average water losses from canal head to outlet at 25%, and from outlet to farm gate at 15%.

Waterlogging, drainage, salinity and flood protection

The increasing diversion of river flows has significantly changed the hydrological balance of the irrigated areas in the past century. Initially, irrigation systems were developed without any provision for drainage. Seepage from irrigation canals and watercourses, and the deep percolation of this water have gradually raised the groundwater table, causing waterlogging and salinity.

It is estimated that about 2.39 million ha had water tables within 1.5 metres of the surface level in June 1989 (which resulted in 4.92 million ha in October 1989, just after the monsoon season), such areas being considered as 'disaster areas' by the government and given high priority for drainage (Figure 8). Since the 1960s, great efforts have been made to provide drainage in the irrigated areas. In 1992, the total drained area was estimated at 5.10 million ha (Figure 9).

FIGURE 8
Extent of waterlogging by province (total: 2.39 million ha in June 1989) (water table depth < 1.5 metres)

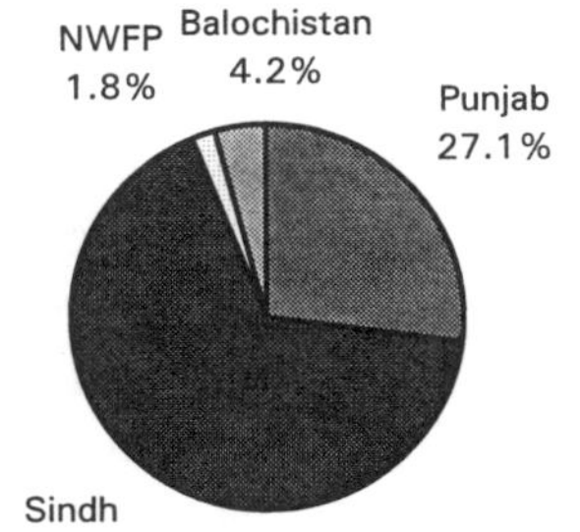

According to the Soil Survey of Pakistan (1985-1990), 1.78 million hectares are considered as severely saline, and 0.18 million hectares as very severely saline (Figure 10), but the survey does not indicate which part is due to irrigation. From 1959 onwards, about 50 Salinity Control and Reclamation Projects (SCARPs) have been initiated to provide a lasting solution to the problem of waterlogging and salinity through subsurface drainage.

FIGURE 9
Drained areas by province (total: 5.1 million ha in 1992)

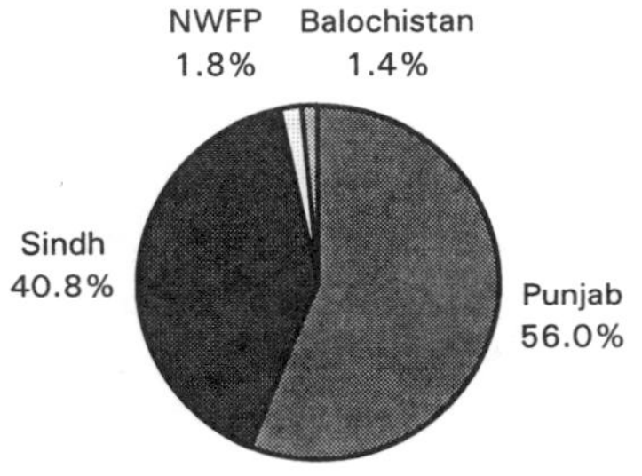

There are about 5 200 km of flood control works, whose maintenance falls under the responsibility of the PID.

INSTITUTIONAL ENVIRONMENT

The Water and Power Development Authority (WAPDA), created in 1958 as a semi-autonomous body, is responsible for planning and execution of schemes pertaining to:

- generation, transmission and distribution of power;
- irrigation, water supply and drainage;
- prevention of waterlogging and reclamation of waterlogged and saline lands;

FIGURE 10
Saline areas, by province, according to the Soil Survey of Pakistan (total: 1.8 million ha in 1994)

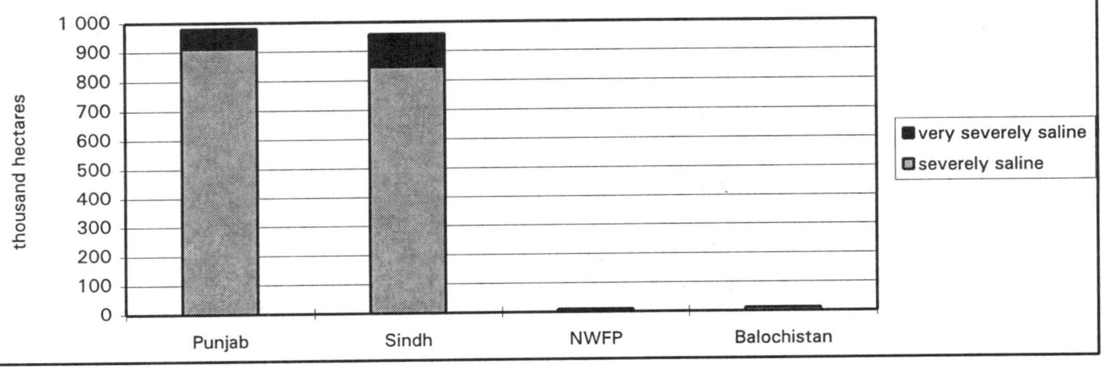

- flood control;
- inland navigation.

In power projects, WAPDA acts as the owner with responsibility for implementation as well as operation and maintenance of the power stations and distribution systems. In other projects, the authority acts as the government agency for the execution of schemes (including drainage schemes), which upon completion are transferred to Provincial Irrigation Departments (PID) for operation and maintenance. However, WAPDA remains responsible for the operation and maintenance of large multipurpose reservoirs and inter-provincial link canals.

The PIDs have historically been responsible for all water-sector activities at the provincial level, including planning, development, and the operation and maintenance of irrigation, drainage, flood control and reclamation works. However, the role of WAPDA has progressively grown in relation to several of these activities. Currently, the main responsibility of the PIDs relates to the operation and maintenance of the irrigation, drainage and flood-control systems. Under the provision of the Irrigation and Drainage Act, the PID is responsible for all aspects of the distribution and use of irrigation supplies including on-farm use; however, in practice, the responsibility ends at the end of the watercourse (*mogha*).

The Provincial Agriculture Departments (PAD) are mainly responsible for organizing input distribution and the extension service to farmers, and notably farm management. The PADs are responsible for the implementation of the on-farm water management programme. Farmers are responsible for operation and maintenance of watercourses and field canals.

The Indus River System Authority (IRSA) has recently been established, with headquarters at Lahore, to provide continuing interaction and a resolution of any disputes among the four provinces on matters relating to sharing the Indus waters. The IRSA is comprised of one representative from each province and a member from the federal Government.

The office of the Chief Engineering Advisor provides advisory services to the Ministry of Water and Power through its Water Engineering and Power Engineering sections, and coordinates the inter-provincial activities related to flood-control works. It acts also as the convenor and co-ordinator of the Dam Safety Council.

The 1967 Land Reform Act established a register of Rights which is a cadastral register for land and water rights.

TRENDS IN WATER RESOURCES MANAGEMENT

Irrigation is central to Pakistan's economy. In the irrigation subsector, measures are expected to be taken to increase water availability, water reliability, equity of water distribution and irrigation efficiency, to reduce waterlogging and salinity, to limit the over-exploitation of the fresh groundwater resources, and to improve cost recovery.

River flows are almost fully utilized, except during the flood period in *Kharif*. The utility of flood water is very marginal unless additional storage is provided in the system to broaden the supply period. According to a World Bank report, Pakistan does not have enough reservoir capacity in its irrigation system to store seasonal waters. Additional storage would also be necessary for providing the flexibility needed by the shift from a supply-based operation system towards a demand-based one. Several sites have already been identified, plans being ready for one of them (the Kalabagh dam).

For most of the multi-purpose reservoirs priority is given to irrigation. The recent increase in thermal generation capacity has reduced the potential conflicts between water releases from reservoirs for hydropower generation and irrigation. Most of the annual storage is required for irrigation, not for hydropower, but conflicts do still arise at times.

Waterlogging and salinity have been identified by the authorities as one of the main issues to be addressed in a near future. The elimination of waterlogging in disaster areas has been set as one of the main targets for the 8th Five-Year Plan (1993-98). Under the Salinity Control and Reclamation Projects, attempts were made to meet the subsurface drainage needs through deep tubewells.

Groundwater extraction by tubewell may cause serious environmental problems and in certain areas, increasing exploitation by tubewells causes water tables to fall at the rate of about 0.3 metres per year.

In a 1994 report, the World Bank proposed changing the Government's approach, and commercializing (in order to later privatize) all water services. The proposed strategy comprises:

- market-determined incentives for irrigation and on-farm drainage;
- establishment of Farmers Organizations;
- drawing up enforceable property rights to water, the first steps being the establishment of communal water rights, which would later be translated into enforceable individual rights;
- establishment on a canal command basis of public utilities responsible for operation and maintenance, cost recovery and future investment;
- legalization of water markets.

The expected results would be: greater transparency in the cost of water leading to a greater efficiency in use, greater equity in distribution, and relief of pressure on groundwater resources.

MAIN SOURCES OF INFORMATION

Economic Adviser's Wing, Finance Division, Government of Pakistan. 1994. Economic survey 1993-94. Islamabad, Pakistan.

Federal Planning Cell. 1990. Water sector investment planning study. Report prepared in cooperation with Sir M. MacDonald and Partners, National engineering services Pakistan Lt, Harza Engineering Co International LP, Associated consulting engineers ACE Ltd, for WAPDA, UNDP and the World Bank. Lahore, Pakistan.

John Mellor Associates, Inc. and Asianics Agro-development International (Pvt) Ltd. 1995. Institutional reforms to accelerate irrigated agriculture. 2 volumes. Islamabad, Pakistan.

Ministry of Food, Agriculture and Cooperatives. 1993. Agricultural statistics of Pakistan 1991-92. Islamabad, Pakistan.

Pakistan National Committee of ICID. 1991. Irrigation and drainage development in Pakistan. Islamabad, Pakistan.

Planning commission, Government of Pakistan. 1994. Eighth Five-Year Plan (1993-98). Islamabad, Pakistan.

The World Bank. 1994a. Pakistan, Irrigation and Drainage: Issues and Options. Report 11884-PAK. Washington DC, USA.

The World Bank. 1994b. Pakistan, A Strategy for Sustainable Agricultural Growth. Report 13092-PAK. Washington DC, USA.

The World Bank. 1995. Pakistan. Balochistan community irrigation and agriculture project. Report 13597-PAK. Washington DC, USA.

Qatar

GEOGRAPHY AND POPULATION

Qatar occupies a peninsula, projecting northwards from the Arabian mainland, on the west coast of the Persian Gulf. Its western coastline joins onto the shores of Saudi Arabia. The peninsula is roughly 180 km long and between 55 and 85 km wide and its total area is estimated at 11 000 km^2. Qatar is stony, sandy and barren and its surface is of low to moderate relief, with the highest elevation of 103 metres above sea level being attained in the southern part, where mesa-type hills and large barchan-type sand dunes serve to break the monotony of an otherwise flat, eroded landscape.

The cultivable land is estimated at 28 000 ha, which is 2.5% of the total area. In 1994, 8 312 ha were estimated to be cultivated as against 2 256 ha in 1980. Of this area of 8 312 ha, 5 987 ha consisted of annual crops (mainly green fodder, vegetables and cereals) and 2 325 ha consisted of permanent crops (mainly dates). The agricultural land is government owned and most farm owners participate only indirectly in the farming process, having permanent positions in other sectors of the economy. Consequently, most farms are run by immigrants.

TABLE 1
Basic statistics and population

Physical areas:			
Area of the country	1995	1 10 000	ha
Cultivable area	1994	28 000	ha
Cultivated area	1994	8 312	ha
- annual crops	1994	5 987	ha
- permanent crops	1994	2 325	ha
Population:			
Total population	1995	551 000	inhabitants
Population density	1995	50	inhab./km^2
Rural population	1995	9	%
Water supply coverage:			
Urban population	1996	100	%
Rural population	1996	100	%

The total population is 551 000 (1995), of which 9% is rural. Between 1980 and 1991 the annual demographic growth rate was estimated at 4.7%. Only about one-fourth of the total population is estimated to be indigenous Qatari population. The capital Doha houses 80% of the total population and the industrial centre of Umm Said, 40 km south, is being extensively developed. Agriculture represents only 1.1% of the country's GDP, which is dominated by oil and its by-products.

CLIMATE AND WATER RESOURCES

Climate

The climate is of the typical arid desert type with a hot, humid summer, with temperatures reaching 44°C. There is some rain in winter, when temperatures range between 10 and 20°C. Average annual rainfall is around 75 mm, although there are wide variations from place to place and from year to year. It varies from 40 mm in the south to over 80 mm in the north. Most rain occus during November to April inclusive, but might be limited to two or three days of heavy rainfall per year which is more destructive than useful. Recorded falls in 24 hours have approached the total annual average.

TABLE 2
Water: sources and use

Renewable water resources:			
Average precipitation		75	mm/yr
		0.825	km^3/yr
Internal renewable water resources		0.051	km^3/yr
Total (actual) renewable water resources	1995	0.053	km^3/yr
Dependency ratio	1995	3.8	%
Total (actual) renewable water resources per inhabitant	1995	96	m^3/yr
Total dam capacity		-	10^6 m^3
Water withdrawal:			
- agricultural	1994	210.6	10^6 m^3/yr
- domestic	1994	65.9	10^6 m^3/yr
- industrial	1994	8.4	10^6 m^3/yr
Total water withdrawal		284.9	10^6 m^3/yr
per inhabitant	1994	528	m^3/yr
as % of total (actual) renewable water resources		538	%
Other water withdrawal	1994	7	10^6 m^3/yr
Average groundwater depletion	1974	42.49	10^6 m^3/yr
Wastewater - Non-conventional water sources:			
Wastewater:			
- produced wastewater	1994	47.6	10^6 m^3/yr
- treated wastewater	1994	25.2	10^6 m^3/yr
- reused treated wastewater	1994	25.2	10^6 m^3/yr
Desalinated water	1995	98.6	10^6 m^3/yr

Water resources

There is practically no permanent surface water - annual surface runoff has been estimated at 1.35 million m^3. Direct and indirect recharge of groundwater from rainwater forms the main natural internal water resources. Two-thirds of the land surface is made up of some 850 contiguous depressions of interior drainage with catchments varying from 0.25 km^2 to 45 km^2 and with a total aggregate area of 6 942 km^2. While direct recharge from rainfall might take place during very rare heavy storms, the major recharge mechanism is an indirect one through runoff from surrounding catchments and the pounding of water in the depression floor. Surface runoff typically represents between 16 and 20% of rainfall. Of the amount reaching the depressions, 70% infiltrates and 30% evaporates.

There are two separate and distinct groundwater regions: the northern half, where groundwater occurs as a freshwater 'floating lens' on brackish and saline water and the southern half where no such lens exists and where water quality is generally brackish with only a thin veneer of freshwater at the top of the water table. Annual groundwater recharge has been estimated at 50.43 million m^3.

The two main aquifers underlying Qatar are recharged in Saudi Arabia. Over most of Qatar the Damman formation does not contain water because of its altitude. It dips lower in south-west Qatar where it contains water, but is also overlain by impervious layers. The artesian aquifer which results from this structure is called the Alat unit of the Damman. Below this aquifer is the Umm er Radhuma, which is similarly artesian. In 1981, the Master Water Resources and Agricultural Development Plan (MWRADP) estimated that in the southern part of Qatar the safe yield of the Alat aquifer is 2 million m^3/year and that of the Umm er Radhuma 10 million m^3/year, based on an estimate of annual flow from Saudi Arabia. However, these safe yields would be substantially reduced if the aquifer were exploited more extensively on the Saudi Arabian side of the border. In the northern and central part the Rus aquifer overlies the Umm el Radhuma aquifer, which is partly an unconfined aquifer,

TABLE 3
Irrigation and drainage

Irrigation potential			-	ha
Irrigation:				
1. Full or partial control irrigation: equipped area		1993	12 520	ha
- surface irrigation			-	ha
- sprinkler irrigation			-	ha
- micro-irrigation			-	ha
% of area irrigated from groundwater		1994	94.2	%
% of area irrigated from surface water		1994	0.0	%
% of area irrigated from non-conventional sources		1994	5.8	%
% of equipped area actually irrigated		1994	66.4	%
2. Spate irrigation area			-	ha
3. Equipped wetland and inland valley bottoms (i.v.b.)			-	ha
Total irrigation (1+2+3)		**1993**	**12 520**	**ha**
- as % of cultivated area			151	%
4. Flood recession cropping area			-	ha
Total water managed area (1+2+3+4)		**1993**	**12 520**	**ha**
- as % of cultivated area			151	%
- increase over last 10 years			-	%
- power irrigated area as % of water managed area			-	%
Full or partial control irrigation schemes:	Criteria			
Large-scale schemes	> 150 ha	1993	1 654	ha
Medium-scale schemes		1993	6 688	ha
Small-scale schemes	< 20 ha	1993	4 178	ha
Total number of households in irrigation		1993	889	
Irrigated crops:				
Total irrigated grain production		1994	3 940	tons
as % of total grain production		1994	100	%
Harvested crops under irrigation (full or partial control)		**1994**	**8 797**	**ha**
- permanent crops: total		1994	2 325	ha
- annual crops: total		1994	6 472	ha
. green fodder (mainly alfalfa)		1994	2 803	ha
. vegetables		1994	2 341	ha
. cereals (mainly barley, some wheat and maize)		1994	1 243	ha
. tubers		1994	55	ha
. pulses		1994	30	ha
Drainage - Environment:				
Drained area			-	ha
as % of cultivated area			-	%
- drained areas in full or partial control irrigated areas			-	ha
- drained areas in equipped wetland and i.v.b			-	ha
- other drained areas			-	ha
- total drained area with subsurface drains			-	ha
- total drained area with surface drains			-	ha
Flood-protected area			-	ha
Area salinized by irrigation			-	ha
Population affected by water-borne diseases			-	inhabitants

recharged by percolating rainfall and return flows from irrigation but losing some water to the sea and some through abstractions. The safe yield of the aquifer system in the northern and central part of Qatar is estimated at 13 million m^3/year from the upper layer and 20 million m^3/year from the lower layer (the latter leading to a depletion in 50 years). In total, the estimated safe yield for the whole of Qatar is 45 million m^3/year.

Another potential source of groundwater is beneath the capital Doha itself. According to the MWRADP, considerable volumes of water leak from pipelines and other sources throughout much of Doha. This leakage, estimated at about 15 million m^3/year, has caused the water table to rise locally, flooding basements as well as shallow excavations.

Non-conventional water sources

In 1995, total desalination capacity was 98.6 million m^3/year. There are two desalination plants in Qatar. The desalination plant in Ras Abu Aboud has a capacity of 52 000 m^3/day or 19.0 million m^3/year. The plant in Ras Abu Fontas has a capacity of 218 000 m^3/day or 79.6 million m^3/year. The addition of two new desalination units in Ras Abu Fontas is expected to raise the total desalination capacity of the country in 1996 to 318 000 m^3/day or 116.1 million m^3/year.

In 1987, treated and reused municipal wastewater was almost 70 000 m^3/day, or 25.2 million m^3/year.

FIGURE 1

Water withdrawal (total: 285 million m^3 in 1994)

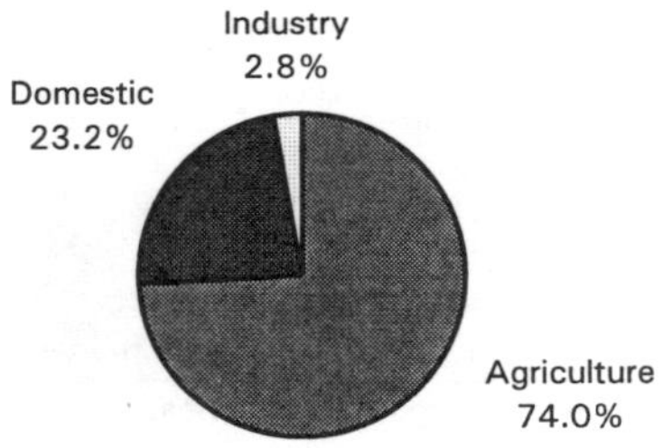

Water withdrawal

In 1994, total water withdrawal was estimated at 292 million m^3, of which 285 million m^3 for agricultural, domestic and industrial purposes (Figure 1). Desalinated water provides the vast majority of Qatar's drinking water (Figure 2). Of the total reused treated wastewater of 25.2 million m^3, 25% is supplied to the municipality of Doha to be used in landscape irrigation on the roads and areas planted with trees. The remaining 75% is conveyed via pipelines to ponds and then repumped to west Doha to be used in irrigation of farms.

FIGURE 2

Origin of water used by sector (total: 285 million m^3 in 1994)

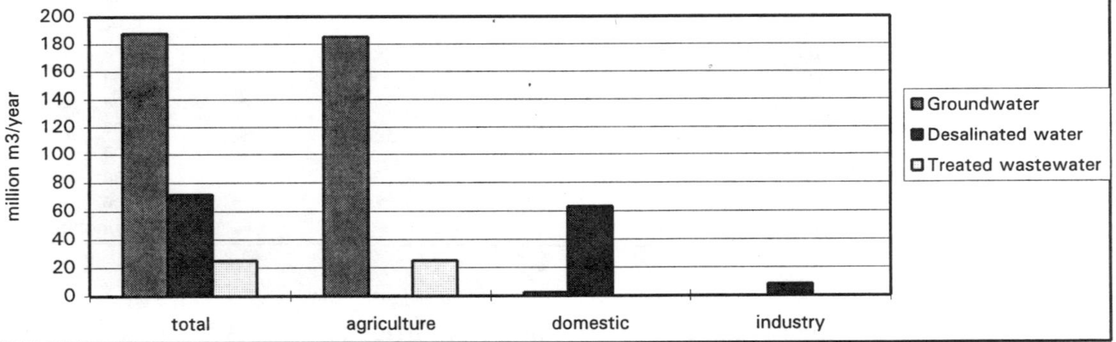

Total groundwater extraction is estimated at 188 million m^3. Given an annual recharge of about 50 million m^3, mining of groundwater is 138 million m^3, which is much more than the considered safe yield of 45 million m^3 (which already includes fossil water).

IRRIGATION AND DRAINAGE DEVELOPMENT

Agriculture is not possible without irrigation. The part of the 28 000 ha of cultivable area that can be considered to assess irrigation potential, depends on the future availability of alternative sources of water, because groundwater is already being depleted at the present rate of abstraction.

In 1993, there were 1 130 registered farms in the country, of which 889 were actually in operation. The area equipped for irrigation was estimated at 12 520 ha, while 8 312 ha was actually cultivated, which is 66% of the equipped area. Most of the water used is groundwater (Figure 3).

FIGURE 3
Origin of irrigation water f/p (total: 12 520 ha in 1993)

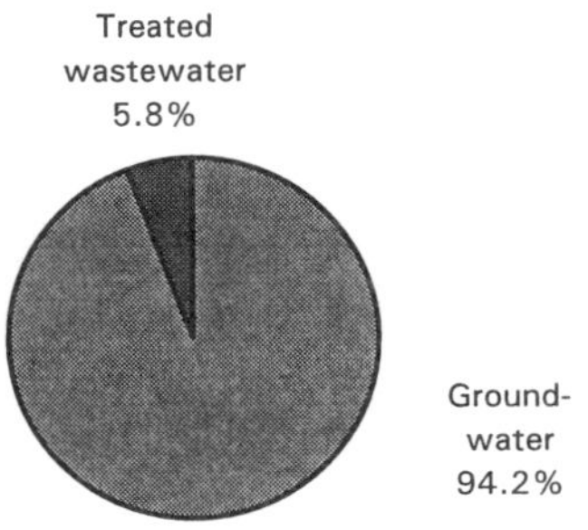

Surface irrigation (basins and furrows) is the most commonly used irrigation technique. Examples of relatively large-scale projects that use modern irrigation techniques are: the Ar Rakiyah project, where 17 centre pivots cover 723 ha, and the Al Mashabiyah project, where 12 000 date palms are irrigated by bubblers and more than 30 ha of vegetables are irrigated by drippers in experimental and private farms. Some examples of the costs of irrigation development are: $US 7 400/ha for a drip irrigation system with buried PVC pipes for vegetables and $US 7 700/ha for a bubbler irrigation system with buried PVC pipes for date palms.

FIGURE 4
Typology of f/p control irrigation schemes (total: 12 520 ha in 1993)

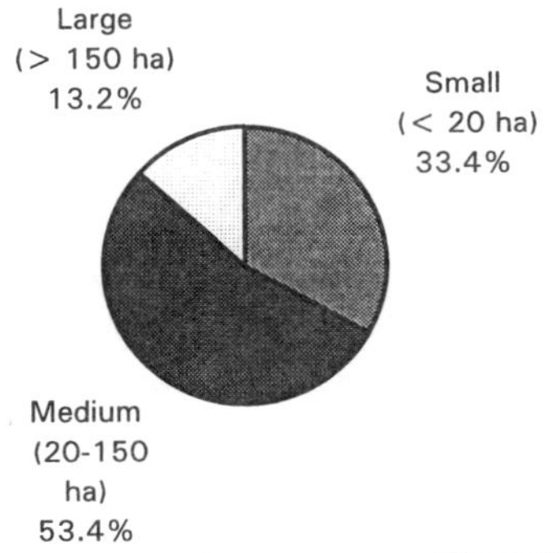

There are no national criteria for distinguishing between the size of the schemes. The classification of registered farms and agricultural projects according to equipped areas assumes that small schemes work areas less than 20 ha, covering 33% of the total equipped area, medium size schemes work areas between 20 and 150 ha, covering 54% of the total area, and large schemes work areas of over 150 ha, covering 13% of the total area (Figure 4).

The major irrigated crops are green fodder, vegetables, fruit trees and cereals (Figure 5). Tomatoes are the main winter vegetable, melons the main summer vegetable. The main fruit trees are dates and citrus. Alfalfa is the main green fodder crop. Barley is the main cereal, with a small quantity of wheat and maize.

Over-extraction of groundwater results in a fast deterioration of the water quality, due to sea water intrusion and upward diffusion of deeper saline water. For this reason, a kind of shifting cultivation takes place within a farm, i.e., as soon as soil salinity has reached intolerable limits, new land is put into cultivation and the degraded land is completely abandoned.

FIGURE 5
Irrigated crops (total: 8 797 ha in 1994)

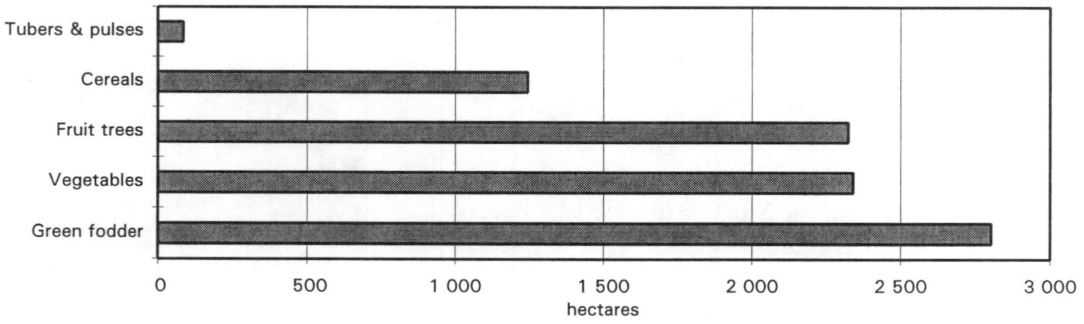

INSTITUTIONAL ENVIRONMENT

The institutions in charge of water resources assessment, development and irrigation are:

- the Ministry of Municipal Affairs and Agriculture together with: the Department of Agricultural and Water Research (DAWR), the Drainage Division and the Environmental Department;
- the Ministry of Electricity and Water together with the Water Networks Department;
- the Supreme Council for Planning;
- the Ministry of Energy and Industry.

Based on the recommendation of the DAWR, Decree No. 1 of 1988 was issued to regulate drilling of wells and groundwater use. The Ministry of Municipal Affairs and Agriculture formed the Farms and Wells Committee to be responsible for implementing the groundwater laws. The articles of the groundwater laws which have been implemented are those connected with granting permits for drilling, alteration and modification of wells. What remains to be done is to put into action the articles concerning water use and specifically the installation of water meters in all wells in the state, not exceeding the specified irrigated area and the allocated amount of water for each farm following the instructions issued by the DAWR regarding conveyance and irrigation systems.

TRENDS IN WATER RESOURCES MANAGEMENT

The main trends in the national strategy and policy for the development of water resources and irrigation consist in fulfilling two objectives:

- the improvement of the present situation of water use. This comprises the redesign of existing farms, rehabilitation of abandoned farms, implementation of modern irrigation systems in all productive farms, strengthening of extension and training and implementation of water laws;

- the enhancement of water resources. This could be achieved by importing fresh water, increasing the quantity of treated wastewater, irrigation with saline water, increasing the natural groundwater recharge rate, development of deep aquifers and monitoring saline water intrusion.

The main constraint facing the implementation of the above-mentioned water resources and irrigation development plan is the high capital cost and the low return from using water in agricultural production.

MAIN SOURCES OF INFORMATION

Civil Engineering Department, Drainage Division. 1996. Treated Sewage Effluent. MMAA [Ministry of Municipal Affairs and Agriculture].

DAWR [Department of Agricultural and Water Research], Agricultural Economics and Statistics Section. 1994. Agricultural Statistics 1993. MMAA. Doha, Qatar.

DAWR, Groundwater Unit. 1995. Groundwater Balance. MMAA.

DAWR. 1982. Brief annual report 1981-82: irrigation and drainage section. Prepared by M.A. Hashim and N.G. Dastane. Ministry of Industry and Agriculture/FAO.

FAO. 1978. Integrated water and land use in Qatar. Rome.

Halcrow-Balfour Ltd. 1981. Master Water Resources and Agricultural Development Plan. Ministry of Industry and Agriculture.

Saudi Arabia

GEOGRAPHY AND POPULATION

Saudi Arabia, with a total area of about 2.15 million km^2, is by far the largest country in the Arabian Peninsula. It is bordered in the north by Jordan, Iraq and Kuwait, in the east by the Persian Gulf with a coastline of 480 km, in the south-east and south by Qatar, the United Arab Emirates, Oman and Yemen, and in the west by the Red Sea with a coastline of 1 750 km.

It can be divided into 4 main physiographic units:

- the Western Mountains, called the Arabian Shield, with a peak at 2 000 metres above sea level and crossed by deep valleys;
- the Central Hills, which run close to the western mountains and lie in the centre of the country. Their elevation ranges between 900 and 1 800 metres above sea level;
- the Desert Regions, which lie to the east of the Central Hills, with elevations ranging between 200 and 900 metres. Sand dunes are commonly found in these deserts;
- the Coastal Regions, which include the coastal strip along the Red Sea with a width of 16 to 65 km. The important part is the Tahama Plain in the south. The plain on the eastern side overlooks the Persian Gulf, is generally wide and includes the region of oases.

TABLE 1
Basic statistics and population

Physical areas:			
Area of the country	1995	214 969 000	ha
Cultivable area	1992	52 684 000	ha
Cultivated area	1992	1 608 000	ha
- annual crops	1992	1 512 500	ha
- permanent crops	1992	95 500	ha
Population:			
Total population	1995	17 880 000	inhabitants
Population density	1995	8	inhab./km^2
Rural population	1995	20	%
Water supply coverage:			
Urban population	1990	100	%
Rural population	1990	74	%

The cultivable area has been estimated at 52.7 million ha, which is almost 25% of the total area. In 1992, the cultivated area was 1 608 000 ha, of which 1 512 500 ha consisted of annual crops and 95 500 ha consisted of permanent crops.

The total population is 17.9 million (1995), of which 20% is rural. In 1992, about 73% were estimated to be Saudi nationals. The natural demographic growth rate of Saudi nationals is probably over 3% and structurally the age of the population is very young, with 62.6% aged under 24 years in 1995, according to World Bank estimates. In 1994, agriculture employed 5.5% of the labour force compared with 6.5% in 1989. The share of GDP accounted for by agriculture, which was 8.8% in 1993, has risen since the early 1980s, mainly as a result of the decline in revenue from the petroleum sector and government efforts to pursue a policy of greater self-sufficiency in agriculture.

TABLE 2
Water: sources and use

Renewable water resources:			
Average precipitation		59	mm/yr
		126.8	km^3/yr
Internal renewable water resources		2.4	km^3/yr
Total (actual) renewable water resources	1995	2.4	km^3/yr
Dependency ratio	1995	0	%
Total (actual) renewable water resources per inhabitant	1995	134	m^3/yr
Total dam capacity	1993	475	$10^6 m^3$
Water withdrawal:			
- agricultural	1992	15 308	$10^6 m^3/yr$
- domestic	1992	1 517	$10^6 m^3/yr$
- industrial	1992	193	$10^6 m^3/yr$
Total water withdrawal		17 018	$10^6 m^3/yr$
per inhabitant	1992	1 040	m^3/yr
as % of total (actual) renewable water resources		709	%
Other water withdrawal		-	$10^6 m^3/yr$
Average groundwater depletion	1990	13 558	$10^6 m^3/yr$
Wastewater - Non-conventional water sources:			
Wastewater:			
- produced wastewater		-	$10^6 m^3/yr$
- treated wastewater	1991	454	$10^6 m^3/yr$
- reused treated wastewater	1992	217	$10^6 m^3/yr$
Desalinated water	1995	714	$10^6 m^3/yr$

CLIMATE AND WATER RESOURCES

Climate

Saudi Arabia falls in the tropical and subtropical desert region. The winds reaching the country are generally dry, and almost all the area is arid. Because of the aridity, and hence the relatively cloudless skies, there are great extremes of temperature, but there are also wide variations between the seasons and regions. In the central region, the summer (May to October) is overwhelmingly hot and dry, with maximum temperatures of over 50°C, while the winter is dry and cool with night-time temperatures close to freezing. There can be severe frost generally and even weeks of snow in the mountains. The western and eastern regions are hot and humid in the summer months, with maximum temperatures around 42°C, while the winters are warm.

In the north, annual rainfall varies between 100 and 200 mm. Further in the south, except near the coast, annual rainfall drops below 100 mm. The higher parts of the west and south do, however, experience appreciable rainfalls and over some small areas 500 mm/year is not uncommon. Total precipitation has been estimated at 126.8 km^3/year, which is equal to 59 mm/year over the whole country.

Water resources

Heavy rainfall sometimes results in flash floods of short duration. River beds are dry for the rest of the time. Part of the surface runoff percolates through the sedimentary layers in the valleys and recharges the groundwater, some is lost by evaporation. The largest quantity of runoff occurs in the western region, which represents 60% of the total runoff although it covers only 10% of the total area of the country. The remaining 40% of the total runoff occurs in the far south of the western coast (Tahama) which covers only 2% of the total area of the country. Total surface water resources have been estimated at 2.2 km^3/year, most of it

TABLE 3
Irrigation and drainage

Irrigation potential		-	ha
Irrigation:			
1. Full or partial control irrigation: equipped area	1992	1 608 000	ha
- surface irrigation	1992	547 000	ha
- sprinkler irrigation	1992	1 029 000	ha
- micro-irrigation	1992	32 000	ha
% of area irrigated from groundwater	1992	95.6	%
% of area irrigated from surface water	1992	3.2	%
% of area irrigated from non-conventional sources	1992	1.2	%
% of equipped area actually irrigated	1992	100	%
2. Spate irrigation area		-	ha
3. Equipped wetland and inland valley bottoms (i.v.b.)		-	ha
Total irrigation (1+2+3)	**1992**	**1 608 000**	**ha**
- as % of cultivated area	1992	100	%
4. Flood recession cropping area		-	ha
Total water managed area (1+2+3+4)	**1992**	**1 608 000**	**ha**
- as % of cultivated area	1992	100	%
- increase over last 10 years	1982-92	170	%
- power irrigated area as % of water managed area	1992	95	%
Full or partial control irrigation schemes: Criteria			
Large-scale schemes > 200 ha	1992	428 000	ha
Medium-scale schemes	1992	730 000	ha
Small-scale schemes < 5 ha	1992	450 000	ha
Total number of households in irrigation	1992	188 370	
Irrigated crops:			
Total irrigated grain production	1992	4 645 424	tons
as % of total grain production	1992	100	%
Harvested crops under irrigation (full or partial control)	**1992**	**1 608 000**	**ha**
- permanent crops: total	1992	95 500	ha
- annual crops: total	1992	1 512 500	ha
. wheat	1992	907 300	ha
. fodder crops	1992	293 100	ha
. other cereals (mainly sorghum, barley)	1992	195 400	ha
. vegetables	1992	109 100	ha
. other annual crops	1992	7 600	ha
Drainage - Environment:			
Drained area	1992	44 000	ha
as % of cultivated area		2.7	%
- drained areas in full or partial control irrigated areas		-	ha
- drained areas in equipped wetland and i.v.b		-	ha
- other drained areas		-	ha
- total drained area with subsurface drains		-	ha
- total drained area with surface drains		-	ha
Flood-protected area		-	ha
Area salinized by irrigation		-	ha
Population affected by water-borne diseases		-	inhabitants

infiltrating to recharge the aquifers. About 1 km^3 recharges the usable aquifers. The total (including fossil) groundwater reserves have been estimated at about 500 km^3, of which 340 km^3 are probably abstractable at an acceptable cost in view of the economic conditions of the country.

Dams

In 1993, approximately 185 dams of various sizes had been constructed for flood control and groundwater recharge with a combined storage capacity of 475 million m^3. About 45 new dams were planned to be built. A major dam, the Bisha dam in the south-west with a capacity of 325 million m^3, was under construction in 1993.

Desalinated water and treated wastewater

Saudi Arabia is the largest producer of desalinated water from the sea. In 1992, there were 18 desalination and power plants in the western coast, with a total capacity of over 0.7 million m^3/day of water and 1 286 MW of electricity, and four plants on in the east coast, with a total capacity of over 1.1 million m^3/day of water and 1 550 MW of electricity. Moreover three plants were under construction and two plants under bidding. In 1992, actual desalinated water production was about 675 million m^3.

Some 265 million m^3 were produced on the western coast, of which over 50% was exported to the city of Jeddah, while 410 million m^3 were produced on the eastern coast, of which over 65% was exported to the city of Riyadh, which is located about 400 km from the sea. The total length of pipelines used for the transmission of desalinated water is about 3 722 km. The capacity of desalinated water reservoirs amounted to 1.85 million m^3. In 1995, desalinated water production was 714 million m^3.

FIGURE 1
Water withdrawal (total: 17 km^3 in 1992)

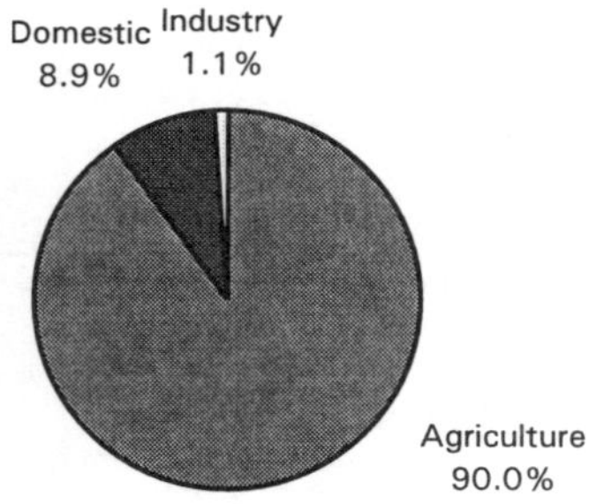

In 1991, there were 22 sewage treatment plants in operation (10 of them being waste stabilization ponds) with a total wastewater treatment of 1.2 million m^3/day or 454 million m^3/year. Five new plants were being designed and the 22 existing were proposed for upgrading/extension in order to have a total flow of 1.8 million m^3/day by the year 2000. In 1992, 217 million m^3 of treated wastewater were reused.

FIGURE 2
Origin of water used by all sectors (total: 16.3 km^3 in 1990)

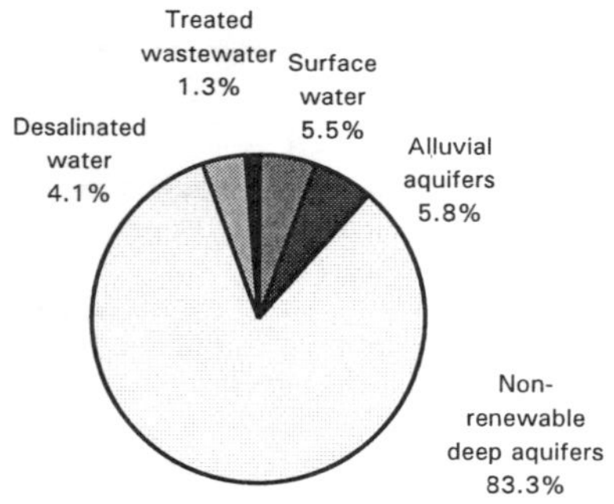

Water withdrawal

In 1992, total water withdrawal was estimated at 17 km^3, of which 90% for agricultural purposes (Figure 1). In 1990, total water withdrawal was estimated at 16.3 km^3. Desalinated water is used for municipal purposes, as it is too saline, even after treatment, for irrigation. Treated wastewater is used to irrigate non-edible crops, for landscape irrigation and for industrial cooling. However, most of the water used (> 13.5 km^3) comes from non-renewable, deep aquifers (Figure 2). At the 1990 rate of abstraction, it is estimated that the usable reserves will last for a maximum of 25 to 30 years. The quality of the abstracted water is likely to deteriorate with time because of the flow from low quality water in the same aquifers towards the core of the depression at the point of use. In 1988, there were 4 667 multi-purpose government wells and 44 080 multi-purpose private wells.

IRRIGATION AND DRAINAGE DEVELOPMENT

The most recent soil surveys (1989) and classifications put the area of land suitable for irrigated agriculture at about 10 million ha. However, as shown above, the limiting factor is water. At present, depletion of non-renewable fossil water is already taking place at a very fast rate.

All agriculture is irrigated and in 1992 the water managed area was estimated at about 1.6 million ha, all equipped for full/partial control irrigation. Surface irrigation is practised on the old agricultural lands, cultivated since before 1975, which represent about 34% of the irrigated area (Figure 3). Sprinkler irrigation is practised on about 64% of the irrigated areas. The central pivot sprinkler system covers practically all the lands cropped with cereals. Normally, pumped groundwater from one deep well supplies one or two central pivots. The irrigation application efficiency of this method is estimated at between 70 and 85%. Vegetables and fruit trees are in general irrigated by drip and bubbler methods respectively. Groundwater is used on almost 96% of the irrigated area, treated wastewater on 1% (Figure 4).

FIGURE 3
Irrigation techniques f/p (total: 1 608 000 ha in 1992)

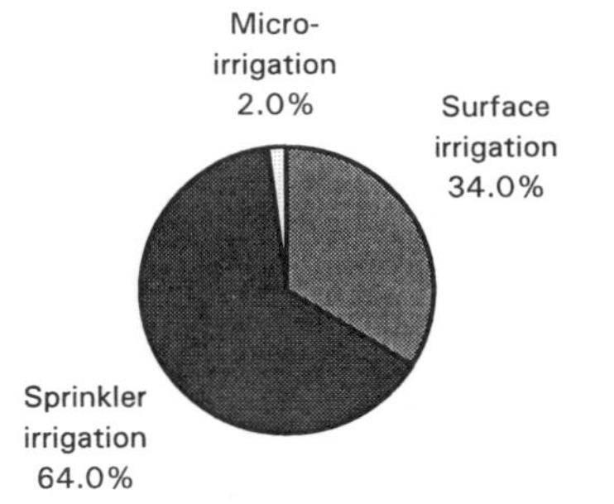

FIGURE 4
Origin of irrigation water f/p
(total: 1 608 000 ha in 1992)

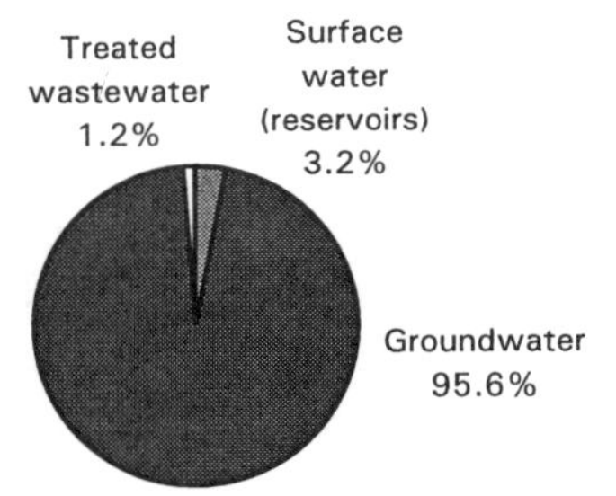

In 1992, 428 000 ha were estimated to be cultivated by 1 070 large farms, with an area of more than 200 ha each. The total area of medium farms (5 - 200 ha) was 730 000 ha, comprising 7 300 farms. Small farms (< 5 ha) covered 450 000 ha, comprising 180 000 farms (Figure 5).

FIGURE 5
Typology of f/p control irrigation schemes
(total: 1 608 000 ha in 1992)

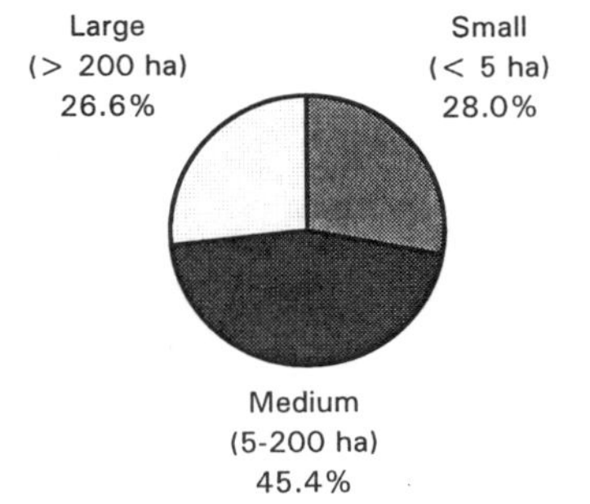

The average cost for irrigation development is about $US 1 093, 372 and 251/ha for micro-irrigation, sprinkler irrigation and surface irrigation systems respectively. Water is free of charge.

The cropped area has more than tripled between 1977 and 1992 (Figure 6). In general, there is only one cropping season.. The major irrigated crop is wheat. In 1988, it consumed almost 40% of the total quantity of irrigation water, while it covered almost 62% of the irrigated area. Other major crops are fodder, other cereals (particularly sorghum and barley), fruit trees and vegetables (Figure 7). Since 1988, self-

FIGURE 6
Evolution of the cropped area

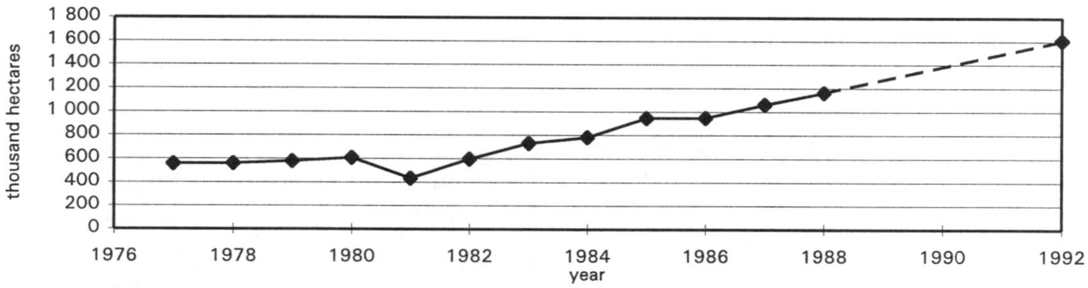

FIGURE 7
Irrigated crops f/p (total: 1 608 000 ha in 1992)

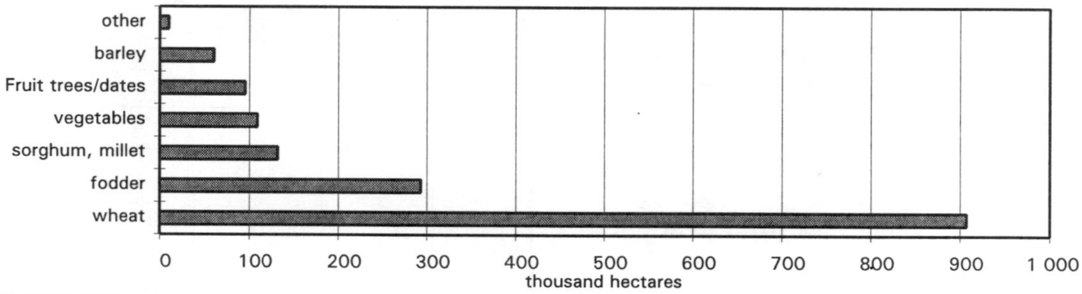

sufficiency in wheat has been reached and part of the production is being exported. In 1992, wheat production was almost 4.1 million tons, while national demand was only about 1.2 million tons. Vegetables, fruits and dates and fodder are also exported.

Waterlogging, drainage and salinity

Waterlogging and drainage problems occur in the central and southern parts of the country, due to the existence of shallow, impervious layers. About 44 000 ha, or 2.7% of the irrigated area, have drainage facilities. The drainage systems mainly consist of open drainage canals. In several projects, such as the Al-Hassa irrigation project in the east, agricultural drainage water is reused for irrigation after blending with fresh groundwater. Old irrigation networks are replaced with concrete-lined canals.

Soil salinity is being noticed in parts of the newly developed areas, due to poor irrigation water quality and thedrainage conditions of some soils.

INSTITUTIONAL ENVIRONMENT

The Ministry of Agriculture and Water (MAW) is in charge of water resources research, drinking water supply, construction, and the operation and maintenance of water resources projects.

The Agency of Water Affairs within MAW has three main programmes:

- water resources development, which includes all activities related to geological and hydrological studies, wastewater reuse investigations, well drilling and dam construction, and the preparation of the national water plan;
- drinking water supply, which includes the construction of drinking water supply networks to various towns and cities that do not have local water authorities or municipalities;
- the operation and maintenance programme. On-farm management is the farmers responsibility, but problems affecting more than one farmer (common irrigation networks, drainage, pest control, etc.) are under the responsibility of the Ministry.

The General Organization for Desalination of MAW is responsible for the construction, operation and maintenance of desalination plants.

The Al-Hassa Irrigation and Drainage Authority (HIDA) is part of MAW and is in charge hydrological studies and data collection to improve the use of water for irrigation. It is also responsible for irrigation water conservation, estimation of crop water requirements, irrigation water distribution to the farms and the operation and maintenance of irrigation and drainage canal systems in the irrigation schemes managed by MAW.

The Ministry of Municipal and Rural Affairs is responsible for the construction, operation and maintenance of drinking water supply networks and for the installation of water connections at house level. It is also responsible for the construction, operation and maintenance of wastewater treatment plants.

TRENDS IN WATER RESOURCES MANAGEMENT

In 1981 there began a change in agricultural cropping patterns by adopting new technologies, exercising extensive and effective agricultural extension, using improved seed varieties with high productivity and providing advanced plant protection services in line with modern agricultural methods.

The government's involvement in the agricultural sector has been extensive. During the 1980s food self-sufficiency, particularly in wheat and dairy products, became a major priority and, with the support of heavy subsidies, the added value in agriculture increased by more than 70% in the period 1985-91. Wheat production was even sufficient to enable Saudi Arabia to become the world's sixth largest wheat exporter. However despite its success, this policy is a threat to the country's water reserves. On economic grounds, the 1991/92 harvest was estimated to have cost the government around $US 480/ton compared with world prices for wheat of $US 100/ton. At present, the national goal is the diversification of agricultural production in order to meet the growing demand for other types of crops and to adjust the wheat production to the level of annual national consumption.

Because of the development of agriculture, which is by far the largest water user, the depletion of fossil groundwater takes place at very fast rates. It is expected that at the present rates of abstraction all the reserves will be used within the next 25 to 30 years. The Ministry of Planning has proposed a target to reduce annual irrigation water use from the current 15.3 km^3 to 14.7 km^3 by the year 2000. Measures to be taken are:

- implementation of effective irrigation schedules at farm level to deliver irrigation water according to actual crop need, which is expected to result in a saving of water of at least 30%;
- replacement of surface irrigation systems by sprinkler irrigation and micro-irrigation systems;
- shifting of some of the fodder and cereals areas from high water consumption zones to lower water consumption zones and cultivation of crops with lower water requirements;
- introduction of water meters at farm level to control the pumping of water.

Extensive pumping of groundwater has resulted in a significant drop in the groundwater level (for example 100 metres in the north-west in the last decade), requiring deeper and larger holes to be drilled and a higher head for pumping which results in a higher production cost. Groundwater quality has also deteriorated to the point where it can no longer be used for municipal supply without expensive treatment. Furthermore, only half the groundwater reserves are located near the areas of demand. The coastal areas suffer increasingly from sea water intrusion.

While Saudi Arabia is already by far the largest producer of desalinated water, future development will have to depend even more on the development of this source and on the reuse of treated wastewater. However, as up to present the desalinated water is still too saline for agricultural use, the problem of the rapid depletion of fossil water is still a long way from being solved.

MAIN SOURCES OF INFORMATION

Abdallah E. Dabbagh and Wali A. Abderrahman. 1996. Management of groundwater resources in Saudi Arabia under various irrigation water use scenarios. In: Arabian Journal of Science and Engineering.

Ali A. Al Jaloud. 1991. Agriculture and fertilizer use in Saudi Arabia. Research Institute of Natural Resources and Environment. Country paper for expert consultation on fertigation/chemigation, Cairo, Egypt, 8-11 September 1991.

Department of Economic Studies and Statistics. 1989. Agricultural Sector Development: Graphical Indicators. Ministry of Agriculture and Water.

Ministry of Agriculture and Water. 1992. Agricultural statistical yearbook.

Munther J. Haddadin, Director of the Regional Office for Integrated Development. 1990. Progress of the Implementation of the Mar del Plata Action Plan in the Region of the Near East and North Africa; Country: Saudi Arabia. UNDTCD [United Nations Department of Technical Cooperation for Development].

National Irrigation Department. 1988. Wastewater reuse for agricultural purposes. Ministry of Agriculture and Water.

Saline Water Conversion Corporation. Annual Report 1992.

Walid A. Abderrahman. 1994. Advanced and modern efficient irrigation technology in the Arab World.

Somalia

GEOGRAPHY AND POPULATION

Somalia, with a total area of 637 660 km^2, has the longest coastline in Africa: in the north on the Gulf of Aden and in the east on the Indian Ocean. It is bordered by Kenya in the south, by Ethiopia in the east and by Djibouti in the north-east.

The cultivable area was estimated at about 8 million ha in 1985, or 13% of the total area. In 1984, it was estimated that about 980 000 ha were cultivated with annual crops, i.e. 12% of the cultivable area. About 18 000 ha consisted of permanent crops in 1993.

TABLE 1
Basic statistics and population

Physical areas:			
Area of the country	1995	63 766 000	ha
Cultivable area	1985	8 150 000	ha
Cultivated area	1984	998 000	ha
- annual crops	1984	980 000	ha
- permanent crops	1993	18 000	ha
Population:			
Total population	1995	9 250 000	inhabitants
Population density	1995	15	inhab./km^2
Rural population	1995	74	%
Water supply coverage:			
Urban population	1990	50	%
Rural population	1990	29	%

The total population is about 9.25 million (1995), of which 74% is rural. Average population density is about 15 inhabitants/km^2. The annual demographic growth rate is approximately 3.1%. Agriculture is the second traditional occupation for most Somalis, after nomadic livestock grazing. Some 70% of the working population was engaged in agriculture in 1991 and this sector accounted for 65% of the country's GDP, including forestry and fisheries. Bananas are the principal cash crop, accounting for 40.3% of export earnings in 1988.

CLIMATE AND WATER RESOURCES

Climate

The climate of Somalia is of the arid to semi-arid type, with an average annual temperature of 27°C. It is hotter and drier in the interior and on the Gulf of Aden, but cooler on the Indian Ocean Coast. Annual rainfall is less than 250 mm in the north, about 400 mm in the south, and 700 mm in the south-west. On average, the country receives 253 mm of rainfall per year. Rainfall distribution is bi-modal. It falls mostly in the *gu* (mid-April to June) and the *der* (October to December) seasons. The country is regularly subject to periods of drought.

Water resources

The total internally produced water resources are estimated at 6 km^3/year, the incoming surface water resources at 9.74 km^3/year. Water resources in Somalia are dominated by surface water.

TABLE 2
Water: sources and use

Renewable water resources:			
Average precipitation		253	mm/yr
		161.3	km^3/yr
Internal renewable water resources		6.0	km^3/yr
Total (actual) renewable water resources	1995	15.74	km^3/yr
Dependency ratio	1995	61.9	%
Total (actual) renewable water resources per inhabitant	1995	1 702	m^3/yr
Total dam capacity		-	10^6 m^3
Water withdrawal:			
- agricultural	1987	786	10^6 m^3/yr
- domestic	1987	24	10^6 m^3/yr
- industrial	1987	0	10^6 m^3/yr
Total water withdrawal		810	10^6 m^3/yr
per inhabitant	1987	99	m^3/yr
as % of total (actual) renewable water resources		5.1	%
Other water withdrawal		-	10^6 m^3/yr
Average groundwater depletion		-	10^6 m^3/yr
Wastewater - Non-conventional water sources:			
Wastewater:			
- produced wastewater		-	10^6 m^3/yr
- treated wastewater		-	10^6 m^3/yr
- reused treated wastewater		-	10^6 m^3/yr
Desalinated water	1990	0.1	10^6 m^3/yr

Along the Gulf of Aden, there is a mountainous zone with rugged relief which is subject to torrential flows, causing considerable erosion. The land slopes down towards the south and the south-flowing watercourses peter out in the sands of the desert.

The rest of the country consists of a plateau, which is crossed by the two main rivers of Somalia, the Shebelli and Juba rivers, originating from the Ethiopian plateau and draining in a south-easterly direction towards the Indian Ocean. Over 90% of the discharge of these rivers originates from runoff in the Ethiopian highlands and there are large variations in discharge from year to year. Within Somalia the discharge decreases rapidly, due to losses by seepage, evaporation, overbank spillage due to low channel capacity and water abstraction. Often the rivers cease to flow in the lower reaches during the early part of the year. Contribution to river flow from inside Somalia occurs only during heavy rainfall.

The contribution of other drainage basins to surface water is generally insignificant. This normally consists of occasional runoff in seasonal watercourses.

Groundwater potential is limited because of the limited potential for recharge. In the northern region, some subsurface flows in the wadis are tapped for small irrigated farms (1-25 ha). However, studies are still needed required for the exploration of groundwater.

Dams

There are no dams on the Shebelli river within Somalia, but off-stream storage exists at Jowhar (200 million m^3), upstream from the greater part of the irrigated lands and downstream of the Jowhar sugar estate. A second off-stream storage reservoir, which would store 130-200 million m^3, is proposed for Duduble, upstream of Jowhar. Another proposed dam is the Baardhere dam on the Juba river, primarily for hydropower, but which should also provide maximum water control and storage in the Juba Valley irrigation projects.

TABLE 3
Irrigation and drainage

Irrigation potential		1985	240 000	ha
Irrigation:				
1. Full or partial control irrigation: equipped area		1984	50 000	ha
- surface irrigation			-	ha
- sprinkler irrigation			-	ha
- micro-irrigation			-	ha
% of area irrigated from groundwater			-	%
% of area irrigated from surface water			-	%
% of area irrigated from non-conventional sources			-	%
% of equipped area actually irrigated			-	%
2. Spate irrigation area		1984	150 000	ha
3. Equipped wetland and inland valley bottoms (i.v.b.)			-	ha
Total irrigation (1+2+3)		**1984**	**200 000**	**ha**
- as % of cultivated area			20	%
4. Flood recession cropping area			-	ha
Total water managed area (1+2+3+4)		**1984**	**200 000**	**ha**
- as % of cultivated area			20	%
- increase over last 10 years			-	%
- power irrigated area as % of water managed area			-	%
Full or partial control irrigation schemes:	Criteria			
Large-scale schemes	> - ha	1984	25 000	ha
Medium-scale schemes		1984	0	ha
Small-scale schemes	< - ha	1984	25 000	ha
Total number of households in irrigation		1984	33 750	
Irrigated crops:				
Total irrigated grain production		1984	150 000	tons
as % of total grain production		1984	30	%
Harvested crops under irrigation			**-**	**ha**
- permanent crops: bananas		1984	3 000	ha
- annual crops: total			-	ha
. maize (mainly spate)		1984	150 000	ha
. sugar cane		1984	9 800	ha
. rice		1984	1 300	ha
.			-	ha
. other annual crops			-	ha
Drainage - Environment:				
Drained area			-	ha
as % of cultivated area			-	%
- drained areas in full or partial control irrigated areas			-	ha
- drained areas in equipped wetland and i.v.b.			-	ha
- other drained areas			-	ha
- total drained area with subsurface drains			-	ha
- total drained area with surface drains			-	ha
Flood-protected area			-	ha
Area salinized by irrigation			-	ha
Population affected by water-borne diseases			-	inhabitants

Water withdrawal

In 1987, total water withdrawal was estimated at 0.87 km^3 (Figure 1). Agricultural water withdrawal, about 0.79 km^3 or 97% of total withdrawal, is mainly for the full or partial control irrigation schemes.

FIGURE 1
Water withdrawal (total: 810 million m^3 in 1987)

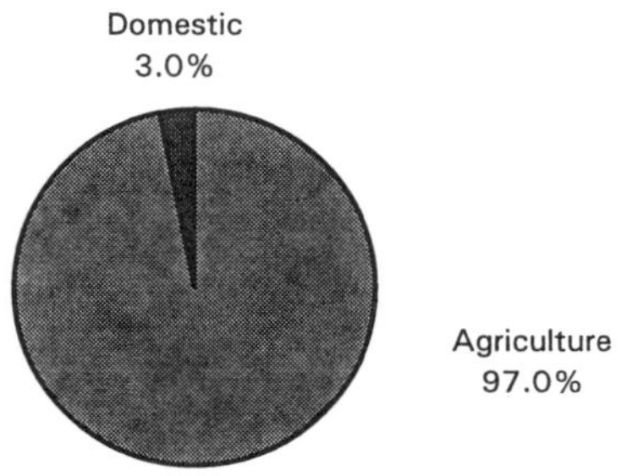

IRRIGATION AND DRAINAGE DEVELOPMENT

Despite the importance of irrigation for the main cash crops in Somalia (bananas and sugar cane), the development of irrigation and drainage systems is very poor. In the main irrigated areas in the Juba and Shebelli valleys, there is no organized system of water allocation and management and there is a salinity problem. Irrigation potential is estimated at 240 000 ha. In 1984, the total water managed area was about 200 000 ha, of which only 50 000 ha had reasonably controlled irrigation, the rest being spate irrigation entirely for maize production (Figure 2).

FIGURE 2
Distribution of the water managed areas
(total: 200 000 ha in 1984)

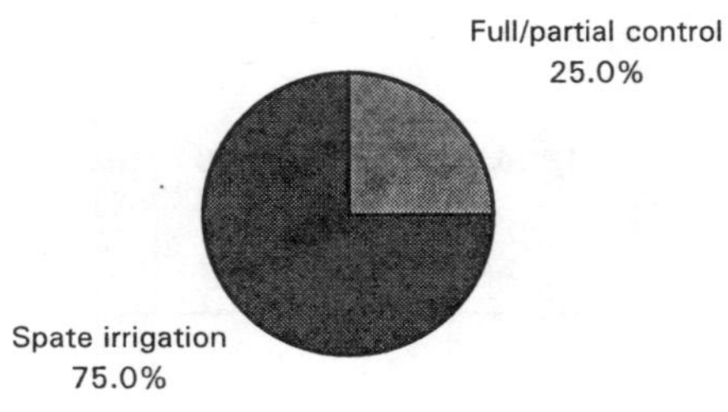

FIGURE 3:
Typology of f/p control irrigation schemes
(total: 50 000 ha in 1984)

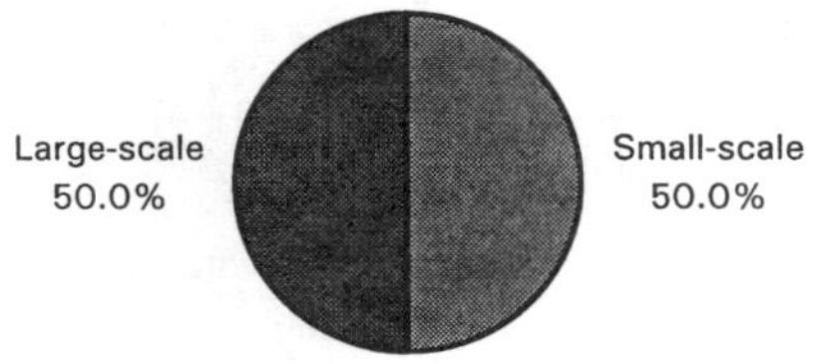

In 1984, as regards full or partial control irrigation schemes, about half were traditional, small-scale schemes and half medium and large private and state schemes (Figure 3), with irrigated farms supporting some 135 000 people. Irrigation is practised mostly along the Shebelli and Juba rivers. For irrigation water management and drainage, services are almost non-existent. The lack of an effective water authority or management system is one reason for the low (20-25%) irrigation efficiency and deteriorating performance. Farmers, individually or in groups, abstract water from rivers or canals regardless of crop rotation or crop water needs. Water use is governed by proximity to the distribution outlet and extraction upstream.

FIGURE 4
Major irrigated crops (1984)

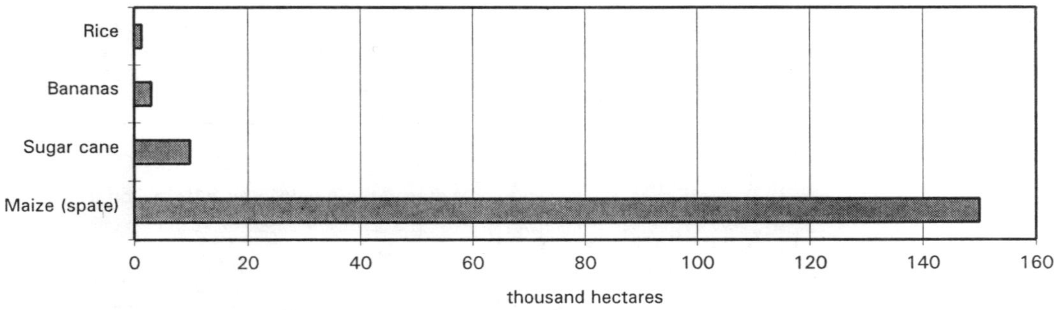

The major irrigated crops are maize (spate), sugar cane (mainly state farms) and bananas (mainly private farms) (Figure 4). Official statistics show, for the period 1970-1986, a slight decrease in area and a static yield for bananas, while there was an increase in area and a considerable drop in yield for sugar cane. This can be attributed mainly to the increasing drainage problems and soil salinity at the Jowhar sugar estate on the Shebelli river. As for maize, an important irrigated and rainfed crop, production increased sharply due to the wide expansion of the planted area (from 102 000 ha in 1971 to 350 000 ha in 1985), while yields remained low. In irrigated areas, maize yields remained low because of the inefficient irrigation system, limited availability of research, the absence of higher yielding varieties, and a shortage of inputs.

INSTITUTIONAL ENVIRONMENT

The main institution in charge of water resources management and development in Somalia is the Ministry of Mineral and Water Resources (MMWR), and its National Water Centre (NWC). The Water Development Agency (WDA) is responsible for operations exploiting groundwater resources for domestic water supply.

TRENDS IN WATER RESOURCES MANAGEMENT

A study carried out by the World Bank in 1987 outlined a proposed strategy for the development of irrigation, drainage and water management systems. In the short term, the plan included: rehabilitation of the existing systems; completion of the Water Master Plan for the Shebelli and Juba rivers; registration of all irrigated farms within a cadastral system; allocation of legal water rights and management of the water allocation system; development of an effective system of water charges; gradual transfer of operation and maintenance responsibility to the holders of water rights; and incentives for adaptive research.

For the long term, the construction of the Duduble reservoir, to control the Shebelli river, and the Baardhere dam on the Juba river were planned. However, even with the completion of the Duduble reservoir, the water availability would still be limited in quantity and quality due to the salinity problem. Upstream 'return-to-river' drainage might contribute to higher salinity further downstream unless outfall drains to the ocean were constructed, in which case the maximum area in the Shebelli valley would be about 50 000 - 60 000 ha, if irrigation efficiency increased to 50%. As far as the Baardhere dam is concerned, it was originally designed to irrigate up to 170 000 ha, but the size of the dam seems to have already been reduced to irrigate 50 000 ha, because of the need to share water with Ethiopia.

The future of irrigation and agriculture in general is primarily related to a return to political stability.

MAIN SOURCES OF INFORMATION

Indian Ocean Newsletters. 1985. Productive sectors of the Economy.

FAO. 1989. A brief description of major drainage basins affecting Somalia. National Water centre, Mogadishu. Field document No. 14. FAO/SOM/85/008, prepared by D. Kammer. FAO, Rome.

World Bank. 1987. Agricultural Sector Survey: Main report and strategy. Report No. 6131-SO. Washington DC.

Sudan

GEOGRAPHY AND POPULATION

Sudan is situated in the north-eastern corner of Africa and is the largest African country, with a total area of about 2.5 million km^2. On the north-east it is bordered by the Red Sea and it shares common borders with nine countries: Eritrea and Ethiopia in the east, Kenya, Uganda and Zaire in the south, the Central African Republic, Chad and Libya in the west and Egypt in the north. Sudan consists of a flat internal plain, lying at about 325 metres above sea level. It is crossed by the Nile river and its tributaries and by a number of mountains. In the south is the Sudd Region, the great wetland which is a maze of channels, lakes and swamps. The most remarkable feature of the Sudd area is its flatness: for 400 km, from south to north, the slope is a mere 0.01% and much of it is still flatter. The soils of the whole area are generally clayish and poor in nutrients. The northern third of the country is covered by a sandy desert with mobile and fixed sand dunes in the north-western part which is considered to be an extension of the eastern outskirts of the Great Desert.

TABLE 1
Basic statistics and population

Physical areas:			
Area of the country	1995	250 581 000	ha
Cultivable area	1995	105 000 000	ha
Cultivated area	1995	7 600 000	ha
- annual crops	1995	7 350 000	ha
- permanent crops	1995	250 000	ha
Population:			
Total population	1995	28 098 000	inhabitants
Population density	1995	11	inhab./km^2
Rural population	1995	75	%
Water supply coverage:			
Urban population	1995	85	%
Rural population	1995	35	%

The cultivable area is estimated to be 105 million ha, or 42% of the total area. The cultivated land is 7.6 million ha, which is 7% of the cultivable area. Only about 3% consists of permanent crops, the remaining area consisting of annual crops.

The population of Sudan is about 28 million (1995), of which 75% is rural. The average population density is about 11 inhabitants/km^2, but there are substantial regional variations and half the population lives on just 15% of the land. The annual demographic growth rate averaged 2.8% between 1985-93. Some 80% of the population works in the agricultural sector. In 1994, agriculture accounted for 37.1% of Sudan's GDP and it provided over 80% of the country's exports.

CLIMATE AND WATER RESOURCES

Climate

The climate of Sudan varies from continental in the northern parts, through savannah in the centre, to equatorial in its most southern parts. Rainfall varies from 20 mm/year in the north to some 1 600 mm/year in the far south. Average annual rainfall is 436 mm.

TABLE 2

Water: sources and use

Renewable water resources:			
Average precipitation		436	mm/yr
		1 092.6	km^3/yr
Internal renewable water resources		35.0	km^3/yr
Total (actual) renewable water resources	1995	88.5	km^3/yr
Dependency ratio	1995	77.3	%
Total (actual) renewable water resources per inhabitant	1995	3 150	m^3/yr
Total dam capacity	1995	8 800	$10^6\ m^3$
Water withdrawal:			
- agricultural	1995	16 800	$10^6\ m^3/yr$
- domestic	1995	800	$10^6\ m^3/yr$
- industrial	1995	200	$10^6\ m^3/yr$
Total water withdrawal		17 800	$10^6\ m^3/yr$
per inhabitant	1995	633	m^3/yr
as % of total (actual) renewable water resources		20.1	%
Other water withdrawal	1995	200	$10^6\ m^3/yr$
Average groundwater depletion		-	$10^6\ m^3/yr$
Wastewater - Non-conventional water sources:			
Wastewater:			
- produced wastewater		-	$10^6\ m^3/yr$
- treated wastewater		-	$10^6\ m^3/yr$
- reused treated wastewater		-	$10^6\ m^3/yr$
Desalinated water	1990	0.4	$10^6\ m^3/yr$

Water resources

Water used in Sudan derives almost exclusively from surface water resources, as groundwater is used in only very limited areas, and mainly for domestic water supply. There are large areas in Sudan where the exploitation of groundwater has been hampered by cost, as the water table is very deep. Internally produced water resources are estimated at 35 km^3/year. Incoming water resources are estimated at 119 km^3/year, resulting in total natural water resources of 154 km^3/year.

Surface water is provided mainly by the Nile river. The main part of Nile is formed by the confluence of the Blue Nile (65%) and the White Nile (23%) in the capital Khartoum and receives, before flowing into Egypt, one more tributary, the Atbara river (12%). Both the Atbara and the Blue Nile rivers originate in the Ethiopian plateaux, while the White Nile originates from the Equatorial Lakes Plateau. The rivers of the Ethiopian catchment are marked by the extreme range in discharge between the peak and low periods, while the flow from the Equatorial Lakes Plateau is more uniform. At its peak the former provides nearly 90% of all water reaching Egypt, the latter only 5%. During the months with low flow the contributions are 30% and 70% respectively. The available average annual flow of the Nile is about 84 km^3 at Aswan at the Sudano-Egyptian border, of which more than 80% between August and October. According to the Nile water agreement between Sudan and Egypt, Sudan's share is 18.5 km^3/year, measured at Aswan, at the border with Egypt. The agreement does not consider possible future reduction in water flowing from upstream countries.

Apart from the Nile system, there are also the seasonal rivers of Gash and Baraka in eastern Sudan. During the rainy period of July-September, the water flow, which is very violent, is drawn off into canals and spread over the land forming a very fertile delta area (spate irrigation). The water balance of Sudan is very complex, due in part to extensive evaporation

TABLE3

Irrigation and drainage

Irrigation potential	1995	2 784 000	ha
Irrigation:			
1. Full or partial control irrigation: equipped area	1995	1 900 000	ha
- surface irrigation		-	ha
- sprinkler irrigation		-	ha
- micro-irrigation		-	ha
% of area irrigated from groundwater	1995	4.0	%
% of area irrigated from surface water	1995	96.0	%
% of area irrigated from non-conventional sources	1995	0.0	%
% of equipped area actually irrigated	1995	63.2	%
2. Spate irrigation area	1984	46 200	ha
3. Equipped wetland and inland valley bottoms (i.v.b.)		-	ha
Total irrigation (1+2+3)	**1995**	**1 946 200**	**ha**
- as % of cultivated area		26	%
4. Flood recession cropping area		-	ha
Total water managed area (1+2+3+4)	**1995**	**1 946 200**	**ha**
- as % of cultivated area		26	%
- increase over last 10 years		-	%
- power irrigated area as % of water managed area	1984	37.3	%
Full or partial control irrigation schemes: Criteria			
Large-scale schemes > - ha		-	ha
Medium-scale schemes		-	ha
Small-scale schemes < - ha		-	ha
Total number of households in irrigation		-	
Irrigated crops:			
Total irrigated grain production	1989	670 562	tons
as % of total grain production	1989	20	%
Harvested crops under irrigation (full or partial control)	**1989**	**1 012 242**	**ha**
- permanent crops: total		-	ha
- annual crops: total	1989	1 012 242	ha
. sorghum	1989	355 320	ha
. cotton	1989	332 640	ha
. wheat	1989	165 060	ha
. groundnut	1989	91 140	ha
. sugar cane	1989	68 082	ha
Drainage - Environment:			
Drained area		-	ha
as % of cultivated area		-	%
- drained areas in full or partial control irrigated areas		-	ha
- drained areas in equipped wetland and i.v.b		-	ha
- other drained areas		-	ha
- total drained area with subsurface drains		-	ha
- total drained area with surface drains		-	ha
Flood-protected area		-	ha
Area salinized by irrigation		-	ha
Population affected by water-borne diseases		-	inhabitants

from the swamps, the best known being the Sudd or Jonglei area on the White Nile in the southern part of Sudan where only half the water entering the region is estimated to flow out of it. Furthermore, from year to year variations in water resources greatly reduce the amount of water actually available for use. In the Table below, elements in the water balance of Sudan are presented, based on information from the Ministry of Irrigation and Water Resources, and including an assessment of the 80%-probability figures.

According to this table, total water 'lost' by evaporation from swamps amounts to 68 km^3/year. It is calculated that the reclamation of part of the swamps, through the construction of the Jonglei Canal that would direct downstream a proportion of the water considered lost each year by spill from the river and evaporation in the swamps, could lead to

Elements in the water balance of Sudan

ELEMENTS	Average (km^3/year)		80% reliable (km^3/year)	
1. Water resources potential:				
- Nationally generated surface water				
. Nilotic	23		15	
. Other	5		3	
- Groundwater recharge	7		5	
Total internal renewable water resources		35		23
Nilotic tributaries at border		117		102
Other international streams		2		1.7
Total natural renewable water resources		154		126.7
2. Currently available water resources:				
Net available surface runoff after losses		86		70.9
Nile water *		20.6		
Other regional streams		1.4		
Internally produced surface runoff		0.7		
Groundwater		0.7		
Total currently available water resources		23.4		

* *This figure takes into account the Sudano-Egyptian agreement of 1959 and corresponds to18.5 km^3/year at the border with Egypt.*

an increase of 10 km^3/year in surface water resources. Sudan and Egypt have agreed to share the additional water equally between them.

Dams

There are four large dams, two on the Blue Nile (the Sennar and the Roseires), one on the Atbara river, and one on the White Nile. The original storage capacity of the first three reservoirs was 5.7 km^3 in total, but by 1990 had been reduced to 3.2 km^3 due to siltation. Apart from the Jebal Aulia dam on the White Nile, which serves mainly to regulate the Nile flow, the other dams serve both as flood control structures and for irrigation.

FIGURE 1
Water withdrawal (total: 17.8 km^3 in 1994)

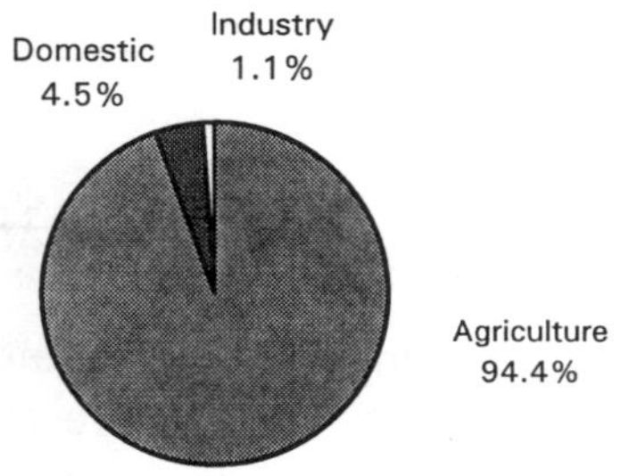

Water withdrawal

In 1995, total water withdrawal was estimated at 17.8 km^3, of which over 94% for agricultural purposes (Figure 1).

Environmental issues

The construction of the Jonglei Canal began in 1978 for a planned total length of 360 km and an average width of 50 metres, but the work stopped in November 1983, after 240 km, because of civil unrest in the region. By that time it also became clear that the 'water losses' in fact create resources in terms of pasture and fisheries and that the canal causes enormous human and environmental problems in the area, where millions of birds come to stay during winter. The canal would have a devastating effect on the region's eco-system and micro-climate, thereby accelerating desertification. If ever restarted, it is likely that the project would have to be revised substantially, taking into consideration the conditions of the local population and the environmental aspects.

IRRIGATION AND DRAINAGE DEVELOPMENT

Agriculture, and especially crop production, is the most significant element in Sudan's economy. Most of the cash crops are produced by irrigation, and irrigated agriculture represents about 50% of total crop production. Sudan has an irrigation potential of about 4.8 million ha considering land resources. Taking into account water resources, the irrigation potential has been estimated at almost 2.8 million ha (without considering possible large scale developments at in the enormous wetland in southern Sudan).

The total water managed area is around 1.95 million ha, or 26% of the cultivated area. Some 46 200 ha is under spate irrigation in the Gash and Tohar deltas; the rest are full or partial control irrigation schemes (Figure 2). All irrigation water comes from surface water. Most schemes are large-scale schemes, which are managed by quasi-public Agricultural Corporations (AC), while small-scale schemes are owned and operated by individuals or cooperatives. The combined Gezira-Managil scheme, located between the Blue and the White Nile, constitutes one of the largest irrigation complexes in the world under single management (about 760 000 ha). It receives its water from the Sennar dam on the Blue Nile and more than 100 000 tenant farmers and their families operate the scheme in partnership with the government and the Sudan Gezira Board, which provides administration, credit and marketing services. Originally planned for the cultivation of cotton, more and more areas are coming under food crop production. Other large schemes are the Rahad Scheme, which receives its water from the Roseires dam on the Blue Nile, and the New Halfa Project (also known as Khashm Al-Girba), located on the Atbara river in the east of the country. The latter project was partly financed by Egypt after the newly constructed High Aswan dam created lake Nasser, which flooded the Sudanese town of Wadi Halfa in 1964. The inhabitants were moved to the new irrigated agricultural lands where they have been growing a variety of crops for over 30 years.

FIGURE 2
Distribution of the water managed areas (total 1 946 200 ha in 1990)

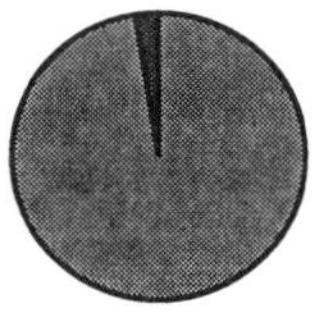

Although irrigated agriculture has been Sudan's greatest economic investment, returns have been far below potential. A study by the World Bank showed that, during the period 1976-1989, yields were low and extremely variable, and cultivated areas suffered gradual decline. Cropping intensity in the Gezira Scheme, dropped from 75% to 57%, as 126 000 ha were taken out of production due to siltation and mismanagement of the canals, leading to reduced availability of water. Because of bad water management, water supply is about 12% below crop requirements at crucial points in the growth cycle, while at the same time as much as 30% of the water delivered is not used by crops. The new government in Sudan reports that since 1990 there has been considerable improvement in agricultural crop production and returns. Major irrigated crops are cash crops (cotton, groundnut, sugar), wheat and sorghum (Figure 3).

FIGURE 3
Irrigated crops (total: 1.01 million ha in 1989)

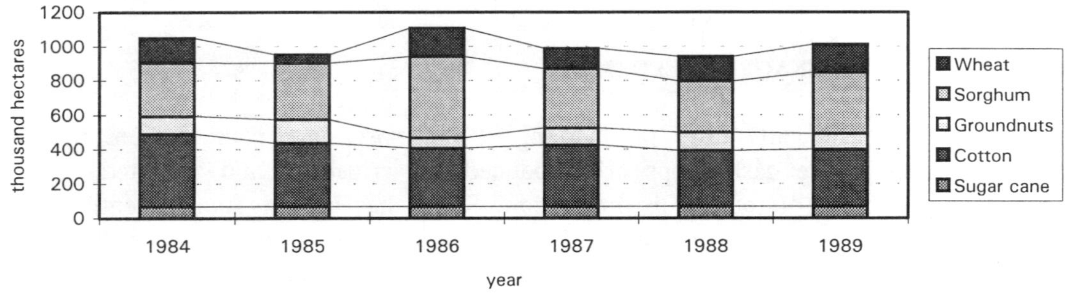

INSTITUTIONAL ENVIRONMENT

Since independence in 1956, Sudan has established a well developed institutional infrastructure in an attempt to make its irrigation sector more efficient. At the top, the National Nile Waters Commission determines the allocation of water to each province. The Ministry of Irrigation and Water Resources (MOI&WR) is responsible for the delivery of irrigation water to the major canals and, jointly with the Agricultural Corporations (AC), for the operation of minor canals. MOI&WR is the sole authority for surface water resources assessment and development in Sudan, whereas the Groundwater Corporation has the same responsibility regarding subterranean water resources. The Hydraulic Research Station (HRS) of MOI&WR deals with the development of theoretical and applied research concerning surface water resources. Regarding the domestic water supply, the responsibility for its management and development is divided between the Urban Waters Corporation and the Rural Waters Corporation.

TRENDS IN WATER RESOURCES MANAGEMENT

Under the reforming policies of the post-1989 government in Sudan, the development of irrigated agriculture has become the crucial. There are ongoing studies for the expansion and rehabilitation of irrigation projects. Also, the construction of new hydraulic structures, such as dams and major irrigation canals, are an essential part of the planned development. Important projects include the raising the height of Roseires dam, the excavation of two major canals for water diversion from the Blue Nile, and the construction of the Merawi dam in the northern part of the country. Further irrigation development is limited by the amount of water allocated by the Nile Water Agreement with Egypt.

Three major constraints to irrigation development in Sudan are:

- the dual management of minor canals by ACs and MOI&WR which has always adversely affected water delivery, making irrigation timing very variable;
- the ineffective process of the annual removal of silt and weed from irrigation canals, which is the responsibility of quasi-public monopolies, causing continuous shrinkage of the actual cultivable area;

- the continuous increase in the cost of development which is aggravated by the continuous devaluation of the local currency.

MAIN SOURCES OF INFORMATION

Abdelsalam A. 1991. Sedimentation in Sudan Multipurpose Reservoirs. Internal Publ. Hydraulic Research Station, Sudan.

Craig, G.M. (ed.). 1991. The agriculture of Sudan. Oxford University Press.

Howell, Paul; Lock, Michael; Cobb, Stephen (eds). 1988. The Jonglei Canal: Impact and opportunity. Cambridge University Press.

World Bank. 1990. Sudan - Reversing the Economic Decline. Country Economic Memorandum.

Syria

GEOGRAPHY AND POPULATION

Syria, with a total area of 185 180 km^2, is bordered in the north by Turkey, in the east and south-east by Iraq, in the south by Jordan, in the south-west by Israel and in the west by Lebanon and the Mediterranean sea. Administratively, the country is divided into 14 governorates, one of which being the capital Damascus.

The country can be divided into 4 physiographic regions:

- the coastal region between the mountains and the sea;
- the mountains and the highlands extending from north to south parallel to the Mediterranean coast;
- the plains, or interior, located east of the highlands and including the plains of Damascus, Homs, Hama, Aleppo, Al-Hassakeh and Dara'a;
- the Badiah and the desert plains in the south-eastern part of the country, bordering Jordan and Iraq.

TABLE 1
Basic statistics and population

Physical areas:			
Area of the country	1995	18 518 000	ha
Cultivable area	1993	5 939 047	ha
Cultivated area	1993	4 938 670	ha
- annual crops	1993	4 266 256	ha
- permanent crops	1993	672 414	ha
Population:			
Total population	1995	14 661 000	inhabitants
Population density	1995	79	inhab./km^2
Rural population	1995	48	%
Water supply coverage:			
Urban population	1993	90	%
Rural population	1993	60	%

In 1993, the cultivable land was estimated at 5.94 million ha, or 32% of the total area of the country. The cultivated land was estimated at 4.94 million ha, which is 83% of the cultivable area. Of this area, 4.27 million ha consisted of annual crops and 0.67 million ha consisted of permanent crops. About 62% of the cultivated area is located in the three northern governorates Aleppo, Al-Reqqa and Al-Hassakeh, representing only 33% of the total area of the country.

The total population is 14.6 million (1995), of which 48% is rural. Actual population growth is 3.3%. In 1993, agriculture employed around 22.5% of the total labour force, accounted for nearly 28% of GDP and 60% of non-oil exports.

CLIMATE AND WATER RESOURCES

Climate

Syria's climate is Mediterranean with continental influence: cool rainy winters and warm dry summers, with relatively short spring and autumn seasons. Large parts of Syria are exposed to high variability in daily temperature.The maximum difference in daily temperature can be as high as 32°C in the interior and about 13°C in the coastal region. Total annual precipitation ranges from 100 to 150 mm in the north-west, 150 to 200 mm from the south

TABLE 2
Water: sources and use

Renewable water resources:			
Average precipitation		252	mm/yr
		46.6	km^3/yr
Internal renewable water resources		7.0	km^3/yr
Total (actual) renewable water resources	1995	26.26	km^3/yr
Dependency ratio	1995	80.3	%
Total (actual) renewable water resources per inhabitant	1995	1 791	m^3/yr
Total dam capacity	1994	15 800	10^6 m^3
Water withdrawal:			
- agricultural	1993	13 600	10^6 m^3/yr
- domestic	1993	530	10^6 m^3/yr
- industrial	1993	280	10^6 m^3/yr
Total water withdrawal		14 410	10^6 m^3/yr
per inhabitant	1993	1 017	m^3/yr
as % of total (actual) renewable water resources		54.9	%
Other water withdrawal		-	10^6 m^3/yr
Average groundwater depletion	1993	1 820	10^6 m^3/yr
Wastewater - Non-conventional water sources:			
Wastewater:			
- produced wastewater	1993	610	10^6 m^3/yr
- treated wastewater	1993	370	10^6 m^3/yr
- reused treated wastewater	1993	370	10^6 m^3/yr
Desalinated water		-	10^6 m^3/yr

towards the central and east-central areas, 300 to 600 mm in the plains and along the foothills in the west, and 800 to 1 000 mm along the coast, increasing to 1 400 mm in the mountains. The Average annual rainfall in the country is 252 mm giving 46.6 km^3.

Water resources

There are 16 main rivers and tributaries in the country, of which 6 main international rivers:

- the Euphrates (Al Furat), which is Syria's the largest river. It comes from Turkey and flows to Iraq. Its total length is 2 330 km, of which 680 km are in Syria;
- the Afrin in the north-western part of the country, which comes from Turkey, crosses Syria and flows back to Turkey;
- the Orontes (El-Assi) in the western part of the country, which comes from Lebanon and flows into Turkey;
- the Yarmouk in the south-western part of the country with sources in Syria and Jordan and which forms the border between these two countries before flowing into the Jordan river;
- the El-Kebir with sources in Syria and Lebanon and which forms the border between them before flowing to the sea.
- the Tigris, which forms the border between Syria and Turkey in the extreme north-eastern part.

The natural average surface runoff to Syria from international rivers is estimated at 28.73 km^3/year. If we include 15.75 km^3 of water entering with the Euphrates, as proposed by Turkey, and 0.43 km^3 of water entering with the Orontes, as agreed with Lebanon, this total becomes 18.11 km^3/year. The Tigris, which is the second most important river in the country, borders the country to the east and has a mean annual flow of 18 km^3. Total natural average outflow from Syria is 31.975 km^3/year, of which an agreement exists for 9.2 km^3, resulting in a total of actual external surface water resources for Syria of 17.91 km^3/year (27.11 km^3/year - 9.2 km^3/year). A breakdown of surface water entering, bordering and leaving the country is given in the Table below (page 230).

TABLE 3
Irrigation and drainage

Irrigation potential	1993	1 250 000	ha
Irrigation:			
1. Full or partial control irrigation: equipped area	1993	1 013 273	ha
- surface irrigation	1993	981 273	ha
- sprinkler irrigation	1993	30 000	ha
- micro-irrigation	1993	2 000	ha
% of area irrigated from groundwater	1993	60.2	%
% of area irrigated from surface water	1993	39.8	%
% of area irrigated from non-conventional sources	1993	0.0	%
% of equipped area actually irrigated		-	%
2. Spate irrigation area		-	ha
3. Equipped wetland and inland valley bottoms (i.v.b.)		-	ha
Total irrigation (1+2+3)	**1993**	**1 013 273**	**ha**
- as % of cultivated area		21	%
4. Flood recession cropping area		-	ha
Total water managed area (1+2+3+4)	**1993**	**1 013 273**	**ha**
- as % of cultivated area		21	%
- increase over last 10 years	1983-93	75	%
- power irrigated area as % of water managed area		-	%
Full or partial control irrigation schemes: Criteria			
Large-scale schemes >20 000 ha		-	ha
Medium-scale schemes		-	ha
Small-scale schemes <2 000 ha		-	ha
Total number of households in irrigation	1994	99 950	
Irrigated crops:			
Total irrigated grain production	1993	2 400 000	tons
as % of total grain production	1993	45	%
Harvested crops under irrigation (full or partial control)	**1993**	**1 204 000**	**ha**
- permanent crops: total	1993	123 925	ha
- annual crops: total	1993	1 080 075	ha
. wheat	1993	550 950	ha
. cotton	1993	200 822	ha
. vegetables	1993	128 719	ha
. maize	1993	62 772	ha
. other annual crops	1993	136 812	ha
Drainage - Environment:			
Drained area	1993	273 030	ha
as % of cultivated area		6	%
- drained areas in full or partial control irrigated areas	1993	273 030	ha
- drained areas in equipped wetland and i.v.b		-	ha
- other drained areas		-	ha
- total drained area with subsurface drains		-	ha
- total drained area with surface drains		-	ha
Flood-protected area		-	ha
Area salinized by irrigation	1989	60 000	ha
Population affected by water-borne diseases		-	inhabitants

Although figures for water resources are very difficult to obtain due to the lack of reliable data, it can be estimated that water resources generated from rain falling within the country amount to 7 km^3/year. Groundwater recharge is about 4.2 km^3/year, of which 2 km^3/year discharges into rivers as spring water. Total groundwater inflow has been estimated at 1.35 km^3/year, of which 1.2 km^3 from Turkey and 0.15 km^3 from Lebanon. Although not quantified, the amount of groundwater flowing into Jordan may be significant

Dams and lakes

There are 141 dams in Syria with a total storage capacity of 15.8 km^3. The largest dam is located at Al-Tabka on the Euphrates. It forms the Al-Assad lake with a storage capacity of 11.2 km^3. Medium-sized dams include the Al-Rastan (225 million m^3),

Major rivers entering, bordering and leaving Syria

Name of river	Inflow into Syria (km^3/year)			Outflow from Syria (km^3/year)		
	from	natural	actual	to	natural	agreement
Euphrates *	Turkey	26.29	15.75	Iraq	30	9
Tributaries of Euphrates	Turkey	1.74	1.74		-	
Afrin	Turkey	0.19	0.19	Turkey	0.25	
Orontes, El Kebir	Lebanon	0.51	0.43		1.2	
Yarmouk		-	-	Jordan	0.4	0.2
Baniyas		-	-	Israel	0.125	
Sub-total		28.73	18.11		31.975	9.2
Bordering Tigris	50% of total	9	9			
Total	**Inflow**	**37.73**	**27.11**	**Outflow**	**31.975**	**9.2**

* : *Turkey has unilaterally promised to secure a minimum flow of 15.75 km3/year at its border with Syria*

the Mouhardeh (50 million m^3) and the Taldo (15.5 million m^3). There are some 20 dams classified as small, the largest of which is the Dara'a, with a storage capacity of 15 million m^3. The majority of these dams are located near Homs and Hama.

Apart from the Al-Assad lake, there are five lakes in Syria, the largest being lake Jabboul near Aleppo with a surface area of about 239 km^2. Lake Qattineh near Homs is the main perennial lake in Syria.

Water withdrawal

Total annual water withdrawal in Syria in 1993 was estimated at 14.41 km^3/year, of which agricultural use accounted for 94% (Figure 1). The assessment of agricultural water use is based on the assumption that the maximum water requirement is about 13 450 m^3/ha per year. The treatment of domestic wastewater is carried out mainly in the towns of Damascus, Aleppo, Homs and Salamieh. The total amount of treated wastewater was estimated at 0.37 km^3/year in 1993, which represents 60% of the total produced volume of 0.61 km^3/year. All treated wastewater is reused.

FIGURE 1
Water withdrawal (total: 14.41 km^3 in 1993)

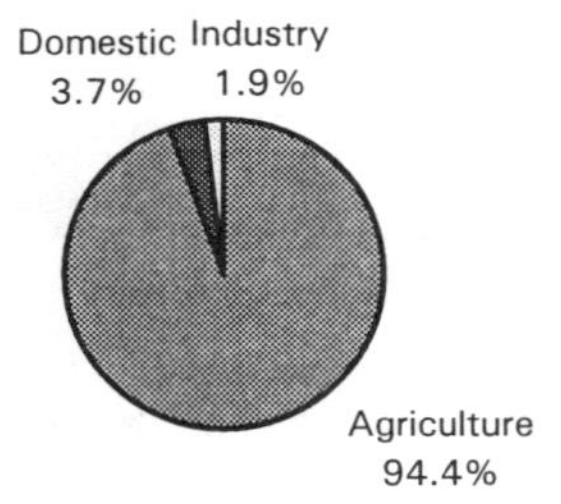

International rivers and agreements

As mentioned above, as far as the Euphrates river is concerned, Turkey unilaterally decided to secure a minimum flow of 500 m^3/s (15.75 km^3/year) to enter Syria, although more has been demanded. According to an agreement between Syria and Iraq (1990), Syria agrees to share the Euphrates water with Iraq on a 58% (Iraq) and 42% (Syria) basis, which corresponds to a flow of 9 km^3/year at the border with Iraq when using the figure of 15.75 km^3/year from Turkey. Up to now, there has been no global agreement between the three countries concerning the Euphrates waters. An agreement was signed in 1955 between Syria and Jordan regarding the allocation of the water of the Yarmouk river, and was further revised in 1987. A recent agreement between Lebanon and Syria on the Orontes river has led to a share of 80 million m^3/year for Lebanon and the remainder for Syria.

IRRIGATION AND DRAINAGE DEVELOPMENT

Estimates on irrigation potential, based solely on soil resources, lead to a figure of around 5.9 million ha, which is roughly equal to the cultivable area. Considering the water resources available at present, irrigation potential is estimated at 1 250 000 ha. The exact figure, however, depends on how Syria reaches agreements with neighbouring countries on the sharing of river waters in the future.

In 1993, the total area equipped for irrigation was estimated at 1 013 273 ha, which is 81% of the above irrigation potential. Irrigation is mainly developed in the north-eastern part of the country and more than one-third of the irrigated areas are located in the Al-Hassakeh governorate.

Surface irrigation is reported to be practised on 981 273 ha (Figure 2). The predominant system of surface irrigation practised is basin irrigation. Traditionally, irrigation in Syria has relied on flooding for cereals, furrow irrigation for vegetables and basin irrigation for fruit trees. Sprinkler irrigation is practised on 30 000 ha, mostly in the Homs, Aleppo and Al-Hassakeh governorates. It has been developing on a wider scale recently, mainly because of groundwater scarcity, which has encouraged farmers to develop water-saving techniques, and because of equipment becoming available on the local market. Micro-irrigation is practised on 2 000 ha.

FIGURE 2
Irrigation techniques f/p (total: 1 013 273 ha)

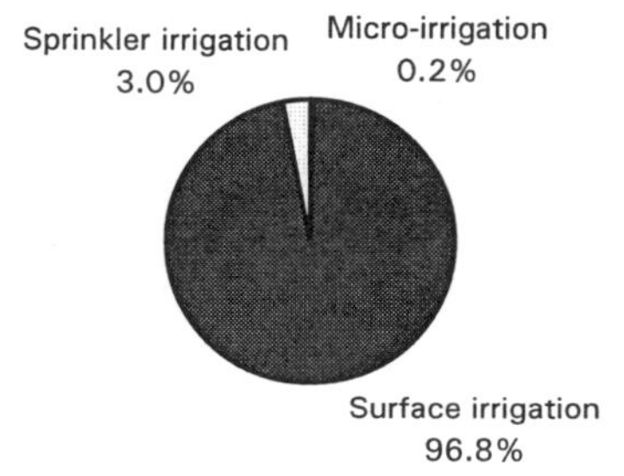

In 1993 it was estimated that 60.2% of the area was irrigated from groundwater and 39.8% from surface water (gravity-fed or pumped from the rivers and lakes) (Figure 3). The use of groundwater for irrigation has been expanding rapidly in the last five years because irrigation from groundwater is cheaper than irrigation by gravity. Since pumped water is free, the only investment expense required is the well and a suitable pump. Farmers obtain credit at preferential rates of interest to purchase fuel and imported pumps at subsidized prices. A second factor explaining the increase in irrigation from groundwater is the large number of farmers with small farm holdings. The latest available data estimate that over 75% of the farmers have holdings of less than 10 ha. On average, it is estimated that a household has 3.5 ha of irrigated land. Since most private farmers want secure and independent access to a supply of water, they often dig their own wells. In 1994, the total number of wells in the country was estimated at 122 276 of which 53 453 were not licensed (Figure 4).

FIGURE 3
Origin of irrigation water f/p (total: 1 013 273 ha)

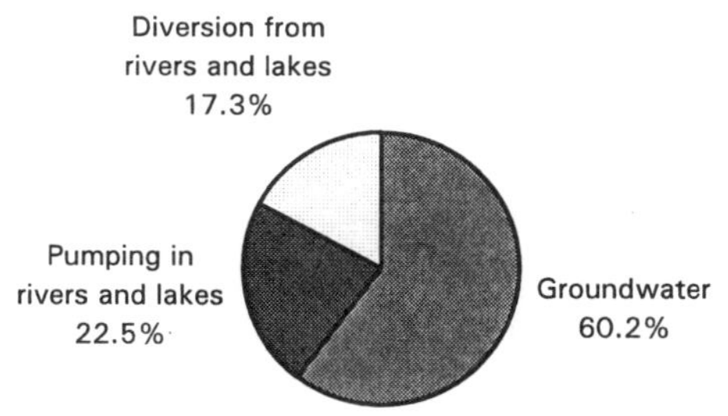

FIGURE 4
Licensed and not licensed wells (total: 122 276 wells in 1994)

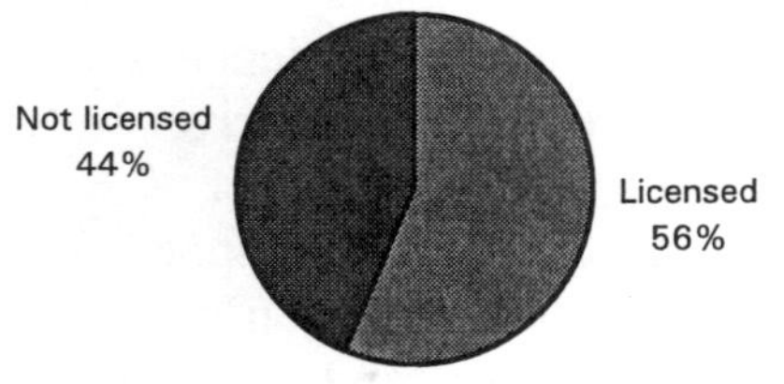

Of the total area of 1 013 273 ha equipped for irrigation, 349 820 ha are involved in government irrigation projects and 78% of this land is also equipped for drainage (see the Table below). It is estimated that 105 000 ha of these government irrigated areas need rehabilitation.

Government projects

Size	Criteria	Total area in ha	as % of total
Large	> 20 000 ha	257 860	74
Medium		47 840	14
Small	< 2 000 ha	44 220	12
Total		349 820	100

There is a wide variation in cropping patterns in the irrigated areas, depending on the water resources available and the agro-climatological conditions. Strategic crops such as wheat and cotton are concentrated in the northern and eastern part of the country. More than 50% of the wheat and cotton produced comes from the Al-Hassakeh governorate, in the north-eastern part of the country.The production of winter vegetables is centred in the coastal region, while summer vegetables are produced mainly in the internal plains, especially in the central and southern regions. In 1993, of the total area equipped for irrigation of 1.01 million ha, about 0.12 million ha were planted with fruit trees and olives, while 0.89 million ha were used for annual crops. The cropping intensity for annual crops reached 121%, leading to a total cropped area of annual crops of about 1.08 million ha of which 0.19 million ha with double cropping (winter and summer), 0.49 million ha planted during the winter only and 0.21 million ha planted during the summer only (Figure 5).

FIGURE 5
Irrigated crops f/p (total 1 204 000 ha in 1993)

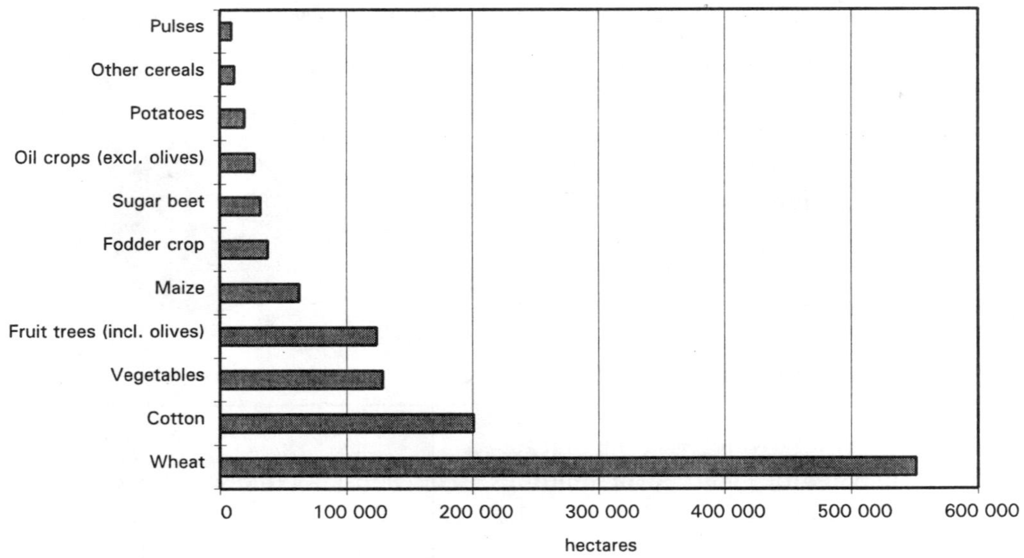

The average yield for irrigated wheat was estimated at 4 tons/ha in 1993, varying between 2.5 and 5.2 tons/ha over the different governorates. The yield for rainfed wheat varied between 1.3 tons/ha for standard varieties and 2 tons/ha for high productivity varieties. The average yield of irrigated cotton was 3.1 tons/ha. Irrigated barley is mainly used as fodder crop with yields reaching 15 tons/ha.

The average cost of surface irrigation development varies between $US 3 800/ha for small, 6 600/ha for medium and 7 600/ha for large schemes. The average operation and maintenance cost varies between $US 60 and 85/ha per year for gravity and $US 155 and 238/ha per year for pumping. However, private farmers using water from government schemes pay only $US 26/ha per year. The current capital cost of installing 1 ha of micro-irrigation is estimated at $US 1 000 for locally manufactured equipment and $US 1 400 for imported equipment. There is no fee charged for irrigation water, only for operation and maintenance.

Drainage is mainly developed in the governorates bordering the Euphrates river. In the Al-Reqqa governorate, for instance, 62% of the irrigated area is drained. About 24% of the total drained area is power drained. The drainage systems are generally mixed systems of surface and subsurface drainage. In 1993, 60 000 ha of irrigated land were estimated to be affected by salinization. Some 5 000 ha in the Euphrates basin have been abandoned due to water logging and salinity problems. In the new irrigation scheme, open drainage systems have been installed on 90% of the irrigated land. Only a small area has been equipped with subsurface drains.

Unit costs for irrigation development have increased considerably in the last two decades and this is one of the reasons why since the 1970s attention has also been given to drainage and irrigation rehabilitation, mainly in the Euphrates valley where irrigation through pumping from the river has developed rapidly since the 1950s. Appreciable progress has been made in restoring large irrigated areas which went out of cultivation due to waterlogging and salinity especially in the lower and middle parts of the Euphrates valley.

INSTITUTIONAL ENVIRONMENT

The Ministry of Irrigation is in charge of irrigation, dams, planning, research, operation and maintenance and pollution control. The Directorate of Irrigation is involved in water resources studies and surveys, water legislation and sharing international waters. There are three other departments under the responsibility of the Ministry of Irrigation: the Euphrates Basin Development Authority, the Euphrates Basin Land Reclamation Authority and the General Company of Major Water Resources Studies.

The Ministry of Irrigation is in charge of groundwater monitoring and the issuing of licenses for groundwater well drilling. In some areas with a high concentration of wells, such as parts of the Aleppo and Salamieh areas, the water table is dropping. The Ministry is exploring means of increasing the recharge of the shallow aquifers.

There are 4 other organizations involved in the water sector in Syria:

- the Ministry of Agriculture and Agrarian Reform, Directorate of Irrigation and Water Uses;
- the Ministry of Housing and Public services, Directorate of Water Supply and Waste Water;

- the State Planning Commission, Section: Irrigation and Agriculture Sector;
- the State Environmental Affairs Commission, Section: Water Environment Safety Sector.

TRENDS IN WATER RESOURCES MANAGEMENT

The agriculture sector is a major source of income, foreign exchange and labour in Syria. The irrigated area produces over 50% of the total value of agricultural production on about 18.6% of the cultivated land. A large part of the wheat production, as well as all major industrial crops including cotton, tobacco and sugar beet are produced on irrigated farms. The development and utilization of water so far has been carried out on an ad hoc basis, mainly responding to various demands. The need to increase food production has resulted in the construction of dams for irrigating lands under their command; and attractive prices for food crops, particularly for wheat recently, have resulted in a rapid increase in wells and the over exploitation of groundwater. In addition, an increase in population and the proliferation of industries in major cities have contributed to further exploitation of the limited water resources.

Under the latest development plans, between 60 and 70% of public investment in agriculture was allocated to irrigation development, in particular for the construction of the Taqba dam and the establishment of the Assad reservoir.

Irrigated agriculture in the Damascus basin has to compete for both land and water with the residential and industrial expansion of the city. Irrigated agriculture has already been seriously reduced in the Damascus basin in recent decades.

Irrigation development to a large extent depends on how Syria reaches agreements with neighbouring countries on the sharing of river waters (Turkey, Lebanon, Jordan and Iraq). Identifying and implementing policies, programs, projects and techniques to improve water use efficiency and to better control surface water and groundwater exploitation are important challenges facing Syrian policy-makers. Future projects have been planned by the government for the development of major schemes: 91 000 ha in the Euphrates basin, 150 000 ha in the Khabour (tributary of the Euphrates) basin with the construction of a reservoir north of Al-Hassakeh, 150 000 ha in the Tigris basin by pumping from the Tigris river and 72 000 ha in the Orontes basin (Al-Ghab).

MAIN SOURCES OF INFORMATION

Central Bureau of Statistics. 1993. Statistical abstract 1992. Damascus, Syria.

Directorate of Irrigation and Water Use. 1993. Water requirements for the agricultural plan 1993-94. Damascus.

ESCWA (Economic and Social Commission for Western Asia) / FAO / MAAR (Ministry of Agriculture and Agrarian Reform). 1995. Evaluation of agricultural policies in the Syrian Arab Republic (policy analysis matrix approach). Damascus, Syria.

ESCWA/FAO/MAAR. 1995. National Farm Data Handbook 1994. United Nations, New York.

FAO. 1993. Irrigation sub-sector review. Mission Report. Rome, Italy.

Ministry of Agriculture and Agrarian Reform. 1993. Annual plan for agricultural production 1993-94. Damascus, Syria.

Ministry of Agriculture and Agrarian Reform. 1993. Balance of land utilization for 1993.

Soumi, George. 1993. Management and development of water resources and their use rationalisation. Report to the Technical Conference of the Arab Agricultural Engineers Federation in Tunis.

UNDP/FAO. 1994. Improved management of water resources for agricultural use (phase II). SYR/90/001. Damascus, Syria.

Tajikistan

Note: *Tajikistan was officially included amongst the countries of FAO's Regional Office for the Near East in March 1996. For the present publication, it has not been possible to perform an in-depth survey of the information on water and irrigation and so the data presented below are probably very incomplete.*

GEOGRAPHY AND POPULATION

Tajikistan is a mountainous, landlocked country in Central Asia, bordered in the west and north-west by Uzbekistan, in the north-east by Kyrgyzstan, in the east by China and in the south by Afghanistan. It became independent in September 1991 and has a total area of 143 100 km². The different regions of the country are separated by high mountain ranges and are often cut off from each other during the winter months. The Pamir mountains in the south-east are part of the Himalayan mountain chain and are among the highest and most inaccessible mountains in the world.

TABLE 1
Basic statistics and population

Physical areas:			
Area of the country	1995	14 310 000	ha
Cultivable area	1992	-	ha
Cultivated area	1992	812 000	ha
- annual crops	1992	756 000	ha
-permanent crops	1992	56 000	ha
Population:			
Total population	1995	6 101 000	inhabitants
Population density	1995	43	inhab./km²
Rural population	1995	68	%
Water supply coverage:			
Urban population		-	%
Rural population	1992	40	%

In 1992, the total cultivated area was estimated at 812 000 ha, or less than 6% of the total area of the country. About 756 000 ha consisted of annual crops and 56 000 ha consisted of permanent crops, of which more than half are vineyards. In 1991, there were 206 collective farms (*kolkoz*), 362 state farms (*sovkholz*) and 19 inter-enterprise farms (*meshkov*). Private plots and land allocated to state farm employees totalled only about 75 000 ha in 1991.

The total population is about 6.1 million (1995), of which 68% is rural. Annual population growth is estimated at 3.1%. Almost 47% of the labour force is engaged in agriculture. The agricultural sector, which has registered negative growth rates since 1989, fell by about 10% in 1991 due to cold weather and heavy rainfall early in the planting season. In 1992, the situation worsened as shortages of key inputs and civil unrest severely hampered production in spite of relatively favourable weather conditions. The agricultural sector as a whole declined by 28% in 1992.

CLIMATE AND WATER RESOURCES

Climate

Tajikistan's mountainous terrain gives rise to a wide range of climates. In those areas where cultivation takes place, which is mainly in the flood plains of the rivers, the climate consists of hot, dry summers and mild, warm winters.

TABLE2
Water: sources and use

Renewable water resources:			
Average precipitation		-	mm/yr
		-	km^3/yr
Internal renewable water resources		61.8	km^3/yr
Total (actual) renewable water resources	1995	39.5	km^3/yr
Dependency ratio	1995	24.5	%
Total (actual) renewable water resources per inhabitant	1995	6 474	m^3/yr
Total dam capacity		-	10^6 m^3
Water withdrawal:			
- agricultural	1989	11 088	10^6 m^3/yr
- domestic	1989	630	10^6 m^3/yr
- industrial	1989	882	10^6 m^3/yr
Total water withdrawal		12 600	10^6 m^3/yr
per inhabitant	1989	2 065	m^3/yr
as % of total (actual) renewable water resources		31.9	%
Other water withdrawal		-	10^6 m^3/yr
Average groundwater depletion		-	10^6 m^3/yr
Wastewater - Non-conventional water sources:			
Wastewater:			
- produced wastewater		-	10^6 m^3/yr
- treated wastewater		-	10^6 m^3/yr
- reused treated wastewater		-	10^6 m^3/yr
Desalinated water		-	10^6 m^3/yr

Water resources

Water is the most abundant natural resource in Tajikistan. There are 8 500 glaciers in the country, covering 6% of the total land area. These glaciers store 455.9 km^3 of water and, together with winter rain and snow, feed the valleys by flowing into 947 streams and rivers for a total length of 28 500 km. Annual total water runoff from the mountains is about 61.8 km^3.

Almost all rivers flow into one of the two major river systems of Central Asia, the Amu Darya and the Syr Darya, which are the main contributors to the Aral Sea. The majority of the rivers in Tajikistan belong to the Amu Darya river basin, including the Pianj, the Vakhsh and the Kafirnigan rivers. The Syr Darya river basin represents only a small area in the north. Although the country covers only 5.7% of the total area of the Aral Sea basin, it produces 54.8 km^3/year of runoff, which is 44% of the total runoff of the Aral Sea basin. About 1 300 lakes make up 1% of the total area of the country.

Dams

Hydroelectricity is the most important source of energy in Tajikistan and its hydropower production comes third in the world, after the United States and Russia. Exploitation of this important resource had been a high priority of the former Soviet Union since the early 1960s, when the construction of several dams on the lower Vakhsh river began, including the Nurek Dam. This dam is situated in a narrow canyon, 300 metres deep and 40 metres wide. In spite of its large size, with a capacity of 10.5 km^3, the Nurek reservoir does not provide adequate flow regulation, since its capacity is about half the annual inflow of the Vakhsh river (20.5 km^3). To provide year to year and seasonal storage to meet the irrigation demand in Uzbekistan, Turkmenistan and Kazakhstan, it was decided in the early 1980s to construct a larger dam (the Rogun Dam), with a reservoir capacity of 13.5 km^3, about 70 km upstream of

TABLE 3
Irrigation and drainage

Irrigation potential		1994	1 570 000	ha
Irrigation:				
1. Full or partial control irrigation: equipped area		1992	718 000	ha
- surface irrigation			-	ha
- sprinkler irrigation			-	ha
- micro-irrigation			-	ha
% of area irrigated from groundwater		1992	8.4	%
% of area irrigated from surface water		1992	91.6	%
% of area irrigated from non-conventional sources		1992	0.0	%
% of equipped area actually irrigated		1992	97.2	%
2. Spate irrigation area			-	ha
3. Equipped wetland and inland valley bottoms (i.v.b.)			-	ha
Total irrigation (1+2+3)		**1992**	**718 000**	**ha**
- as % of cultivated area			88	%
4. Flood recession cropping area			-	ha
Total water managed area (1+2+3+4)		**1992**	**718 000**	**ha**
- as % of cultivated area			88	%
- increase over last 10 years			-	%
- power irrigated area as % of water managed area		1992	59.9	%
Full or partial control irrigation schemes:	Criteria			
Large-scale schemes	> - ha	1992	670 000	ha
Medium-scale schemes		1992	0	ha
Small-scale schemes	< - ha	1992	48 000	ha
Total number of households in irrigation			-	
Irrigated crops:				
Total irrigated grain production			-	tons
as % of total grain production			-	%
Harvested crops under irrigation (full or partial control)		**1992**	**698 000**	**ha**
- permanent crops: total			-	ha
- annual crops: total			-	ha
. cotton		1992	296 000	ha
. fodder crops			-	ha
. vegetables			-	ha
. potatoes			-	ha
. other annual crops			-	ha
Drainage - Environment:				
Drained area			-	ha
as % of cultivated area			-	%
- drained areas in full or partial control irrigated areas			-	ha
- drained areas in equipped wetland and i.v.b.			-	ha
- other drained areas			-	ha
- total drained area with subsurface drains			-	ha
- total drained area with surface drains			-	ha
Flood-protected area			-	ha
Area salinized by irrigation			-	ha
Population affected by water-borne diseases			-	inhabitants

the Nurek Dam. Construction of this dam started in 1988 and was planned to be completed in 1997-2002. However, progress has been slow, mainly because of financial problems. Furthermore, in May 1993 a flood caused severe damage and destroyed much of the infrastructure.

Water withdrawal

Total water withdrawal was 12.6 km^3 in 1989, of which 88% for agricultural purposes (Figure 1). About 91% of the total water used is surface water, mainly from the Amu Darya river system. An inadequate supply of clean drinking water is a serious health hazard in Tajikistan. Even before the civil war in 1992, only 65% of the total population had access to piped water and in many towns the distribution systems were in a poor condition. In rural areas, only 40% of the population had piped water, while the remainder obtained water from open sources. Most water is of poor quality as a result of contamination from pesticides, fertilizers, industrial waste, inadequate sewage treatment and open distribution systems.

FIGURE 1
Water withdrawal (total: 12.6 km^3 in 1989)

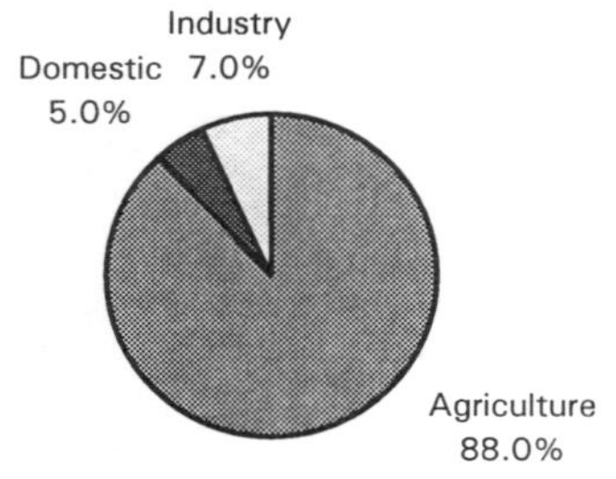

Water sharing agreements

Water allocation in Central Asia was initially governed by decrees from Moscow. More recently, the allocation was confirmed by agreements between riparian republics. According to the agreement with the downstream riparians of Uzbekistan, Turkmenistan and Kazahstan, Tajikistan is entitled to 8% of the Syr Darya's annual natural flows and 13% of those of the Amu Darya.

Environmental issues

One of the most serious policy dilemmas facing the countries in Central Asia concerns the environmental problems of the Aral Sea basin, as a result of the lack of environmental considerations in agricultural practises, industrial development and waste management.

IRRIGATION AND DRAINAGE DEVELOPMENT

Irrigation potential is estimated at about 1.57 million ha. Water availability is a limiting factor due to water sharing agreements with the downstream riparian countries. In 1992, the area equipped for irrigation was 718 000 ha, of which 698 000 ha were actually irrigated. The growth rate of irrigated land has been considerably lower than in the other states of the Aral Sea basin, primarily due to the difficulty of building irrigation systems under mountainous conditions. Tunnels of 12 km long have been dug through the mountains to bring water from the mountains to the valleys. In total, some 43 000 km of irrigation canals have been built.

FIGURE 2
Origin of irrigation water f/p
(total: 718 000 ha in 1992)

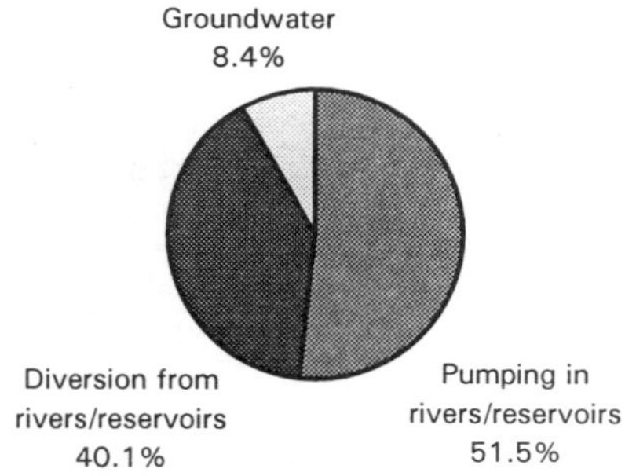

Approximately 60 000 ha are irrigated from groundwater and 370 000 ha are served by pumps to take water from the rivers, with lifts ranging from 10 to 400 metres (Figure 2). Almost all irrigation is surface irrigation using the furrow method. Investments in modern water saving techniques, such as sprinkler irrigation and micro-irrigation, are still considered to be quite expensive (between $US 6 000 and 7 000/ha). According to a recent World Bank Study, water can be saved under the existing system through better intake controls, by lining canals to reduce water loss and by improving drainage which will also reduce salinity levels.

More efficient water use can also be achieved through better on-farm water management, land levelling and changes in the cropping pattern.

About 670 000 ha, or over 90% of the irrigation schemes, are large- to medium-scale irrigation projects (Figure 3).

FIGURE 3
Typology of f/p control irrigation schemes (total: 718 000 ha in 1992)

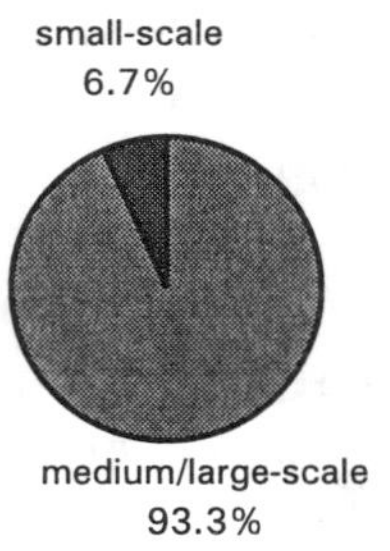

The major irrigated crop is cotton, with an area of about 296 000 ha in 1992. Per hectare yields of cotton used to be the highest in Central Asia but have fallen considerably because of poor weather and lack of fertilizes, pesticides and fuel. In 1992, much cotton remained unharvested because of civil unrest. Production in 1993 was 38% below 1991 levels. Other major irrigated crops are fodder crops (grains and green fodder), potatoes, vegetables, melons and fruit.

A serious problem in irrigated agriculture is the lack of adequate drainage systems and the associated waterlogging and salinization, in total affecting approximately 180 000 ha. According to the Soviet Union Environmental Survey of 1989, 15% of Tajikistan's total cropland, or about 122 000 ha, were salinized and 15 000 ha of irrigated cropland were completely unusable because of high saline concentrations. Soil erosion is another effect of intensive irrigation and, according to the survey of 1989, has affected almost 53% of all irrigated cropland.

INSTITUTIONAL ENVIRONMENT

The Ministry of Water Management is responsible for managing and maintaining the irrigation systems in Tajikistan. Irrigation project design is carried out by the Tajik Giprovodkhoz and new system construction by the Tajik Vadstroi.

Coordination of water sharing between the riparian countries is overseen by the Inter-Governmental Coordination Committee and implemented by two river basin commissions (BVO), one for the Amu Darya and one for the Syr Darya. The BVOs control all river and canal offtakes that affect more than one country, manage inter-republic and inter-sectoral allocations and monitor water use and quality.

The State Committee for Nature Protection (Goskompriroda) is responsible for establishing and implementing environmental policy, coordinating the environmental activities of other public and private entities, and for environmental monitoring. It has two separate research divisions: the Tajik Research Centre for the Protection of Water Resources and the Research Laboratory for the Protection of Nature.

The National Hydrometeorological (Hydromet) Service collects, analyses and disseminates hydrometeorological data to forecast floods and allocate water efficiently for irrigation, hydropower and other uses. It also monitors the glaciers and takes part in the activities to manage the region's river basins.

TRENDS IN WATER RESOURCES MANAGEMENT

Tajikistan's economy is mainly dependent upon agriculture and therefore the development of sustainable land use practices is of prime importance. According to the World Bank Study, for irrigated areas this includes: improving yields, improving on-farm water management to minimize water waste, avoid waterlogging and salinization, and develop environmentally sound practices for using fertilizers, pesticides and other agricultural chemicals. On rainfed areas and pasture land, management practices are required to maintain productivity and avoid land degradation.

There appears to be substantial room for increasing agricultural output from existing crops in the short run by using improved crop varieties, integrated pest control, better production techniques, and by improving on-farm water management.

A significant degree of cooperation among the republics will be crucial in finding a solution to the Aral sea crisis.

MAIN SOURCE OF INFORMATION

World Bank. 1994. Tajikistan: A World Bank Country Study. Washington D.C.

Tunisia

GEOGRAPHY AND POPULATION

Tunisia, situated in the north of Africa, has a total area of 193 610 km^2, of which 11 160 km^2, almost 7%, consist of lakes and *shats* (salty depressions). It is bordered in the north and north-east by the Mediterranean Sea, in the south-east by Libya and in the south and west by Algeria.

The country can be divided into four physiographic regions:

- The North-western Mountains, which are located at the eastern end of the two mountain ranges, the Atlas El-Talli and the Desert Atlas, which extend from Morocco through Algeria and reach a peak at 1 500 metres in Tunisia. This area is crossed by permanent rivers;
- The Southern Mountains, which slope towards the east to the Coastal Plains and towards the west to the Desert Plains and are covered by sand dunes;
- The Coastal Plains which run close to the Mediterranean Sea as wide plains;
- Desert Plain. This plain forms the northern boundary of the Great Desert (Sahara). A number of *shats* exist in this plain, the largest one being the Shat El-Jarid with an area of 5 000 km^2 at a minimum elevation of 15 metres below sea level.

The cultivable area is estimated at 8.7 million ha, which is about half the total area of the country. In 1993, the cultivated area was estimated at 4.25 million ha, of which 2.18 million ha consisted of annual crops and 2.07 million ha of permanent crops.

TABLE 1
Basic statistics and population

Physical areas:		
Area of the country	1995	16 361 000 ha
Cultivable area	1991	8 700 000 ha
Cultivated area	1993	4 254 600 ha
- annual crops	1993	2 180 800 ha
- permanent crops	1993	2 073 800 ha
Population:		
Total population	1995	8 896 000 inhabitants
Population density	1995	54 inhab./km^2
Rural population	1995	43 %
Water supply coverage:		
Urban population	1992	91 %
Rural population	1992	65 %

The total population is 8.9 million (1995) and 43% is rural as against 60% 25 years ago. The annual demographic growth rate is less than 2%. Most of the towns and the people are in the northern and central coastal zones. The rural population lives mainly in 4 500 villages. Agriculture employed 30% of the labour force in 1993. There are about 380 000 farm holdings, with an average area of 14 ha. During the period 1989-93 GDP increased about 5% per year. The importance of the agricultural sector in the economy decreased from 1960 to 1994: in 1960 it accounted for 24% of the country's GDP, while in 1994 this figure had fallen to 16%.

CLIMATE AND WATER RESOURCES

Climate

In the region close to the sea a semi-arid climate prevails, while in the south the climate is of the desert type. The average annual temperature rises from 17°C in the north to over 21°C in

TABLE 2
Water: sources and use

Renewable water resources:			
Average precipitation		207	mm/yr
		33.9	km^3/yr
Internal renewable water resources		3.52	km^3/yr
Total (actual) renewable water resources	1995	4.12	km^3/yr
Dependency ratio	1995	14.6	%
Total (actual) renewable water resources per inhabitant	1995	463	m^3/yr
Total dam capacity	1991	1 510	$10^6\ m^3$
Water withdrawal:			
- agricultural	1990	2 728	$10^6\ m^3/yr$
- domestic	1990	261	$10^6\ m^3/yr$
- industrial	1990	86	$10^6\ m^3/yr$
Total water withdrawal		3 075	$10^6\ m^3/yr$
per inhabitant	1990	382	m^3/yr
as % of total (actual) renewable water resources		74.6	%
Other water withdrawal		-	$10^6\ m^3/yr$
Average groundwater depletion	1990	160	$10^6\ m^3/yr$
Wastewater - Non-conventional water sources:			
Wastewater:			
- produced wastewater		-	$10^6\ m^3/yr$
- treated wastewater	1993	96	$10^6\ m^3/yr$
- reused treated wastewater	1993	20	$10^6\ m^3/yr$
Desalinated water	1990	8.3	$10^6\ m^3/yr$

the south. Average annual rainfall is estimated at 207 mm, but almost half the country receives less than 100 mm/year. Only in the extreme north does rainfall reach up to 1 500 mm/year. About 80% of the rain falls between October and March.

Water resources

The hydrographic system is dense in the north where the Medjerda wadi is the most important water course. This is also the zone where the principal irrigation development and flood protection works have been carried out.

Surface water resources have been estimated at 2.91 km^3/year, of which 2.31 km^3 are produced internally. About 1.5 km^3/year are exploitable at present through reservoirs. It will be possible in the future to exploit another 0.6 km^3/year, but the remaining part could only be used by means of large water conservation works and groundwater recharge systems. At present, there are 18 large dams and 22 hillside dams.

Internal renewable groundwater resources have been estimated at 1.21 km^3/year. At present, there are 83 000 open wells and 1 830 tubewells. Two categories of groundwater resources can be distinguished in function of the depth:

- when the water table is above 50 metres, groundwater is defined as phreatic and can be used for private exploitation (with some restrictions). The potential has been estimated at 669 million m^3/year;
- below 50 metres of depth, the groundwater has been reserved for public exploitation. Potential has been estimated at 1 170 million m^3/year, of which 630 million m^3 is fossil water.

TABLE 3
Irrigation and drainage

Irrigation potential		1991	563 000	ha
Irrigation:				
1. Full or partial control irrigation: equipped area		1991	355 000	ha
- surface irrigation		1991	294 000	ha
- sprinkler irrigation		1991	55 000	ha
- micro-irrigation		1991	6 000	ha
% of area irrigated from groundwater		1991	60.7	%
% of area irrigated from surface water		1991	37.3	%
% of area irrigated from non-conventional sources		1991	2.0	%
% of equipped area actually irrigated		1991	90.7	%
2. Spate irrigation area		1991	30 000	ha
3. Equipped wetland and inland valley bottoms (i.v.b.)			-	ha
Total irrigation (1+2+3)		**1991**	**385 000**	**ha**
- as % of cultivated area			9	%
4. Flood recession cropping area			-	ha
Total water managed area (1+2+3+4)		**1991**	**385 000**	**ha**
- as % of cultivated area			9	%
- increase over last 10 years			-	%
- power irrigated area as % of water managed area		1991	67.5	%
Full or partial control irrigation schemes:	Criteria			
Large-scale schemes	> - ha		-	ha
Medium-scale schemes			-	ha
Small-scale schemes	< - ha		-	ha
Total number of households in irrigation		1991	141 000	
Irrigated crops:				
Total irrigated grain production		1991	82 500	tons
as % of total grain production		1991	4	%
Harvested crops under irrigation (full or partial control)		**1992**	**308 000**	**ha**
- permanent crops: total		1992	130 000	ha
- annual crops: total		1992	178 000	ha
. vegetables		1992	108 000	ha
. wheat		1992	33 000	ha
. other annual crops		1992	37 000	ha
Drainage - Environment:				
Drained area		1991	162 000	ha
as % of cultivated area			4	%
- drained areas in full or partial control irrigated areas		1991	162 000	ha
- drained areas in equipped wetland and i.v.b			-	ha
- other drained areas			-	ha
- total drained area with subsurface drains		1991	162 000	ha
- total drained area with surface drains			-	ha
Flood-protected area			-	ha
Area salinized by irrigation			-	ha
Population affected by water-borne diseases			-	inhabitants

Wastewater

For the last 20 years, reuse of treated wastewater has taken place. In 1993, 96 million m^3 was treated, of which 20 million m^3 was reused. It is planned to increase the treatment to 200 million m^3/year by the year 2000.

Water withdrawal

In 1990, water withdrawal was estimated at about 3.1 km^3/year, of which almost 89% for agricultural purposes (Figure 1). However, the amount of water withdrawn depends to a large extent on the quantity and the distribution of the precipitation. In particular, irrigation water withdrawal varies in function of the rainfall and of the area actually irrigated within the public irrigation network. Of the total of 3.1 km^3 of water used annually, only 1.9 km^3 are estimated to be actually used.

In 1992, the rural population with access to good drinking water within a distance of 3 km was estimated at 65%, while 91% of the urban population was connected to the drinking water supply network.

FIGURE 1
Water withdrawal (total: 3.1 km^3 in 1990)

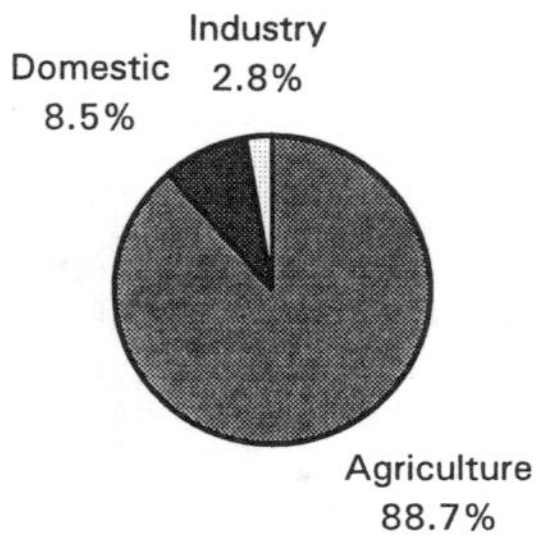

IRRIGATION AND DRAINAGE DEVELOPMENT

The irrigation potential has been estimated at 563 000 ha, based on land and water resources. This area can be divided in to 402 000 ha with possibilities for full or partial control irrigation development and 161 000 ha of spate irrigation.

FIGURE 2
Distribution of the water managed areas (total: 385 000 ha in 1991)

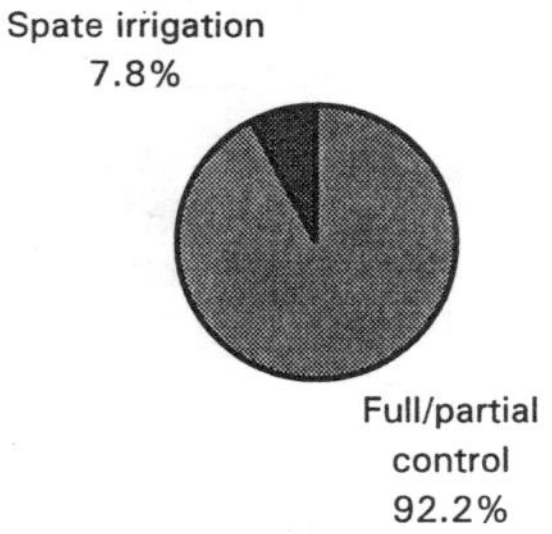

In 1991, the water managed area was estimated at 385 000 ha, including 355 000 ha of full or partial control irrigation schemes and 30 000 ha of spate irrigation (Figure 2). The average annual growth of irrigation development is about 2%, which means that at this rate full potential will have been achieved by the year 2010.

FIGURE 3
Kind of irrigation in the f/p control schemes (total: 355 000 ha in 1991)

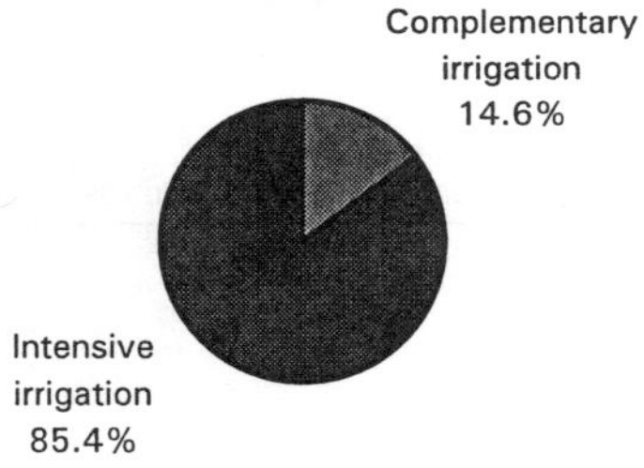

The area equipped for full or partial control irrigation can be divided into 52 000 ha of complementary irrigation and 303 000 ha of so-called intensive irrigation (Figure 3). Surface irrigation techniques are mainly practised on the schemes with complementary irrigation and some sprinkler irrigation is used on cereals in case of a severe rainfall deficit and on fruit trees during the early years of their development. The intensive irrigation schemes can be divided into (Figure 4):

- 140 000 ha of small schemes developed individually by the farmers (70 000 households) through private investments and irrigated by open wells or by pumping of water in the wadis;
- 163 000 ha of schemes with a communal irrigation network, irrigated by tubewells in the case of medium schemes (60 000 ha; 47 000 households) and by dams in the case of large schemes (103 000 ha; 24 000 households). These schemes have been developed with public investment.

The communal irrigation water distribution systems are in general modern. There are prefabricated canals or low pressure pipes for surface irrigation and high pressure pipes for sprinkler irrigation (Figure 5). Surface irrigation remains the most important irrigation technique (Figure 6). At present, about 6 000 ha are irrigated by treated wastewater, the remaining part by surface water (dams, pumping in wadis) and by groundwater (open wells, tube-wells) (Figure 7).

The evolution of the area equipped for full or partial control intensive irrigation is shown in Figure 8. In 1991, the actually irrigated area was 270 000 ha, which is about 89% of the equipped intensive irrigation area.

Fruit trees (dates, oranges, grapes, etc.) and vegetables (tomatoes, melons, water melons, potatoes, etc.) are cultivated on about 70% of the full or partial control schemes actually irrigated. However, recent government encouragement for the development of dairy livestock has led to an increase in the irrigation of cereals and fodder crops. The irrigation sector contributes about 30-35% to the agricultural value of the country. In 1992, about 308 000 ha were actually irrigated and harvested, which is 80% of the water managed area of 385 000 ha (Figure 9).

The cost of irrigation development for the public schemes varies between $US 6 000 and 7 000/ha depending on the size of the scheme. The annual operation and maintenance costs are about $US 130/ha, of which 25-30% is spent on energy.

Large-scale public irrigation schemes are managed by the state while medium-scale public irrigation schemes are managed by users associations (AIC or Associations d'intérêt collectif). The services of the state or AIC are in charge of the operation and maintenance of the irrigation network as well as of the distribution of water to the farmers, applying a water charge according to volume. However irrigation water is still subsidized by the state for up to 20-30% of its real exploitation cost.

FIGURE 4
Typology of the f/p control intensive irrigation schemes (total: 303 000 ha in 1991)

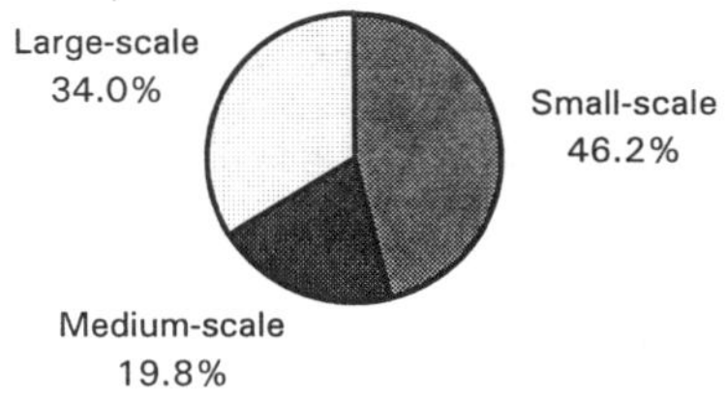

FIGURE 5
Water distribution systems in the f/p control intensive irrigation schemes (total: 303 000 ha in 1991)

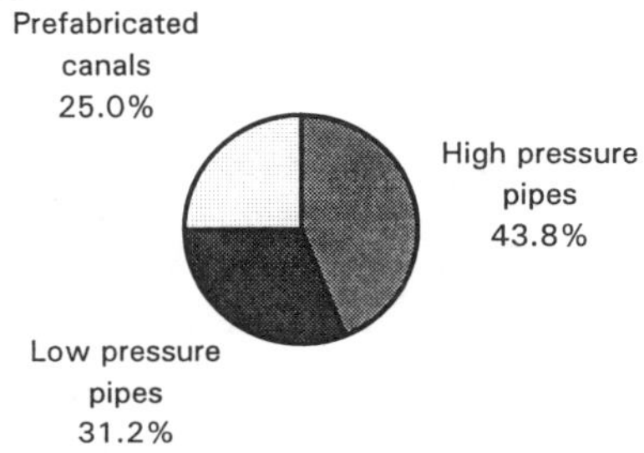

FIGURE 6
Irrigation techniques in the f/p control intensive irrigation schemes (total: 303 000 ha in 1991)

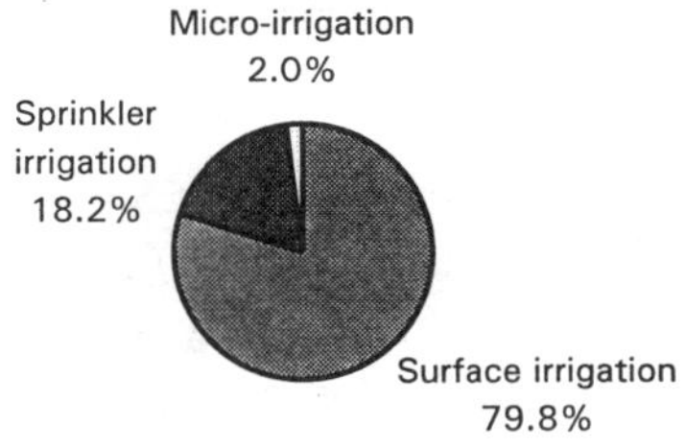

FIGURE 7
Origin of irrigation water in the f/p control intensive irrigation schemes (total: 303 000 ha in 1991)

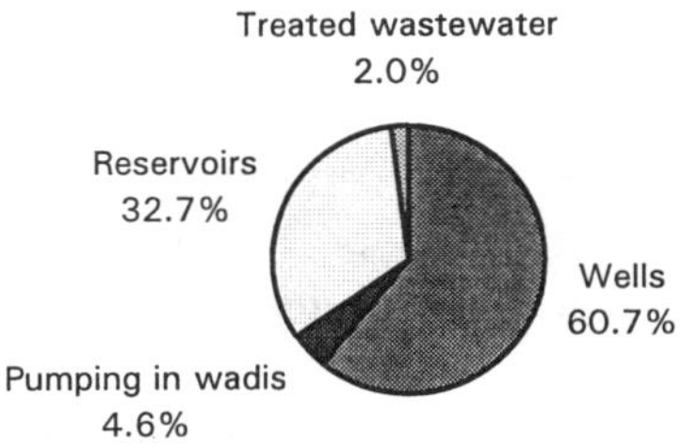

Irrigation water is quite saline (1.5-4.0 g/l), but the degree of salinization of the irrigated soils is not yet a serious problem due to the low intensification. A monitoring system has been set up for all the schemes with a high risk of salinization. Subsurface drainage is not very developed (162 000 ha) and is limited to soils with a high water table (schemes in the north).

FIGURE 8
Evolution of the area equipped for 'intensive' irrigation

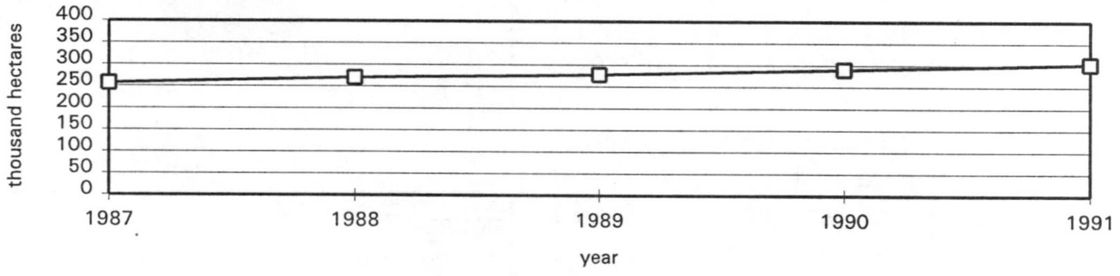

FIGURE 9
Irrigated crops (total: 308 000 ha in 1992)

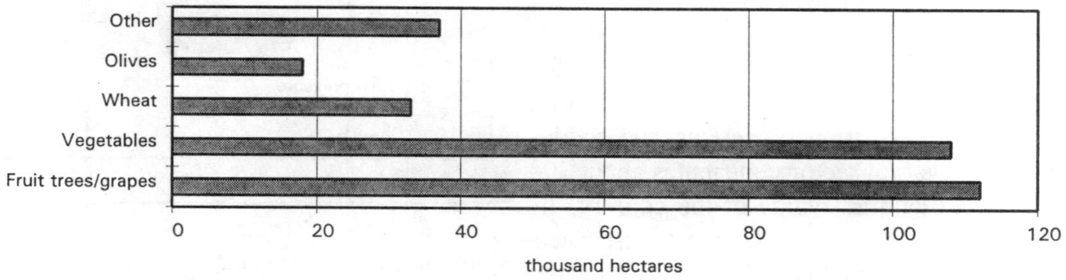

INSTITUTIONAL ENVIRONMENT

The Ministry of Agriculture is the main institution involved in the water sector. The main General Directorates in charge of water are:

- The General Directorate of Water Resources (DGRE or Direction générale des ressources en eau), which is in charge of the monitoring and evaluation of water resources;

- The General Directorate of Large Hydraulic Works (DGGTH or Direction générale des grands travaux hydrauliques), the main activity of which is the construction of dams;

- The General Directorate of Hydraulic Studies and Works (DGETH or Direction générale des études et travaux hydrauliques), the main activities of which are: general hydraulics studies, construction of hillside dams, development of large-scale schemes and management of the dams;

- The General Directorate of Rural Engineering (DGGR or Direction générale du génie rural), which is responsible for irrigation, rural equipment and for drinking water supply to the rural population.

The regional agricultural development offices (CDRA or Commissariat régional de développement agricole), linked to the Ministry of Agriculture, are the institutions responsible at regional level for the development of public irrigation schemes. The management of these schemes is shared between CDRA and AIC.

The water code, promulgated in 1975, determines all interventions in the water sector.

TRENDS IN WATER RESOURCES MANAGEMENT

The strategy for the future use of water, adopted in 1990, aims at the development of 90% of surface water resources and 100% of groundwater resources by the year 2010, by means of the construction of 21 dams, 235 hillside dams and 610 deep tube-wells.

This improvement in the control of water resources is associated with the agricultural development of the irrigation schemes. The objectives of the country are at present to encourage the adoption of water saving techniques, to favour the reuse of treated wastewater for irrigation, to expand the irrigated areas and to set up the related activities necessary for agricultural development and a better use of irrigation schemes.

MAIN SOURCES OF INFORMATION

Agrar, Coyne and Bellier, GKW, Ministère de l'agriculture. 1992. Economie d'eau 2000.

Direction générale des ressources en eau, Ministère de l'agriculture. 1991. Commission de réflexion sur le développement des ressources en eau de surface.

Direction générale du génie rural, Ministère de l'agriculture. 1992. Rapport du Comité Irrigation. - VIII Plan.

Direction générale du génie rural, Ministère de l'agriculture. 1994. La gestion de l'eau en Tunisie.

Direction générale du génie rural, Ministère de l'agriculture. 1994. Tarification de l'eau d'irrigation 1992 et 1993.

GERSAR and SOTUETEC. 1987. Gestion et entretien des périmètres irrigués.

Hamza, M. and Khanfir, R. 1992. Evolution du potentiel et de l'exploitation des ressources en eau souterraines.

Ministère de l'agriculture. 1992. Annuaire des statistiques agricoles.

Ministère de l'agriculture. 1993. Résultats de l'enquête-suivi de la campagne agricole 1992/93.

Ministère de l'environnement et de l'aménagement du territoire. 1992. Rapport national de la CNUCED.

Office national de l'assainissement. 1994. Rapport d'activités 1993.

Turkey

GEOGRAPHY AND POPULATION

Turkey, with a total area of 779 45 km^2, lies between Europe and Asia and is surrounded by the Black Sea, Bulgaria and Greece in the north, the Aegean Sea in the west, the Mediterranean Sea, Syria and Iraq in the south, Iran in the east and Armenia and Georgia in the north-east.

About 28 million ha, or 36% of the total area, are classified as cultivable. In 1991, the total cultivated area was estimated at 20.5 million ha, of which 17.5 million ha consisted of annual crops and 3.0 million ha consisted of permanent crops, mainly vineyards, fruit trees and olives.

The total population is about 61.9 million (1995), of which 31% is rural, with an annual demographic growth estimated at 2%. Agriculture employed 44.5% of the total labour force in 1994 and accounted for nearly 16% of GDP in 1993. In 1992, agriculture provided 14.9% of the total export revenue.

TABLE 1
Basic statistics and population

Physical areas:			
Area of the country	1995	77 945 000	ha
Cultivable area	1994	28 054 000	ha
Cultivated area	1991	20 480 000	ha
- annual crops	1991	17 450 000	ha
- permanent crops	1991	3 030 000	ha
Population:			
Total population	1995	61 945 000	inhabitants
Population density	1995	79	inhab./km^2
Rural population	1995	31	%
Water supply coverage:			
Urban population	1994	98	%
Rural population	1994	85	%

CLIMATE AND WATER RESOURCES

Climate

Turkey has a subtropical, semi-arid climate with extremes in temperatures. In the east, summers are hot and dry, winters are cold, rainy and snowy. Along the coastal area, a Mediterranean climate is dominant with long, hot, dry summers and short, mild, rainy winters. Rainfall shows great differences from one region to another. Average annual rainfall is 643 mm, ranging from 250 mm in the south-east to over 3 000 mm in the north-east Black Sea area. About 70% of the rain falls in the winter and spring seasons. Average runoff is about 37%.

Surface water resources

The most important hydrological basins in Turkey have been grouped in a table (page 252).

Of the total surface runoff of the country, estimated at 192.8 km^3/year, almost one-fourth comes from the Euphrates (Firat) and the Tigris (Dicle) rivers, that both have their sources in the eastern part of the country. Turkey contributes about 90% of the total annual flow of the Euphrates, while the remaining part originates in Syria and nothing is added further downstream in Iraq. Turkey contributes 38% directly to the main Tigris river and another 11% to its tributaries, that join the main stream of the Tigris further

TABLE 2
Water: sources and use

Renewable water resources:			
Average precipitation		643	mm/yr
		501.2	km^3/yr
Internal renewable water resources		196.0	km^3/yr
Total (actual) renewable water resources	1995	183.76	km^3/yr
Dependency ratio	1995	1.8	%
Total (actual) renewable water resources per inhabitant	1995	2 967	m^3/yr
Total dam capacity	1992	206 000	10^6 m^3
Water withdrawal:			
- agricultural	1992	22 900	10^6 m^3/yr
- domestic	1992	5 200	10^6 m^3/yr
- industrial	1992	3 500	10^6 m^3/yr
Total water withdrawal		31 600	10^6 m^3/yr
per inhabitant	1992	541	m^3/yr
as % of total (actual) renewable water resources		17.2	%
Other water withdrawal		-	10^6 m^3/yr
Average groundwater depletion		-	10^6 m^3/yr
Wastewater - Non-conventional water sources:			
Wastewater:			
- produced wastewater	1994	2 400	10^6 m^3/yr
- treated wastewater	1994	100	10^6 m^3/yr
- reused treated wastewater		-	10^6 m^3/yr
Desalinated water	1990	0.5	10^6 m^3/yr

Major hydrological basins in Turkey

Names of the basins	Total number of basins	Total flow (km^3/year)	Draining to:
Tigris, Euphrates *	2	52.94	Iraq/Syria/Iran (Persian Gulf)
Orontes, Ceyhan, Seyhan, Eastern Mediterranean, Antalya, Western Mediterranean	6	47.42	Mediterranean Sea
Greater Menderes, Smaller Menderes, Gediz, North Aegean, Meric	5	9.59	Aegean Sea
Marmara, Susurluk	2	13.76	Sea of Marmara
Sakarya, Western Black sea, Kizilirmak, Yesilirmak, Eastern Black Sea, Coruh	6	49.81	Black Sea (Coruh first to Georgia with 6.30 km^3/year)
Aras	1	4.63	Armenia/Azerbaijan/Iran (Caspian Sea)
Burdur Lakes Area, Akarcay, Konya Closed Basin, Van Lake	4	7.90	Interior
TOTAL	26	186.05	

* ***Euphrates*** *: Figures of average runoff vary between 26.29 and 31.61 km^3/year. The latter figure was used in this table.*
Tigris *: Figures of average runoff vary between 18.00 and 21.33 km^3/year. The latter figure was used in this table.*

downstream in Iraq. In general, the streams vary greatly in their flow from season to season and year to year. For example, the Euphrates' annual flow at the border with Syria ranged from 15.3 km^3 in 1961 to 42.7 km^3 in 1963.

Average annual surface runoff entering Turkey from other countries is estimated at about 1.8 km^3 (the Orontes, the Tunca). Another 5.8 km^3/year comes from Bulgaria through the Meriç river which forms the border between Turkey and Greece. Average annual surface runoff leaving the country is estimated at 60.4 km^3, while almost 8 km^3 flows in the interior basins.

TABLE 3
Irrigation and drainage

Irrigation potential		1993	8 500 000	ha
Irrigation:				
1. Full or partial control irrigation: equipped area		1994	4 070 746	ha
- surface irrigation		1993	3 806 511	ha
- sprinkler irrigation		1993	263 849	ha
- micro-irrigation		1993	386	ha
% of area irrigated from groundwater		1994	16.5	%
% of area irrigated from surface water		1994	83.5	%
% of area irrigated from non-conventional sources		1994	0.0	%
% of equipped area actually irrigated		1994	74	%
2. Spate irrigation area			-	ha
3. Equipped wetland and inland valley bottoms (i.v.b.)		1994	115 164	ha
Total irrigation (1+2+3)		**1994**	**4 185 910**	**ha**
- as % of cultivated area			20	%
4. Flood recession cropping area			-	ha
Total water managed area (1+2+3+4)		**1994**	**4 185 910**	**ha**
- as % of cultivated area			20	%
- increase over last 10 years			-	%
- power irrigated area as % of water managed area		1994	5.3	%
Full or partial control irrigation schemes:	Criteria			
Large-scale schemes	> 1 000 ha	1994	1 805 390	ha
Medium-scale schemes				ha
Small-scale schemes	< 1 000 ha	1994	2 265 356	ha
Total number of households in irrigation			-	
Irrigated crops:				
Total irrigated grain production			-	tons
as % of total grain production			-	%
Harvested crops under irrigation (full or partial control)			-	**ha**
- permanent crops: total			-	ha
- annual crops: total			-	ha
.			-	ha
.			-	ha
.			-	ha
.			-	ha
. other annual crops			-	ha
Drainage - Environment:				
Drained area		1994	3 143 000	ha
as % of cultivated area			15	%
- drained areas in full or partial control irrigated areas			-	ha
- drained areas in equipped wetland and i.v.b			-	ha
- other drained areas			-	ha
- total drained area with subsurface drains			-	ha
- total drained area with surface drains			-	ha
Flood-protected area		1994	799 000	ha
Area salinized by irrigation			-	ha
Population affected by water-borne diseases			-	inhabitants

Groundwater resources

A figure of 12.2 km^3/year for groundwater is given for the year 1994. It probably represents identified development potential, and a figure of 20.0 km^3/year for yearly groundwater recharge is probably a more realistic estimate. Groundwater flows to other countries are not known. However, the sources of the Khabour river, situated in Syria, with a runoff of 1.2 km^3/year, have their origin in groundwater coming from Turkey.

Dams

By the end of 1991, the construction of 164 large dams, mostly rockfill or earthfill dams, and 765 small dams had been completed and put into service for water supply, irrigation, hydropower and flood control. Total dam capacity is about 206 km^3. In addition, in 1992, 78 large dams and 172 small dams were under construction. The Ataturk Dam on the Euphrates (Firat) river in the south-eastern part of the country, with a total storage capacity of 48.5 km^3, is one of the 10 largest dams in the world. In the beginning of 1990, the filling of the reservoir behind the dam started and was completed in 1992. The surface area of the reservoir is about 817 km^2.

FIGURE 1
Water withdrawal (total: 31.6 km^3 in 1992)

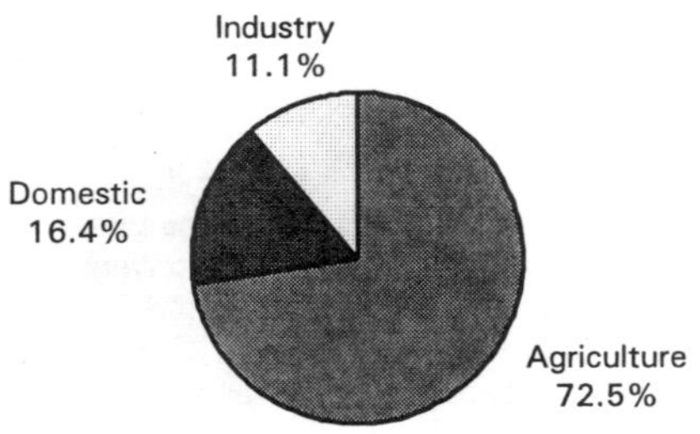

Water withdrawal

It is estimated that 95 km^3 of the total surface runoff of 192.8 km^3/year could be technically developed for economic use. In 1992, total annual water withdrawal was 31.6 km^3, of which agricultural use accounted for 72% (Figure 1). About 24% of total water withdrawal, or 7.6 km^3, was estimated to be groundwater, of which 3.7 km^3 for agricultural purposes. Almost 98% of the urban and 85% of the rural populations have access to safe drinking water. The treatment of domestic wastewater is estimated at 0.1 km^3/year.

International rivers and agreements

The total length of the boundary between Turkey and its surrounding countries is 2 753 km, of which 615 km is formed by rivers: 238 km with Bulgaria and Greece, 76 km with Syria, 58 km with Iraq and Iran, 243 km with Armenia and Georgia. In 1927, Turkey and the USSR signed a 'Treaty on the Beneficial Uses of Boundary Waters', in which they agreed to share water on a fifty-fifty basis. A Joint Boundary Water Commission was established (although without legal identity) to control the use of the frontier water. In 1973 the two governments signed an additional 'Treaty on the Joint Construction of the Arpacay or Ahuryan Storage Dam'. In a similar vein, Turkey and Greece after the Treaty of Lausanne (1923) signed several protocols regarding the control and management of the Meric river which forms the border between Greece and Turkey.

The protocol concerning the Euphrates and the Tigris rivers dates back to 1946 when Turkey and Iraq agreed that the rivers' control and management depended to a large extent on the regulations of flow in Turkish source areas. Turkey, at that time, agreed to begin monitoring the two rivers and to share related data with Iraq. In 1980 Turkey and Iraq further specified the nature of the earlier protocol by establishing a Joint Technical Committee on Regional Waters. After a bilateral agreement in 1982, Syria joined the committee. Turkey has unilaterally guaranteed to allow 15.75 km^3/year (500 m^3/s) of water across the border to Syria, but no formal agreement has been obtained so far on the sharing of the Euphrates water.

IRRIGATION AND DRAINAGE DEVELOPMENT

Out of the cultivable area of 28 million ha, almost 26 million ha are classified as suitable for irrigation. Considering the water resources this area is reduced to 12.5 million ha. When moreover economic considerations are brought in, the official estimated irrigation potential of

Turkey is 8.5 million ha, of which 93% would be from surface water resources and 7% from groundwater.

FIGURE 2
Distribution of the water managed areas (total: 4 185 910 ha in 1994)

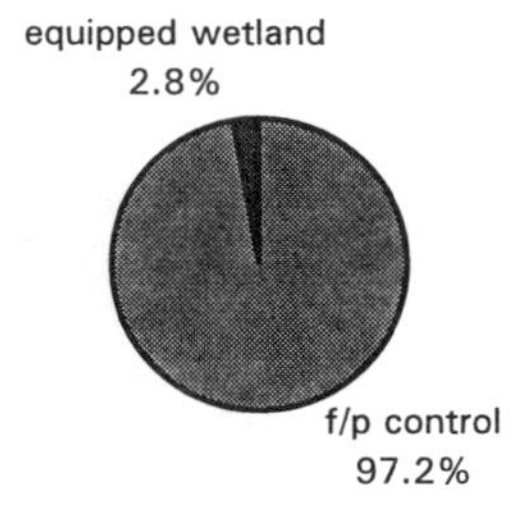

Irrigation development in Turkey is carried out by the public sector, represented by DSI (State Hydraulic Works) and GDRS (General Directorate of Rural Services), and by the private sector (farmers and groups of farmers). As at January 1994, the total water managed area was estimated at almost 4.2 million ha, or 49% of the irrigation potential, of which 115 164 ha was equipped wetland (Figure 2). Over 3.1 million ha have so far been developed by the public sector, compared with less than 0.5 million ha in 1965 (Figure 3). The total water managed area can be classified in four categories as shown in the table below.

FIGURE 3
Evolution of irrigation development by the public sector

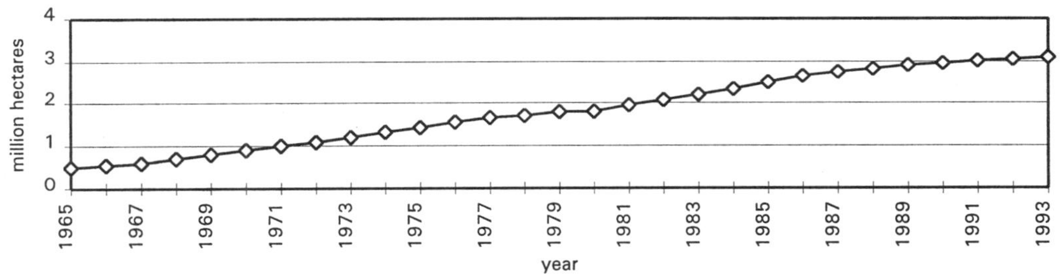

On public schemes, the national average of the irrigation ratio (the part of the equipped area actually irrigated) varies between 64 and 72% with wide regional and annual fluctuations (28-82% and 34-76% respectively).

Irrigation management (1994)

Water managed area (ha)	As % of total	Equipped by	Operated by
1 342 651	32	DSI	DSI
462 739	11	DSI	GDRS/Other agencies *
1 300 520	31	GDRS	GDRS/Other agencies *
1 080 000	26	Private farmers	Private (groups of) farmers
4 185 910	100	TOTAL	

* *Other agencies: irrigation cooperatives, heads of villages, municipalities, some state organizations*

Of the total area developed by the public sector for full and partial control irrigation 20% is irrigated from groundwater, while for the area developed by the farmers the figure is only 7% (Figure 4).

FIGURE 4
Origin of irrigation water f/p (total: 4 070 746 ha in 1994)

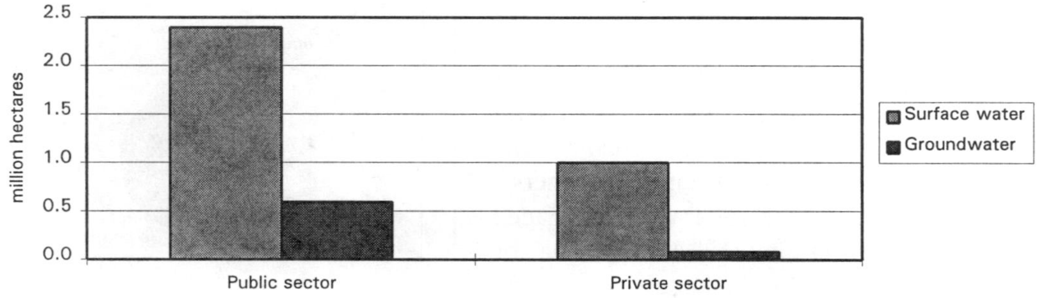

Almost 94% of the total area is irrigated using surface irrigation methods (furrow, basin, border, wild flooding). The remaining part is under sprinkler irrigation (mainly hand-move) and some micro-irrigation, mainly in the Aegean and Mediterranean regions (Figure 5). The conventional (hand-move) sprinkler irrigation is common all over Turkey among the farmers and an estimated 200 000 ha are irrigated using this method. On DSI schemes, 63 849 ha are irrigated by sprinklers, mainly for sugar beet, cereals, beans, alfalfa, cotton, sunflowers, water melons and vegetables. Micro-irrigation is practised on 386 ha of DSI schemes, mainly for citrusfruits, vineyards, vegetables, strawberries and watermelons.

FIGURE 5
Irrigation techniques f/p
(total: 4 070 746 ha in 1994)

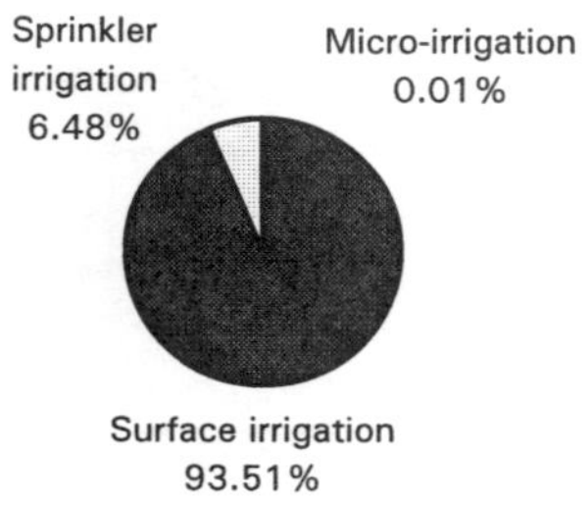

During the last three decades GDRS has carried out a lot of on-farm water development works, for example the reclamation of saline and alkaline soils on 803 000 ha and open drains on 3 143 000 ha. In 1992, about 223 000 ha were estimated to be power drained. The flood protected area amounted to almost 800 000 ha in 1994. It was estimated in 1992, that of the total area operated by DSI, about 41 000 ha was salinized by irrigation.

The cost of irrigation development varies between $US 1 750/ha for small schemes (< 1 000 ha) and $US 3 000/ha for large schemes (> 1 000 ha, with 70% being > 10 000 ha) (Figure 6). The average cost of operation and maintenance varies between $US 50/ha for small schemes and $US 70/ha for large schemes (excluding dams). The cost of drainage development (the average for sub-surface and open drains) was estimated at $US 274/ha in 1994. Water rates are charged on a cropped area basis, with different rates for each crop. During 1990-92 the average water charges for large schemes were estimated at $US 35/ha.

FIGURE 6
Size of irrigation schemes f/p
(total: 4 070 746 ha in 1994)

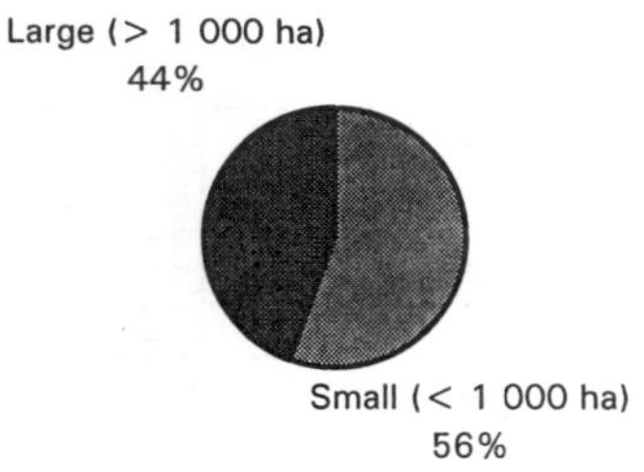

Data for major crops and their production are not available for the whole country. In 1991, 60% of the total area equipped and operated by DSI (over 1.34 million ha), or 810 000 ha, was cropped (Figure 7).

FIGURE 7
Irrigated crops on areas managed by DSI (total: 810 000 ha in 1991)

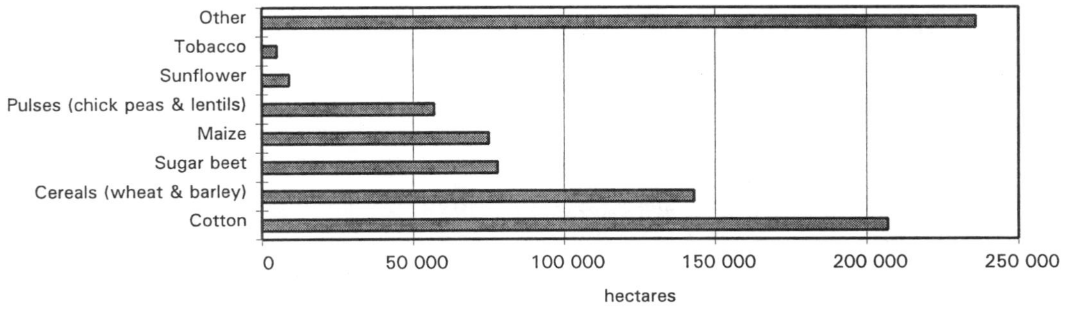

The average yield for irrigated cereals (wheat and barley) was 4.1 t/ha as against 2.1 t/ha for rainfed crops. For irrigated pulses (chick peas and lentils) the average yield was 2.5 t/ha, for cotton 2.9, for sunflowers 2.3, for maize 7.9, for sugar beet 55.2 and for tobacco 1.1 t/ha.

INSTITUTIONAL ENVIRONMENT

There are two governmental organizations involved in major irrigation and drainage development:

- The State Hydraulic Works (DSI) was established in 1954 as a legal entity and brought under the aegis of the Ministry of Public Works and Settlement and is responsible for the planning, design, construction and operation of water resources development for various purposes like irrigation, flood control, swamp reclamation, hydropower development, navigation and water supply to cities with over 100 000 inhabitants.

- The General Directorate of Rural Services (GDRS) was established in 1984 by incorporating the soil conservation and irrigation organization, the rural settlement organization and the rural roads, water and electricity organization into one organization. It is responsible for the development of small-scale irrigation schemes and small reservoirs, rural roads and water supply to rural areas. It is also responsible for land consolidation and the on-farm development of all irrigation projects, including the projects developed by DSI. It was formerly under the Ministry of Agriculture and Rural Affairs, but now falls under the Prime Minister's Office.

TRENDS IN WATER RESOURCES MANAGEMENT

In several areas, problems are emerging as urban activities encroach onto agricultural lands. There is an increasing interest in using the land as a vehicle for the treatment and disposal of the wastewater from agri-business and urban activities. In particular there is current concern about the use of polluted water resources to irrigate agricultural lands, especially in western Turkey, which has been experiencing water shortages on a regular basis in recent years.

A regional agricultural development project, the South East Anatolian Project (GAP), is planned for the lower Euphrates river and the Tigris river basin within the boundaries of Turkey. The project involves the integrated development of irrigated agriculture and agro-industry, and supporting services, including communications, health and education. It includes 13 major projects of which 7 are in the Euphrates basin and 6 in the Tigris basin. After full development it will include 22 dams, 19 hydroelectric power plants and the irrigation of almost 1.7 million hectares. The water obtained from the Ataturk dam on the Euphrates is carried to the Harran Plain by the Sanli Urfa tunnel system, which is the largest tunnel system in the world in view of its length and flow rate. The waters of the river pass through tunnels which are 26.4 km in length and 7.62 m in diameter with an estimated flow of about 328 m^3/s, which is one-third of the total flow of the Euphrates.

However, the downstream riparian states, Syria and Iraq, do not agree with water abstraction for irrigation. Especially as far as the Euphrates river is concerned, problems regarding sharing water might arise between these countries. Turkey has unilaterally guaranteed to allow 15.75 km^3/year of water from the Euphrates river across the border to Syria, but no formal agreement has been obtained so far. Although fewer water problems exist on the Tigris river, mainly because more than 50% of the water originates downstream of Turkey (as against less than 10% of the Euphrates water), here also the use of the water resources has to be carefully planned.

According to the different scenarios established, full irrigation development by the countries in the Euphrates-Tigris river basin would lead to water shortages and solutions will have to be found at basin level through regional cooperation.

MAIN SOURCES OF INFORMATION

Adem Ilbeyi and Bulent Sonmez. 1995. Water management and irrigation practices in Turkey. Country paper presented at the TCDC Regional Workshop on improved water management technologies for sustainable agriculture in arid climates, Cairo, 25-29 March 1995.

DSI (State Hydraulic Works). 1992. Dams and hydroelectric power plants in Turkey.

E. Esat Kiratlioglu. 1995. Land and water resources management in Turkey. Article published in MEDIT No 1/95.

FAO, IAP-WASAD. 1993. National Action Programme for the Republic of Turkey.

Kulga, Dincer and Cakmak, Cuma. 1994. The role of DSI in water and sustainable agricultural development. IAP/WASAD/NAP/Gvt.

Nedeco/Dapta/Su-Yapi/Temelsu. 1991. Irrigation master plan. DSI.

Osman Tekinel, Riza Kanber, Bulent Ozekic. 1992. Water resources planning and development in Turkey. In Proceedings of the situation of agriculture in Mediterranean countries, organized by CIHEAM/CCE-DGI. Adana, 3-9 September 1992.

State Institute of Statistics (SIS). 1995. Statistical Yearbook 1994 of Turkey.

Suleyman Kodal and Erkan Benli. 1993. Research and development on irrigation and drainage technologies in Turkey. In Proceedings of the Technical Consultation on integrated rural water management, Rome, 15-19 March 1993.

World Bank. 1993. Irrigation management and investment review. Report No. 11589-TU.

Turkmenistan

Note: *Turkmenistan was officially included amongst the countries of FAO's Regional Office for the Near East in March 1996. For the present publication, it has not been possible to perform an in-depth survey of the information on water and irrigation and so the data presented below are probably very incomplete.*

GEOGRAPHY AND POPULATION

Turkmenistan is a landlocked country in Central Asia, bordered in the west by the Caspian Sea, in the north-west by Kazakhstan, in the north-east by Uzbekistan, in the south-east by Afghanistan and in the south-west by Iran. It formally declared its independence from the USSR in October 1991. The total area of the country is 488 100 km^2. The Kara Kum desert comprises 80% of the total area of the country. In the south-west, along the border with Iran, lies the Kopetdag mountain chain with its highest peak at 2 919 metres. The Kougitangtau mountain with a peak at 3 137 metres is located in the east at the border with Uzbekistan.

TABLE 1
Basic statistics and population

Physical areas:			
Area of the country	1995	48 810 000	ha
Cultivable area		-	ha
Cultivated area	1993	1 480 000	ha
- annual crops	1993	1 400 000	ha
- permanent crops	1993	80 000	ha
Population:			
Total population	1995	4 099 000	inhabitants
Population density	1995	8	inhab./km^2
Rural population	1995	55	%
Water supply coverage:			
Urban population		-	%
Rural population		-	%

In 1993, the total cultivated area was estimated at 1 480 000 ha, which is 3% of the total area of the country. About 95% consisted of annual crops, and 5% of permanent crops, including vineyards, pistachio nuts, figs and olives.

The total population is about 4.1 million (1995), of which 55% is rural. Annual population growth was estimated at almost 2% in 1995, while it was 2.5% during the period 1979-89. In 1994, 44% of the total labour force was engaged in agriculture, including forestry. Agriculture accounts for almost 50% of the country's GDP.

CLIMATE AND WATER RESOURCES

Climate

The climate of Turkmenistan is of the subtropical desert type. In the Kara Kum desert, rainfall is so rare, that people remember single rainstorms that occurred years ago.

Water resources

The country has few internal renewable water resources, estimated at 1 km^3/year. The Amu Darya is the largest incoming river with an annual flow estimated at 70 km^3. It enters the country in the south-east after running along the border between Afghanistan and Turkmenistan over a length of about 60 km. It flows to the north-east and runs along the border between Turkmenistan and Uzbekistan for about 200 km before entering Uzbekistan.

TABLE 2
Water: sources and use

Renewable water resources:			
Average precipitation		-	mm/yr
		-	km³/yr
Internal renewable water resources		1.0	km³/yr
Total (actual) renewable water resources	1995	71.0	km³/yr
Dependency ratio	1995	98.6	%
Total (actual) renewable water resources per inhabitant	1995	17 321	m³/yr
Total dam capacity		-	10⁶ m³
Water withdrawal:			
- agricultural	1989	20 748	10⁶ m³/yr
- domestic	1989	228	10⁶ m³/yr
- industrial	1989	1 824	10⁶ m³/yr
Total water withdrawal		22 800	10⁶ m³/yr
per inhabitant	1989	-	m³/yr
as % of total (actual) renewable water resources		32.1	%
Other water withdrawal		-	10⁶ m³/yr
Average groundwater depletion		-	10⁶ m³/yr
Wastewater - Non-conventional water sources:			
Wastewater:		-	
- produced wastewater		-	10⁶ m³/yr
- treated wastewater		-	10⁶ m³/yr
- reused treated wastewater		-	10⁶ m³/yr
Desalinated water		-	10⁶ m³/yr

Several smaller rivers (the Murgap, the Tejen, the Atrek) descend from the Kopetdag and other mountains in the south. The central and western regions have no significant natural waterways.

The Kara Kum canal, the longest irrigation canal in the world, 1 300 km long, was constructed during the 1950s. Its inlet at the Amu Darya river is located just at the point where the river enters the country. It brings water to the oases in the south and to the capital Asghabat.

FIGURE 1
Water withdrawal (total: 22.8 km³ in 1989)

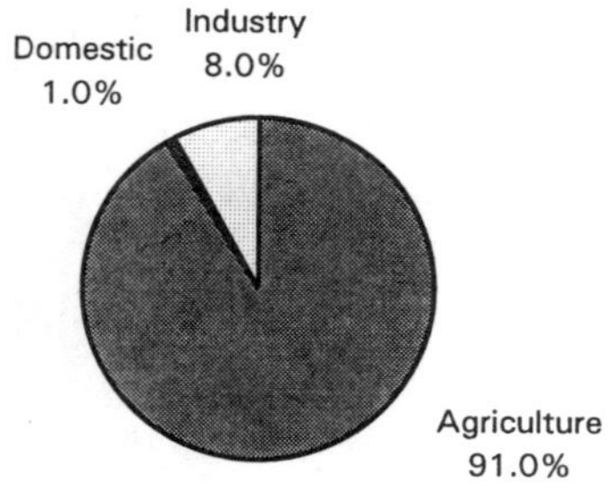

Total water withdrawal was 22.8 km³ in 1989, of which 91% for agricultural purposes (Figure 1). Contamination of surface water and groundwater with agricultural chemicals and pesticides is a serious problem. An inadequate supply of clean drinking water is a serious health hazard in Turkmenistan.

IRRIGATION AND DRAINAGE DEVELOPMENT

Turkmenistan is largely a desert country with nomadic cattle raising and intensive agriculture in the oases, irrigated with water from the Kara Kum canal in the south and the Tashauz canal in the north. Irrigation is estimated to cover around 1 300 000 ha, which is 88% of the cultivated area. About half of the irrigated area is estimated to be planted with cotton, making Turkmenistan the world's tenth largest cotton producer. Other irrigated crops are cereals (mainly wheat), vegetables and fruit.

TABLE 3
Irrigation and drainage

Irrigation potential		-	ha
Irrigation:			
1. Full or partial control irrigation: equipped area	1993	1 300 000	ha
- surface irrigation		-	ha
- sprinkler irrigation		-	ha
- micro-irrigation		-	ha
% of area irrigated from groundwater		-	%
% of area irrigated from surface water		-	%
% of area irrigated from non-conventional sources		-	%
% of equipped area actually irrigated		-	%
2. Spate irrigation area		-	ha
3. Equipped wetland and inland valley bottoms (i.v.b.)		-	ha
Total irrigation (1+2+3)	**1993**	**1 300 000**	**ha**
- as % of cultivated area		88	%
4. Flood recession cropping area		-	ha
Total water managed area (1+2+3+4)	**1993**	**1 300 000**	**ha**
- as % of cultivated area		88	%
- increase over last 10 years		-	%
- power irrigated area as % of water managed area		-	%
Full or partial control irrigation schemes:	Criteria		
Large-scale schemes	> - ha	-	ha
Medium-scale schemes		-	ha
Small-scale schemes	< - ha	-	ha
Total number of households in irrigation		-	
Irrigated crops:			
Total irrigated grain production		-	tons
as % of total grain production		-	%
Harvested crops under irrigation (full or partial control)		-	**ha**
- permanent crops: total		-	ha
- annual crops: total		-	ha
.		-	ha
.		-	ha
.		-	ha
.		-	ha
. other annual crops		-	ha
Drainage - Environment:			
Drained area		-	ha
as % of cultivated area		-	%
- drained areas in full or partial control irrigated areas		-	ha
- drained areas in equipped wetland and i.v.b		-	ha
- other drained areas		-	ha
- total drained area with subsurface drains		-	ha
- total drained area with surface drains		-	ha
Flood-protected area		-	ha
Area salinized by irrigation	1994	598 000	ha
Population affected by water-borne diseases		-	inhabitants

The unprotected banks of the Kara Kum canal have caused massive waterlogging and salinization of the surrounding land. It is estimated that some 46% of the cultivated land is now suffering from severe salinization. The land is also heavily polluted because of the overuse of pesticides and fertilizers.

INSTITUTIONAL ENVIRONMENT

No information available.

TRENDS IN WATER RESOURCES MANAGEMENT

One of the most serious policy dilemmas facing countries in the central Asian region concerns the environmental problems of the Aral Sea basin as a result of the absence of environmental considerations in agricultural practices, industrial development and waste management.

The special problems related to the Aral Sea desiccation are addressed through a potable water and a regional water management programme. Turkmenistan is one of the beneficiaries of the Central Asian Environmental Plan which seeks to ensure potable drinking water and improve water management throughout the area. It is also concerned with cooperation and water policy reforms to solve the Aral Sea disaster.

MAIN SOURCES OF INFORMATION

The following information was available on Internet in August 1996:

Embassy of Turkmenistan in the USA. 1995. Turkmenistan at a glance: Turkmenistan fact sheet.

United States Government. 1995. Turkmenistan: country fact sheet. CIA.

USAID. 1995. Turkmenistan: country profile.

United Arab Emirates

GEOGRAPHY AND POPULATION

The United Arab Emirates (UAE) is a federation of seven emirates: Abu Dhabi, Dubai, Sharjah, Ras Al Khaimah, Fujairah, Umm ul Quwain and Ajman. It is situated in the eastern corner of the Arabian peninsula and is bordered in the north by the Persian Gulf, in the east by the Gulf of Oman and Oman and in the south and west by Saudi Arabia. Six of the seven Emirates lie on the coast of the Persian Gulf, while the seventh, Fujairah, is situated on the eastern coast of the peninsula and has direct access to the Gulf of Oman.

The total area of the United Arab Emirates is about 83 600 km^2, of which 77 700 km^2 is mainland surface area, where the population lives. The total area of the many -and generally uninhabited- islands is about 5 900 km^2. Over 90% of the land is desert. In 1994, the agricultural area was estimated at 72 374 ha, of which 66 682 ha were available for agricultural production, while the remaining 5 962 ha were occupied by farm buildings and surrounding wasteland. In 1993, the total cultivated area was estimated at 54 512 ha, of which 21 683 ha consisted of annual crops and 32 829 ha of permanent crops.

TABLE 1
Basic statistics and population

Physical areas:			
Area of the country	1995	8 360 000	ha
Cultivable area	1993	66 682	ha
Cultivated area	1993	54 512	ha
- annual crops	1993	21 683	ha
- permanent crops	1993	32 829	ha
Population:			
Total population	1995	1 904 000	inhabitants
Population density	1995	23	inhab./km^2
Rural population	1995	16	%
Water supply coverage:			
Urban population	1995	100	%
Rural population	1995	100	%

The total population is about 1.9 million according to UN estimates (1995), of which only 16% is rural. According to the 1995 national census of the Ministry of Planning, the population was about 2.4 million, including both nationals and foreigners. Unofficial estimates suggest that over 80% of the population is made up of non-nationals. The table below shows the distribution of the area and the population, including resident foreigners, over the different Emirates in 1993.

Mainland area and population by emirates
(estimate by the Ministry of Planning, 1993)

Emirate	Mainland area (excl. islands) (km^2)	As % of total area	No. of inhabitants including foreigners	As % of total population	Population density (inh./km^2)
Abu Dhabi	67 340.0	86.7	871 000	41.8	12.9
Dubai	3 885.0	5.0	548 000	26.3	141.1
Sharjah	2 590.0	3.3	342 000	16.4	132.0
Ras Al Khaimah	1 683.5	2.2	141 000	6.8	83.7
Fujairah	1 165.5	1.5	68 000	3.3	58.3
Umm Al Quwain	777.0	1.0	30 000	1.4	38.6
Ajman	259.0	0.3	83 000	4.0	320.5
Total	77 700.0	100.0	2 083 000	100.0	26.8

TABLE 2
Water: sources and use

Renewable water resources:			
Average precipitation		100	mm/yr
		8.36	km^3/yr
Internal renewable water resources		0.15	km^3/yr
Total (actual) renewable water resources	1995	0.15	km^3/yr
Dependency ratio	1995	0	%
Total (actual) renewable water resources per inhabitant	1995	79	m^3/yr
Total dam capacity	1995	80	$10^6\ m^3$
Water withdrawal:			
- agricultural	1995	1 408	$10^6\ m^3/yr$
- domestic	1995	500	$10^6\ m^3/yr$
- industrial	1995	200	$10^6\ m^3/yr$
Total water withdrawal		2 108	$10^6\ m^3/yr$
per inhabitant	1995	1 107	m^3/yr
as % of total (actual) renewable water resources		1 405	%
Other water withdrawal		-	$10^6\ m^3/yr$
Average groundwater depletion	1995	1 495	$10^6\ m^3/yr$
Wastewater - Non-conventional water sources:			
Wastewater:			
- produced wastewater	1995	500	$10^6\ m^3/yr$
- treated wastewater	1995	108	$10^6\ m^3/yr$
- reused treated wastewater	1995	108	$10^6\ m^3/yr$
Desalinated water	1995	385	$10^6\ m^3/yr$

By far the largest emirate is Abu Dhabi and Abu Dhabi City is the capital of both the emirate and the whole country. It also has the largest population numerically, but at the same time the lowest population density among the emirates. Dubai, which has the highest population density, is considered the business capital and the most important port in the country. Over two-thirds of the total population are concentrated in these two emirates. Average annual population growth has been estimated at over 3.5%, including both nationals and foreigners. The male population accounted for over 66% of the total population in 1995, mainly because of male immigrant labour force.

In 1994, agriculture employed an estimated 9% of the labour force and accounted for less than 2% of the country's GDP. The main source of income is the revenue from oil exports. The last five-year plan concentrated on the manufacturing sector, in order to attain a balance in growth with other industries and a diversification of sources of income. Besides development planning at the federal level, each emirate follows an independent development strategy. The federal government is the driving force behind the economy through large public spending.

CLIMATE AND WATER RESOURCES

Climate

The climate is arid with very high summer temperatures. The coastal area, where the bulk of the population lives, has a hot and humid climate in the summer (May-October) with temperatures and relative humidity reaching 46°C and 100% respectively. Winters are generally mild with temperatures between 14 and 23°C. The interior desert region has hot summers with temperatures rising to about 50°C and cool winters during which the lowest temperature may fall to 4°C.

TABLE 3
Irrigation and drainage

Irrigation potential	1993	66 682	ha
Irrigation:			
1. Full or partial control irrigation: equipped area	1993	66 682	ha
- surface irrigation	1993	25 382	ha
- sprinkler irrigation	1993	3 748	ha
- micro-irrigation	1993	37 552	ha
% of area irrigated from groundwater	1993	100	%
% of area irrigated from surface water	1993	0	%
% of area irrigated from non-conventional sources	1993	0	%
% of equipped area actually irrigated	1993	81.7	%
2. Spate irrigation area		-	ha
3. Equipped wetland and inland valley bottoms (i.v.b.)		-	ha
Total irrigation (1+2+3)	**1993**	**66 682**	**ha**
- as % of cultivated area		122	%
4. Flood recession cropping area		-	ha
Total water managed area (1+2+3+4)	**1993**	**66 682**	**ha**
- as % of cultivated area		122	%
- increase over last 10 years		-	%
- power irrigated area as % of water managed area		-	%
Full or partial control irrigation schemes: Criteria			
Large-scale schemes > 5 ha	1993	0	ha
Medium-scale schemes	1993	0	ha
Small-scale schemes < 5 ha	1993	66 682	ha
Total number of households in irrigation	1993	21 194	
Irrigated crops:			
Total irrigated grain production	1993	1 052	tons
as % of total grain production	1993	100	%
Harvested crops under irrigation (full or partial control)	**1993**	**54 512**	**ha**
- permanent crops: total	1993	32 829	ha
- annual crops: total	1993	21 683	ha
. vegetables	1993	12 927	ha
. fodder crops (alfalfa, green fodder)	1993	7 808	ha
. wheat	1993	567	ha
. potatoes	1993	174	ha
. other annual crops	1993	207	ha
Drainage - Environment:			
Drained area		-	ha
as % of cultivated area		-	%
- drained areas in full or partial control irrigated areas		-	ha
- drained areas in equipped wetland and i.v.b		-	ha
- other drained areas		-	ha
- total drained area with subsurface drains		-	ha
- total drained area with surface drains		-	ha
Flood-protected area		-	ha
Area salinized by irrigation		-	ha
Population affected by water-borne diseases		-	inhabitants

Mean annual rainfall is about 100 mm, ranging from less than 40 mm around Liwa in the southern desert to 160 mm in the north-eastern mountains. Almost 90% of rainfall occurs during the winter (October-March) and the wettest month is February. Spring and summer witness only occasional concentrated heavy rainfall. The rainfall distribution is highly variable over space and time. Rainfall tends to be more reliable in the north-eastern mountain region.

Water resources

The total annual surface runoff produced from rain is about 150 million m^3, but there are no perennial streams. The average annual groundwater recharge is about 120 million m^3, most of which comes from infiltration from the river beds. The total groundwater abstraction during the year 1995 is estimated at 1 615 million m^3. This means that groundwater depletion probably amounts to almost 1 500 million m^3/year. However, this figure does not consider the possible annual recharge of groundwater entering from neighbouring countries (for example from the Eastern Arabia Aquifer), as no figures are available. In any case, the over-extraction of groundwater resources has led to a lowering of the water table by more than one metre on average during the last two decades, while sea water intrusion is increasing in the coastal areas.

Dams

To increase the groundwater recharge a number of dams have been built at various locations in the country. There are about 35 dams and embankments of various dimension having a total storage capacity of 80 million m^3. While most of these dams are basically built for recharge purposes, they also provide protection against damage caused by flash floods.

Desalinated water and treated wastewater

The first desalination plant was installed in Abu Dhabi in 1976 with a total capacity of 250 m^3/day. Because of a rapid increase in domestic and industrial water demand more plants were installed, particularly in Abu Dhabi and Dubai. At present there are 35 desalination plants in the UAE, with a total installed capacity of 1 922 m^3/day or 700 million m^3/year, while total actual production is 385 million m^3/year.

FIGURE 1

Water withdrawal (total: 2 108 million m^3 in 1995)

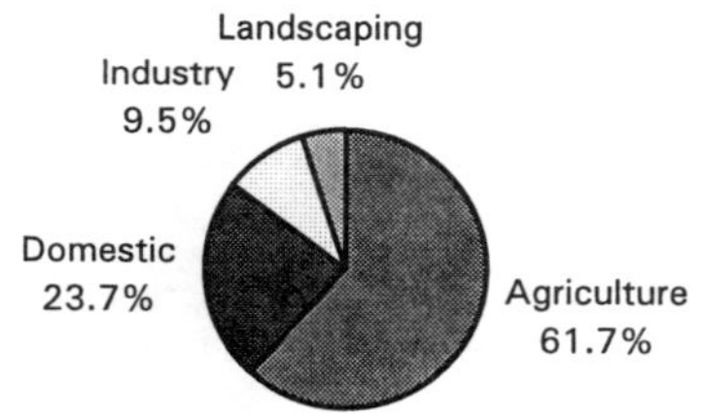

It is estimated that about 500 million m^3 of wastewater were produced in the urban areas during 1995, of which 108 million m^3 were treated and reused.

Water withdrawal

Total water withdrawal was estimated at 2 108 million m^3 in 1995 (Figure 1). Over 76% of the total water withdrawal was groundwater. Agricultural water withdrawal for crops was estimated at about 1 300 million m^3/year (all from groundwater), while landscape irrigation used 108 million m^3 (all treated wastewater). Total water withdrawal for domestic

FIGURE 2

Origin of water used by sector (total: 2 108 million m^3 in 1995)

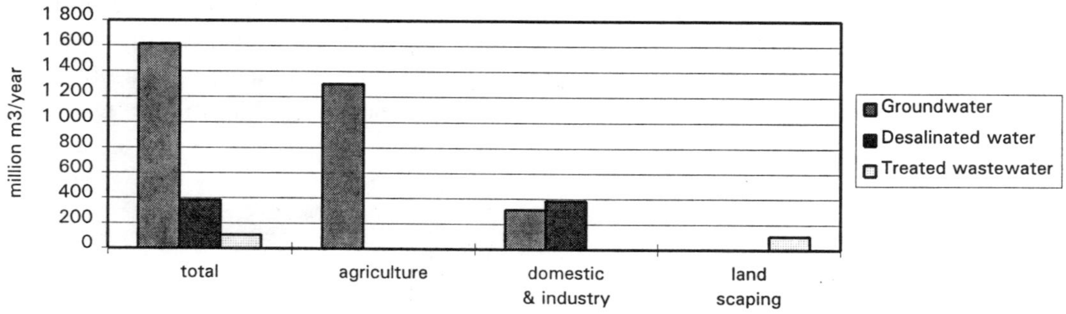

and industrial purposes was estimated at 700 million m^3, of which 385 million m^3, or 55%, consisted of desalinated water and the remaining part of groundwater (Figure 2).

IRRIGATION AND DRAINAGE DEVELOPMENT

The UAE has very limited potential for agricultural development since over 90% of the land is desert, there are no perennial surface water resources and rainfall is meagre and erratic. However, in spite of the harsh weather conditions and soil and water constraints, remarkable progress has been made in the agricultural sector, particularly during the last two decades. The cultivable area increased from 15 050 ha in 1977 to 66 682 ha in 1994. The land is usually developed by transporting suitable soil to areas where water is available and for this reason the term 'cultivable land' is somewhat relative and may change over time. For the same reason the irrigation potential can be estimated at 66 682 ha but may change over time. The main agricultural areas are located in the north-east (Ras Al Khaimah), in the east along the coast from Kalba to Dibba (Fujairah), in the south-east (Al Ain/Abu Dhabi) and in the central region (Dhaid/Abu Dhabi).

In 1994, the total water managed area was 66 682 ha (equal to the cultivable area), of which 41 300 ha were equipped with modern irrigation systems (sprinkler irrigation and micro-irrigation), while on 25 382 ha surface irrigation (basin and furrow) was practised (Figure 3). All irrigation water is groundwater.

FIGURE 3
Irrigation techniques f/p (total: 66 682 ha)

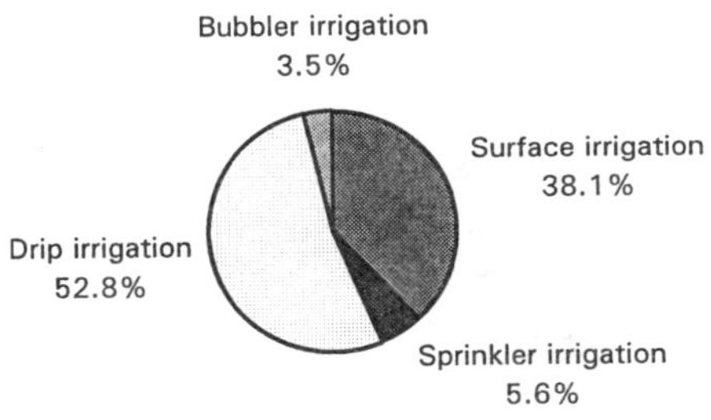

Prior to the introduction of modern irrigation systems all the agricultural land was irrigated by traditional flood and furrow methods. Extensive research was carried out during the period 1976-81 to select suitable irrigation systems and a pilot farm was established in 1983 to introduce sprinkler irrigation and micro-irrigation systems and a subsidy was given to the farmers. These irrigation systems are believed to have saved about 60% of the irrigation water.

Apart from the government's experimental farms, nurseries, afforestation schemes and public gardens, all the agricultural land is owned and developed by private owners. There are 21 194 farm holdings. The government provides subsidies, extension services and other incentives to the farmers. All crops are irrigated and the main crops grown are dates, fruits, vegetables and fodder crops (Figure 4). In 1994, the actually irrigated area was 54 512 ha, or 82% of the equipped area. Every year a part of the total equipped area is left fallow, a kind of shifting cultivation being practised.

The cost of irrigation development has decreased considerably during recent years, mainly due to the local production of pipes and irrigation equipment. The average cost of irrigation development is about $US 3 250/ha and the annual operation and maintenance cost is about $US 400/ha. There are no irrigation water charges levied by the government, but the farmers pay for the drilling of boreholes on their farms and the pumping of the groundwater.

FIGURE 4
Irrigated crops f/p (total: 54 512 ha)

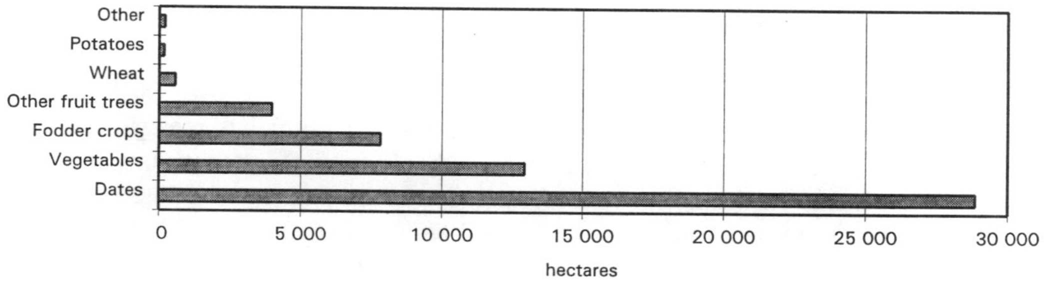

Over-abstraction of groundwater leads to a rapid fall in the water table, especially near the coast, resulting in increased salinity of the soil and water. Encroachment of sea water had already been reported in 1982, when it apparently penetrated as far as 20 km inland in the northern emirates. As a result, several farms are going out of production.

INSTITUTIONAL ENVIRONMENT

There are five main institutions involved in water resources management:

- The Ministry of Agriculture and Fisheries is responsible for the development and management of the agricultural water supply. The water and soil department of the Ministry is in charge of promoting irrigated agriculture and for the planning, investigation and management of groundwater resources, the investigation of quality and salinization of soil due to irrigation, the construction of dams for flood control and groundwater recharge, the operation and maintenance of the hydro-meteorological network, the operation of laboratories, designing of the irrigation networks, etc.

- The Ministry of Electricity and Water is responsible for the drinking water supply, which includes well drilling, the installation and operation of desalination plants, the operation and maintenance of well fields, water supply and planning.

- The General Water Resources Authority is the federal authority responsible for water management and coordination between the other agencies. It is also responsible for formulating the rules and regulations for matters relating to water in the country, including the registration of the water well drilling companies and licenses for drilling.

- Local government water departments and authorities especially in the emirates of Abu Dhabi, Dubai and Sharjah are independently responsible for the supply of drinking water and all water affairs in their respective emirates.

- The Federal Environmental Agency has the power to control and regulate water pollution.

TRENDS IN WATER RESOURCES MANAGEMENT

Groundwater has been affected adversely, both qualitatively and quantitatively, due to over-abstraction. At present, all water used to irrigate agricultural products is groundwater, while treated wastewater is used for landscaping purposes. Future irrigation development using groundwater is very limited and attempts are being made to alleviate the problem of water scarcity by constructing desalination plants and dams, the latter mainly for the recharge of groundwater.

Local food production currently satisfies about one-fourth of UAE's food demand. The emirates are now self-sufficient in some winter vegetables and, in fact, a surplus is produced at certain times of the year. Overall, the government will continue to encourage agriculture, but is also aware that complete self-sufficiency in food is impossible.

Within the national strategy for water management, priority is given to sustainable and economically viable agricultural products and to research on the growth of salt tolerant crops. Utilizing all the possible options, the ultimate aim is to maintain the present level of growth if further development is obstructed due to water scarcity.

MAIN SOURCES OF INFORMATION

Ministry of Agriculture and Fisheries, Department of Statistics. 1993. Statistical Bulletin.

Ministry of Agriculture and Fisheries, Department of Soil and Water. Annual Hydrological Reports 1982-1993.

Ministry of Agriculture and Fisheries, Department of Soil and Water. 1993. Meteorological Yearbook No 3.

Ministry of Planning. Annual Statistical Abstracts 1992-1993.

Mohammed Saqr Al Asam. 1995. United Arab Emirates Water Resources Use in Agriculture and Conservation.

Water and Electricity Department, Abu Dhabi. 1995. Development of Desalination Plants in the United Arab Emirates.

Yemen

GEOGRAPHY AND POPULATION

Yemen, with a total area estimated at 527 970 km^2, is located on the south-western edge of the Arabian Peninsula. Apart from the mainland it includes many islands, the largest of which are Socotra in the Arabian Sea to the far east of the country and Kamaran in the Red Sea. The country is bordered by Saudi Arabia in the north, Oman in the east, the Arabian Sea and the Gulf of Aden in the south, and the Red Sea in the west. A large part of the boundary between Yemen and Saudi Arabia has not yet been defined officially.

The present Republic of Yemen was born in 1990, as a result of the unification of the former Yemen Arab Republic (YAR) and the People's Democratic Republic of Yemen (PDRY). These two parts are sometimes still referred to respectively as the northern and southern part of the country. For administration purposes, the country is divided into 17 governorates.

TABLE 1
Basic statistics and population

Physical areas:			
Area of the country	1995	57 797 000	ha
Cultivable area	1994	3 617 753	ha
Cultivated area	1994	1 052 786	ha
- annual crops	1994	852 786	ha
- permanent crops	1994	200 000	ha
Population:			
Total population	1995	14 501 000	inhabitants
Population density	1995	27	inhab./km^2
Rural population	1995	66	%
Water supply coverage:			
Urban population	1990	65	%
Rural population	1990	48	%

The cultivable land is estimated at about 3.62 million ha, which is 7% of the total area. In 1994, the total cultivated area was 1.05 million ha, or 29% of the cultivable area, of which 0.85 million ha consisted of annual crops and 0.20 million ha consisted of permanent crops.

The total population is 14.5 million (1995), of which 66% is rural. The average population density is about 27 inhabitants/km^2, but in the western part of the country the density can reach up to 300 inhabitants/km^2 (Ibb province) while in the three eastern provinces of the country the density is less than 5 inhabitants/km^2. This is closely related to the physical environment. By far the largest part of the population lives in the Yemen Mountain area in the western part of the country, where rainfall is still significant, although not high in many locations. The hostile environment of the desert and eastern upland areas is reflected by low population density. The average demographic growth rate is estimated at 3.7%, which is very high. In 1990, agriculture accounted for 20% of the country's GDP and employed 62% of the labour force.

CLIMATE AND WATER RESOURCES

Climate

Yemen has a predominantly semi-arid to arid climate, with rainy seasons during spring and summer, and high temperatures prevail throughout the year in low-altitude zones.

TABLE 2
Water: sources and use

Renewable water resources:			
Average precipitation		170	mm/yr
		89.8	km^3/yr
Internal renewable water resources		4.1	km^3/yr
Total (actual) renewable water resources	1995	4.1	km^3/yr
Dependency ratio	1995	0	%
Total (actual) renewable water resources per inhabitant	1995	283	m^3/yr
Total dam capacity	1995	180	$10^6\ m^3$
Water withdrawal:			
- agricultural	1990	2 700	$10^6\ m^3/yr$
- domestic	1990	201	$10^6\ m^3/yr$
- industrial	1990	31	$10^6\ m^3/yr$
Total water withdrawal		2 932	$10^6\ m^3/yr$
per inhabitant	1990	251	m^3/yr
as % of total (actual) renewable water resources		71.5	%
Other water withdrawal		-	$10^6\ m^3/yr$
Average groundwater depletion	1994	700	$10^6\ m^3/yr$
Wastewater - Non-conventional water sources:			
Wastewater:			
- produced wastewater	1992	37	$10^6\ m^3/yr$
- treated wastewater	1990	20	$10^6\ m^3/yr$
- reused treated wastewater		-	$10^6\ m^3/yr$
Desalinated water	1989	10	$10^6\ m^3/yr$

In Aden mean figures of 25°C (January) to 32°C (June) occur, but maximun temperatures over 38°C are quite common, combined with a very high relative air humidity. Mean annual precipitation on the mainland gives a volume of 93.6 km^3. The average annual rainfall ranges from less than 50 mm in the coastal areas and the deserts to 200-400 mm on the slopes of the highlands and more than 1 000 mm on the western slopes of the mountains.

The many different landscapes of Yemen can be grouped into five main geographical/ climatological regions:

- The Coastal Plains: The Plains are located in the west and south-west and are flat to slightly sloping with maximum elevations of only a few hundred meters above sea level. They have a hot climate with generally low to very low rainfall (< 50 mm/year). Nevertheless, the Plains contain important agricultural zones, due to the numerous wadis that drain the adjoining mountainous and hilly hinterland.

- The Yemen Mountain Massif: This massif constitutes a high zone of very irregular and dissected topography, with elevations ranging from a few hundred metres to 3 760 m above sea level. Accordingly, the climate varies from hot at lower elevations to cool at the highest altitudes. The western and southern slopes are the steepest and enjoy moderate to rather high rainfall, on average 300-500 mm/year, but in some places even more than 1 000 mm/year. The eastern slopes show a comparatively smoother topography and average rainfall decreases rapidly from west to east.

- The Eastern Plateau Region: This region covers the eastern half of the country. Elevations decrease from 1 200-1 800 m at the major watershed lines to 900 m on the northern desert border and to sea level on the coast. The climate in general is hot and dry, with average annual rainfall below 100 mm, except in the higher parts. Nevertheless, floods following rare rainfall may be devastating.

TABLE 3
Irrigation and drainage

Irrigation potential		-	ha
Irrigation:			
1. Full or partial control irrigation: equipped area	1994	383 200	ha
- surface irrigation	1994	382 450	ha
- sprinkler irrigation	1994	350	ha
- micro-irrigation	1994	400	ha
% of area irrigated from groundwater	1994	100	%
% of area irrigated from surface water	1994	0	%
% of area irrigated from non-conventional sources	1994	0	%
% of equipped area actually irrigated		-	%
2. Spate irrigation area	1994	98 320	ha
3. Equipped wetland and inland valley bottoms (i.v.b.)		-	ha
Total irrigation (1+2+3)	**1994**	**481 520**	**ha**
- as % of cultivated area		46	%
4. Flood recession cropping area		-	ha
Total water managed area (1+2+3+4)	**1994**	**481 520**	**ha**
- as % of cultivated area		46	%
- increase over last 10 years		-	%
- power irrigated area as % of water managed area		-	%
Full or partial control irrigation schemes:	Criteria		
Large-scale schemes	> 5 ha	-	ha
Medium-scale schemes		-	ha
Small-scale schemes	< 2 ha	-	ha
Total number of households in irrigation		-	
Irrigated crops:			
Total irrigated grain production	1994	156 000	tons
as % of total grain production	1994	20	%
Harvested crops under irrigation (full or partial control)		-	**ha**
- permanent crops: mainly qat		-	ha
- annual crops: total		-	ha
. cereals (sorghum, maize, wheat, barley)	1994	49 110	ha
. vegetables		-	ha
. pulses	1994	26 230	ha
. sesame	1994	20 410	ha
. other annual crops		-	ha
Drainage - Environment:			
Drained area		-	ha
as % of cultivated area		-	%
- drained areas in full or partial control irrigated areas		-	ha
- drained areas in equipped wetland and i.v.b		-	ha
- other drained areas		-	ha
- total drained area with subsurface drains		-	ha
- total drained area with surface drains		-	ha
Flood-protected area	1994	70 000	ha
Area salinized by irrigation		-	ha
Population affected by water-borne diseases		-	inhabitants

- The Desert: Between the Yemen Mountain Massif and the Eastern Plateau lies the Ramlat as Sabatayn, a sand desert. Rainfall and vegetation are nearly absent, except along its margins where rivers bring water from adjacent mountain and upland zones. In the north lies the Rub Al Khali desert, which extends far into Saudi Arabia and is approximately 500 000 km^2 in area. This sand desert is one of the most desolate parts of the world.

- The Islands: The most important of all the islands is Socotra, where more exuberant flora and fauna can be found than in any other region in Yemen.

Water resources

Yemen can be subdivided into four major drainage basins, regrouping numerous smaller wadis:

- the Red Sea basin
- the Gulf of Aden basin
- the Arabian Sea basin
- the Rub Al Khali interior basin

The floods of the wadis in Yemen are generally characterized by abruptly rising peaks that rapidly recede. In between the irregular floods the wadis are either dry or carry only minor base flows.

Surface water resources have been estimated at 2 000 million m^3/year, but this quantity corresponds to the runoff from major rivers and does not include the runoff produced within the smaller catchments. Renewable groundwater resources have been estimated at 1 525 million m^3/year, a large part probably coming from infiltration in the river beds. A major groundwater aquifer was recently discovered in the eastern part of the country with an estimated storage of 10 km^3. This aquifer is still under study and it is not known whether the groundwater is rechargeable or whether it is all fossil water.

The surface runoff to the sea measured in some major wadis is estimated at 270 million m^3/year, the groundwater outflow to the sea at 280 million m^3/year. There might be some groundwater flowing into Saudi Arabia, but no data are available. The existence of surface drainage crossing into Saudi Arabia suggests that some sharing of surface flows could be possible, but details are not known.

Dams

The total dam capacity is estimated at 0.18 km^3. In general, the dams are built for irrigation and domestic purposes, but at the same time they contribute to groundwater recharge. There are also many flood control dams which are not intended to store water, but to divert the spate floods immediately to the adjacent irrigation network (spate irrigation).

Water withdrawal

In 1990 total water withdrawal was estimated at 2 932 million m^3/year, of which over 90% for agricultural purposes (Figure 1). Most of the water used was groundwater (from wells and springs), resulting in groundwater depletion as withdrawal exceeds the annual groundwater recharge. The rates of decline of the groundwater levels is alarmingly high in many zones, especially in the Yemen Highlands, where decline of between 2 and 6 m/year is commonly observed. In coastal zones this leads to the incidence of salt water intrusion. Spring-fed irrigation has reduced significantly as

FIGURE 1
Water withdrawal (total: 2 932 million m^3 in 1990)

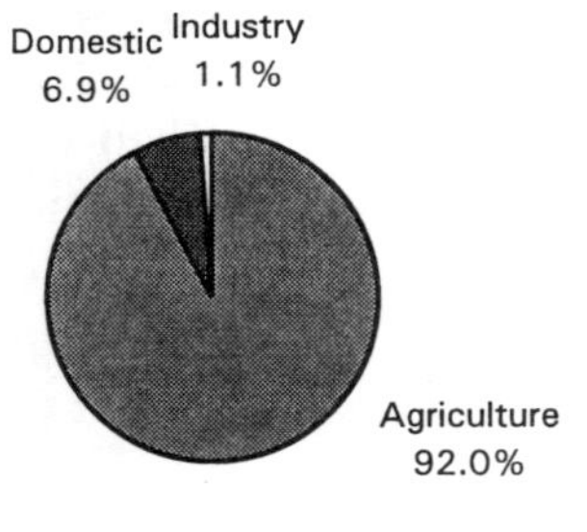

groundwater tables have dropped. The quantity of desalinated water was estimated at 10 million m^3/year in 1989, contributing to the water supply of Aden.

Irrigation and drainage development

In 1994, the total water managed area was estimated at 481 520 ha. A global figure for irrigation potential is not available. About 48 000 ha have been identified for further irrigation development, mostly in the coastal plains and in Wadi Hadramaut.

IGURE 2
istribution of the water managed areas
total: 481 520 ha in 1994)

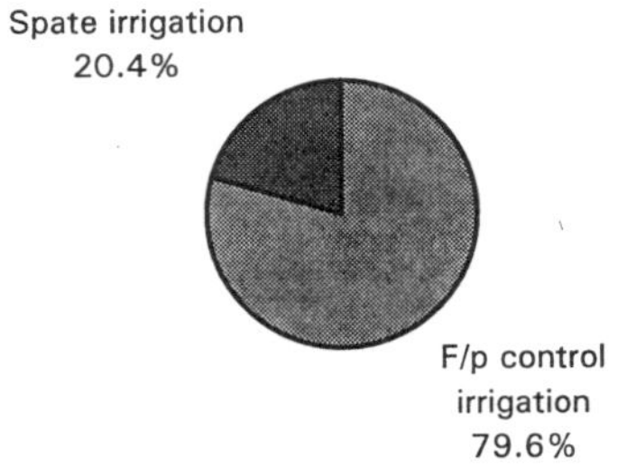

FIGURE 3
Origin of irrigation water f/p (total: 383 200 ha in 1994)

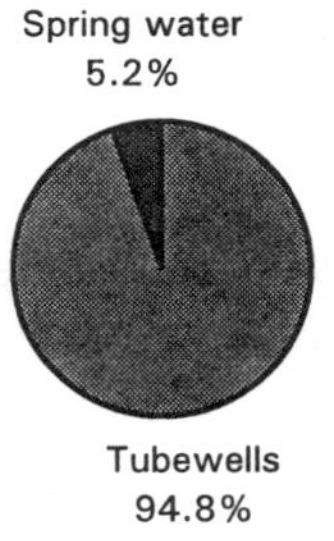

Two main types of water management can be distinguished (Figure 2):

- Full/partial control irrigation: This concerns an area of 383 200 ha, all irrigated from groundwater, of which 363 200 ha from tubewells and 20 000 ha from spring water (Figure 3). In general, new, deeper tubewells replace those which have gone out of production because of declining water tables.

- Spate irrigation: This covers an area of 98 320 ha. Traditionally, farmers in the vicinity of wadis relied on simple earth built diversion systems and irrigation networks. With small to medium spates, these temporary embankments can be effective; with large spates, they are often swept away. In order to give better control of the spate flows, a series of public sector investments, involving the construction of permanent diversion weirs and canal distribution structures, have been made in the main wadis since the early 1970s. Most of these systems, however, have experienced maintenance and water distribution problems because scheme designs conflicted with traditional water rights.

On the remaining cultivated area of 571 266 ha, water harvesting is practised, based on collecting and retaining overland flow in zones where soils permit agriculture. The receiving zone is always smaller than the zone where overland flow is produced, thus a multiplier effect is produced which permits agricultural production in low precipitation zones. The numerous constructed mountain terraces, also called 'the hanging gardens of Yemen', collect and retain rain and overland flow in a similar way.

Overall irrigation efficiency is low, between 35 and 45%, depending on field levelling and the water conveyance system used. Sprinkler irrigation and micro-irrigation are found on a limited number of farms and in pilot projects, using water from tubewells and springs. Almost all irrigation is surface irrigation (Figure 4). It is thought that efficiency could be increased to 60% by lining the canals and installing pipe distribution for surface irrigation, and to over 80% by adopting sprinkler irrigation and micro-irrigation techniques.

FIGURE 4
Irrigation techniques f/p (total: 383 200 ha in 1994)

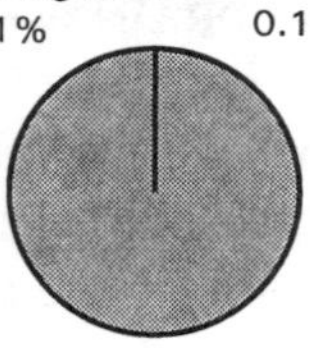

Farm size, including both rainfed and irrigated agriculture, is very small in general: 37% of the farms have less than 0.5 ha, 72% of the farms less than 2 ha, while only 4% of the farms have more than 10 ha.

Very rough estimates of the cost of irrigation development using groundwater from tube-wells lead to figures between $US 250/ha for large schemes (10 ha) and $US 450/ha for small schemes (2 ha) in the lowlands, while in the highlands the cost might vary between $US 575/ha and $US 1 300/ha respectively. The differences in cost between the lowlands and the highlands are mainly due to differences in the characteristics of the aquifers. Operation and maintenance costs vary between $US 175/ha per year in the lowlands and $US 350/ha per year in the highlands.

According to the Constitution, flowing and underground water are defined as 'res communis'. However, a landowner has 'precedence' for water taken from a well on his land. In spring-irrigated areas water can be attached to land in the form of 'turns', which give rights to divert the canal into the field for a fixed period of time. The 'turn' can however be detached from the land and sold or rented separately. This landowner 'precedence' has permitted the private development of deep tubewell extraction, which is in some ways in conflict with Islamic principles. Islamic and customary law has no precedent for dealing with a new technology that allows landowners to extract (and sell) unlimited quantities of water from deep aquifers, and modern law has not yet regulated it either.

The major irrigated cash crops are cotton (12 270 ha), coffee (8 060 ha), sesame (20 410 ha), tobacco and *qat*. As far as the production of *qat* is concerned, no official figures are available, but there have been estimates that at least one-fourth of the irrigated land is under qat production. Main cereal crops are sorghum, maize, wheat and barley, with a total irrigated area estimated at 49 110 ha. Production of vegetables, potatoes and fruits have increased significantly. Pulses also retain an important place. However, no figures on the area covered by all irrigated crops are available for the country as a whole. In 1994, the yield of irrigated wheat was 3.75 tons/ha as against 0.71 to 1.70 tons/ha for rainfed wheat. The yield of irrigated barley was 4.84 tons/ha as against 0.77 to 1.30 ton/ha for rainfed barley.

The flood protection area has been estimated at between 70 000 and 80 000 ha. Salinization due to irrigation exists in several regions, but no figures are available. No drainage systems are reported to exist.

INSTITUTIONAL ENVIRONMENT

The Ministry of Agriculture and Water Resources (MAWR) is responsible for formulating policies for water resources, for food security and for crops, livestock and forestry production, and for coordinating public investment and services in the sector. The General Directorate of Water Resources is located within the Ministry with four general departments: water resources; irrigation and maintenance of water installations; farm mechanization and land reclamation; irrigation studies. Most field services are provided to farmers through decentralized Regional Development Agencies (RDA), supported by technical services at national level. However, the division of responsibility between MAWR, the Agricultural Research and Extension Authority (AREA) and the RDAs with respect to water management is unclear.

Responsibility for coordinating rural water supplies lies within the Water Supply Department of the Ministry of Water and Electricity (MWE).

The General Department of Hydrology is located within the Ministry of Oil and Mineral Resources (MOMR).

TRENDS IN WATER RESOURCES MANAGEMENT

The successful and sustainable exploitation of the water resources in Yemen is threatened. The most serious and obvious problem is the rapid depletion of groundwater resources. Almost all the important groundwater systems in Yemen are being over-exploited at an alarming rate. The socio-economic consequences of groundwater resources depletion are dramatic since groundwater will become too expensive for use in agriculture and, as a result, regional agricultural economies based on groundwater irrigation are doomed to collapse if the water resources are not adequately controlled. The groundwater stocks may be further reduced by groundwater salinization (in coastal areas) and groundwater pollution (in urban areas and areas of intensive agriculture). Environmental degradation occurs, for example in areas where springs have dried up. The scarcity of water leads to ever-increasing competition which, if uncontrolled, might lead to socio-economic problems.

There is an increasing awareness in Yemen of groundwater depletion. The Government of Yemen has committed itself to a sustainable use of the water resources, which was reiterated in an official statement issued at the UN Conference on Environment and Development of 1992 in Rio de Janeiro.

Water resources management in the country suffers because there is no unified central decision-making organization. Several authorities are dealing with water related affairs with minimum integration and coordination. To solve this problem, a Presidential Decree for the establishment of the National Water Resources Authority (NWRA) was issued in October 1995, providing for the merger of the General Directorate of Water Resources of MAWR, the General Department of Hydrology of MOMR and the Technical Secretariat of the previously existing High Water Council. The main duties of the authority will be:

- to prepare water resources policies and strategies;
- to formulate water legislation and regulations along with their enforcement;
- to undertake water resources studies, evaluation and planning;

- to carry out management at basin level, as traditional centralized management has proved to be a failure.

Measures to be implemented at field level may include the introduction of water-saving techniques (improving irrigation efficiencies, imposing a water tariff, etc.), groundwater licensing and enforcement of pollution control regulations.

MAIN SOURCES OF INFORMATION

Central Statistics Office. 1995. Population and housing census, December 1994. Ministry of Planning and Development.

Consulting Engineering Services Private Ltd. 1991. Land and water resources and irrigation development study. New Delhi, India.

General Department of Agricultural Statistics. Agricultural Statistics Yearbooks 1988-1992. Ministry of Agriculture and Water Resources.

General Department of Agricultural Statistics. 1995. Agricultural Statistics Pamphlet 1994. Ministry of Agriculture and Water Resources.

Ministry of Oil and Mineral Resources & TNO Institute of Applied Geoscience (Netherlands). 1995. The water resources of Yemen: a summary and digest of available information. Report compiled by Van der Gun, J.A.M. and Abdul Aziz Ahmed.

Noman, Abdullah Ahmed. 1989. Issues and problems of agricultural development/agricultural holdings. Al Farabi, Beirut.

TS/HWC-UNDP/DESD. 1995. Final reports. Volume III: Surface water resources; Volume IV: Groundwater resources; Volume VI: Water supply, wastewater and sanitation.

World Bank. 1993. Republic of Yemen, Agricultural sector study: Strategy for sustainable agricultural production. Report No 11126-YEM.

Zarina, M., Ismail, F., Abdullah, M., Alshab, M., Shasan. 1991. The first national population policy conference. Central Statistics Office, Ministry of Planning and Development.

General bibliography

Abbot, J.S (chairman). 1988. Micro irrigation world wide usage. Working group on micro irrigation. ICID Bulletin 37(1), pp. 1-12.

ACSAD (Arab Centre for the Studies of Arid zones and Dry lands)/UNESCO/IHE (International Institute Hydraulic and Environmental Engineering). 1988. Water resources assessment in the Arab region. Paris-Delft-Damascus. 418 p.

ACSAD/UNESCO/IHP (International Hydrological Programme). 1995. Groundwater protection in the Arab region. UNESCO, Paris-Cairo. 162 p.

Alexandratos, N. (Editor). 1995. World Agriculture: Towards 2010, an FAO study. J. Wiley and Son, Chichester GB and FAO Rome. 488 p.

Beaumont, P., Blake, G.H. and Wagstaff. 1988. The Middle East: a geographical study. David Fulton Publishers, London. 523 p.

Biswas, A.K. (editor). 1994. International waters of the Middle East: from Euphrates-Tigris to Nile. Water Resources Management Series. Oxford University Press, Bombay/New Delhi. 221 p.

Donkers, Henk. 1994. De witte olie: Water en duurzame ontwikkeling in het Midden-Oosten. Novib, Den Haag. Uitgeverij Jan van Arkel, Utrecht. 176 p.

Economist Intelligence Unit (Various years). Country profiles (All countries). Economist Intelligence Unit, London.

ESCWA/FAO. 1994. Land and water policies in the Near East region: case studies on Egypt, Jordan and Pakistan. New York. 88 p.

Europa Publications Limited. 1993. The Middle East and North Africa 1994. Fortieth edition. Europa Publications Limited, London. 967 p.

FAO. 1995. FAO Production Year Book - 1994. Vol. 48. FAO Statistics Series N° 125. Rome. 243 p.

FAO. 1995. Irrigation in Africa in figures/L'irrigation en Afrique en chiffres. Water Report N° 7, FAO/AGLW, Rome. 336 p.

FAO. 1995. Irrigation potential in Africa: a basin approach. FAO/AGL, Rome (in preparation).

Gleick, P.H. (editor). 1993. Water in Crisis: A Guide tot the world's freshwater resources. Oxford University Press for Pacific Institute, New York. 497 p.

League of Arab States, AOAD (Arab Organization for Agricultural Development). 1994. Arab agricultural statistics. Yearbook, Volume No 14. AOAD, Khartoum. 547 p.

Organization of the Islamic Conference (OIC). 1994. Information on OIC Countries: 3-Agriculture.

Rogers, P. and Lydon, P. (editors). 1994. Water in the Arab world: Perspectives and prognoses. Harvard University Press, USA. 369 p.

Rashid, A. 1994. The resurgence of Central Asia: Islam or Nationalism? Oxford University Press, Karachi. 278 p.

UNESCO/ROSTAS. 1995. Rainfall water management in the Arab region. State of the Art Report. UNESCO, Cairo. 152 p.

United Nations. 1982. Groundwater in the Eastern Mediterranean and Western Asia. UN-DTCD, Natural Resources Water Series No. 9. United Nations, New York. 230 p.

USAID, Bureau for the Near East. 1993. Water resources action plan for the Near East. USAID, Washington D.C. 184 p.

Wangnick, K. 1990. IDA Worldwide desalting plants inventory. Report N° 11, May, Wangnick Consulting, Guarrenburg, Germany in Gleick (1993).

WHO/UNICEF. 1993. Water supply and sanitation sector monitoring report 1993. UNICEF, New York; WHO, Geneva. 57 p.

World Bank. 1994. A strategy for managing water in the Middle East and North Africa. Prepared by Berkoff, J. IBRD/World Bank, Washington D.C. 72 p.

World Development Report 1992: Development and Environment. 1992. Published for the World Bank. Oxford University Press, New York. 308 p.

WRI (World Resources Institute)/UNEP (United Nations Environmental Programme)/UNDP (United Nations Development Programme)/WB (World Bank). 1996. World Resources 1996-97. A guide to the global environment. The Urban Environment. Oxford University Press, New York and Oxford. 365 p.

World Water/WHO. 1987. The international drinking water supply and sanitation directory.

AQUASTAT

For more information on the AQUASTAT programme, or to obtain specific country profiles, contact:

FAO Regional Office for the Near East
P. O. Box 2223
Cairo, Egypt
Facsimile: + 20 2 3495981
E-mail : FAO-RNE@field.fao.org

or:

Chief, Water Resources, Development and Management Service
Land and Water Development Division (AGL)
FAO, Viale delle Terme di Caracalla,
00100 Rome, Italy
Facsimile: + 39 6 5225 6275
E-mail : Land-and-Water@fao.org

Information on AQUASTAT is also available on Internet:

http://www.fao.org/waicent/faoinfo/agricult/agl/aglw/aquastat/aquastat.htm

or:

gopher://gopher.fao.org Then choose 'Information from the Food and Agriculture Organization' and then 'Aquastat'.

FAO TECHNICAL PAPERS

WATER REPORTS

Availability: March 1997

Ar	– Arabic	Multil –	Multilingual
C	– Chinese	*	Out of print
E	– English	**	In preparation
F	– French		
P	– Portuguese		
S	– Spanish		

The FAO Technical Papers are available through the authorized FAO Sales Agents or directly from Distribution and Sales Section, FAO, Viale delle Terme di Caracalla, 00100 Rome, Italy.